AUSTRALIAN MATHEMATICAL SOCIETY LECTURE SERIES

Editor-in-Chief: Professor J.H. Loxton, School of Mathematics, Physics, Computing and Electronics, Macquarie University, NSW 2109, Australia

Editors: Professor C.C. Heyde, School of Mathematical Sciences, Australian National University, Canberra, ACT 0200, Australia

Associate Professor W.D. Neumann, Department of Mathematics, University of Melbourne, Parkville, Victoria 3052, Australia

Associate Professor C.E.M Pearce, Department of Applied Mathematics, University of Adelaide, SA 5005, Australia

Integral:
An Easy Approach after
Kurzweil and Henstock

Lee Peng Yee

The National Institute of Education, Singapore

Rudolf Výborný

The University of Queensland

CAMBRIDGE UNIVERSITY PRESS
Cambridge, New York, Melbourne, Madrid, Cape Town,
Singapore, São Paulo, Delhi, Tokyo, Mexico City

Cambridge University Press
The Edinburgh Building, Cambridge CB2 8RU, UK

Published in the United States of America by Cambridge University Press, New York

www.cambridge.org
Information on this title: www.cambridge.org/9780521779685

First published 2000

A catalogue record for this publication is available from the British Library

ISBN 978-0-521-77968-5 Paperback

Contents

vi   *Contents*

Preface

Presenting the theory of the integral to non-specialists is an old and everlasting problem. At most universities the Riemann integral is taught in introductory courses, even to future mathematicians. The reason for this is that the Riemann integral has an intuitive appeal and basic theorems are easy to prove. This, however, is all that can be said in its favour. This theory is not powerful enough for applications and when it comes to deeper results they are not any easier to prove than the corresponding results in more modern theories. It is true that Riemann with his approach to integration advanced mathematics significantly but that was almost a century and a half ago. We feel the time is now ripe to start teaching more comprehensive theories of integration at all levels.

The theory of integration employed by professional mathematicians was created by Henri Lebesgue at the beginning of the twentieth century. It could hardly be criticized and the mathematical community is happy with it. Unfortunately experience shows that, perhaps because of its abstract character, it is deemed to be difficult by beginners and non-mathematicians. It is not popular with physicists and engineers. The Lebesgue theory does not cover non-absolutely convergent integrals and there is a need then to consider improper integrals. It is an additional and important advantage of the theory expounded in this book that it includes all improper integrals.

In 1957 Jaroslav Kurzweil gave a new definition of the integral, which in some respects is more general than Lebesgue's. Ralph Henstock developped the theory further and started to advocate its use at the elementary level. The Kurzweil–Henstock theory preserves the intuitive appeal of the Riemann definition but has the power of the Lebesgue theory. The aim of this book is to present the Kurzweil–Henstock theory. We wish to give this powerful tool to non-mathematicians and undergraduates and we advocate the widest possible use of **one** integral at all levels. We believe that the desirability of teaching one integral at all levels was also part of the motivation for R. Henstock to develop the theory.

Both authors have taught the Kurzweil–Henstock integral at various levels and various universities, first of all at our home institutions, the National Institute of Education in Singapore and University of Queensland and also at Universität Erlangen-Nürnberg, the University of Canterbury, Northwest Normal University in Lanzhou and the University of the Philippines. We express our gratitude to the Mathematics Departments of these institutions for their understanding of our desire to teach a 'new' integral and support of our research. Our experience is positive at all levels and in the introductory courses, once the students grasped the concept of δ-fine partitions, they found the theory as easy as, or perhaps one should say no more difficult than, the Riemann theory.

Several books have appeared since the inception of the Kurzweil–Henstock theory. Most of these aim at the advanced or graduate level. This is so with the books which the inventors themselves wrote, [15], [16], [18] and [21]. Other books at the same level are Gordon's [12], Pfeffer's [37] and Lee's [23]. The book by DePree and Swartz [8] does contain an introduction to Kurzweil–Henstock theory, but we in contrast cover more material and concentrate solely on integration. J. Mawhin's Introduction à l'Analyse [28] contains the Kurzweil–Henstock integral; obviously it is in French. The book by McLeod [30] is closest to us in its spirit but we use very different and more systematic notation, which we feel is important at the elementary level. We also consider some topics in greater detail, relate the KH-integral to other integrals and give a range of applications including Fourier series.

We hope that our book will be useful at various levels. The first section of Chapter 1 and Chapter 2, with perhaps some omissions, can serve as a first (serious) course on integration. Later sections of Chapter 1 contain a fairly complete account of the Riemann integral but require more mathematical maturity and are **not** intended for a beginner or a non-mathematician. To indicate that these sections are not meant for the first reading they are typeset in a smaller font. We have expounded the Riemann theory to provide easily available comparison for someone who desires it. For instance, the non-integrable derivative of Example 1.4.5 gives an opportunity to appreciate the Fundamental Theorem 2.6.2 but it is far more difficult than the proof of the Fundamental Theorem itself. Chapters 3 and 6 together with some topics from Chapter 7 can form the basis of a course which could be given instead of a first course in Lebesgue theory. Chapters 4 and 5 are not elementary; they give the most general convergence theorems for the Kurzweil–Henstock integral. Exercises are provided at the end of each Chapter. Exercises containing additional information which is worth reading even if one does not intend to work them out in detail are marked by ⓒ; exercises which are not easy are marked by ⓘ .

Finally we wish to acknowledge help when writing this book. We thank the editor of this series, John Loxton, for his friendly attitude and invaluable advice. We are grateful to David Tranah and particularly to Roger Astley from CUP for the care and expertise with which they have published our work. In writing we had advice on computer typesetting and presentation from our friends and colleagues. We specifically mention Anthony Miller from CSIRO in Adelaide, Ding Chuan Song from the Northwest Normal University in China, Chew Tuan Seng from the National University of Singapore, and Peter Adams, Keith Matthews and Ken Smith from the University of Queensland. Peter Adams also produced all figures in this book.

January 1999   Lee Peng Yee
                                Rudolf Výborný

List of Symbols

Symbol	Description	Page
$\mathcal{F} \int_a^b f$	Riesz integral	139
$\mathcal{M} \int_A^B f$	McShane integral	29
$\mathcal{SL} \int_A^B f$	SL-integral	154
$(A) \sum_1^\infty a_n$	Abel sum	286
$\sum_\pi f$	Riemann sum	4
$\sum_\pi f(\xi)(v-u)$	Riemann sum	4
$\sum_\pi f(y)\lvert I \rvert$	Riemann sum	4
$\sum_\pi {}_a^c f$ or $\sum {}_a^c f$	Riemann sum	37
$S(D)$	Darboux upper sum	6
$s(D)$	Darboux lower sum	6
$\pi \ll \delta$	π is δ-fine	23
$\Re z$	real part of the complex number z	69
$\Im z$	imaginary part of the complex number z	69
$\mathrm{Var}_a^b F$	variation of F on $[a, b]$	84
f^N	truncated function	92
$m(S)$	measure of S	120
$m_n(S)$	measure of S in $\mathbb{R}^n$	228
$A \triangle B$	symmetric difference	122
$\mathcal{E}$	a function with small Riemann sums	134, 223
AC^*	see Definition 5.3.1	179
AC	see Definition 5.3.4	182
VB^*	see Definition 5.3.7	183
ACG^*	see Definition 5.4.1	187
ACG	see Definition 5.4.3	188
VBG^*	see Definition 5.4.4	188
$\lvert x \rvert_u$	maximum norm of x in $\overline{\mathbb{R}}^n$	204
$\lvert x \rvert_2$	Euclidean norm of x in $\overline{\mathbb{R}}^n$	204
$C(a, h)$	cube centred at a	204

1
Introduction

1.1 Historical remarks

The history of the integral is both long and interesting. A monograph could easily be devoted to it. Here we make only a few remarks in order to set our topic into a proper historical perspective and refer the interested reader to several excellent books; see Hawkins [13], Medvedev [33], Pesin [36] and van Dalen and Monna [7] for instance. The roots of integration can be traced to Archimedes but the real story of integration starts with Newton and Leibniz. Even today, if $F : [a, b] \mapsto \mathbb{R}$ and $F'(x) = f(x)$ for every $x \in [a, b]$ we say that $F(b) - F(a)$ is the definite *Newton's integral* of f from a to b, in symbols

$$F(b) - F(a) = \mathcal{N} \int_a^b f$$

or briefly

$$F(b) - F(a) = \int_a^b f.$$

We also refer to the function F as the Newton indefinite integral of f. The Newton definition today looks much more solid than the Leibniz definition of an integral as a sum of infinitely many infinitesimal quantities. This is because the concept of derivative is firmly entrenched in our mind as a solidly defined mathematical entity. In Newton's time, however, the concepts of limit and derivative were somewhat nebulous. Despite the logical shortcomings of the beginning of calculus the early masters of calculus, e.g. the Bernoulli brothers and Euler, were able to make wonderful discoveries with the new-found tool. Of all the various definitions that would survive a modern critical scrutiny, by far the simplest and most intuitive is that which was given at the beginning of

1

the modern era by Cauchy (1789–1857) and completed and fully investi-
gated by Riemann (1826–1866). In fact, it is the Riemann theory that is
still today taught at universities to physicists, engineers and others who
need to know integration. A brief account of some finer points of Riemann integration is given in Sections 1.3—1.5. This we do because the
main topic of this book is indebted to Riemann, and we wanted to give
the reader an opportunity to compare results in Riemann integration
with the theory expounded in this book. However, Sections 1.3—1.5 require some mathematical maturity and are not intended for a student's
first reading of the book and are typeset in a smaller font. Apart from
Section 1.2 containing notation, the rest of the book is independent of
Chapter 1.

Among non-specialists there is an almost universal identification of
the integral with the Riemann integral and this is surprising for two reasons. Firstly the Riemann integral, despite its wide use and its intuitive
appeal, has serious shortcomings, as we shall see later. Secondly over
eighty years ago Lebesgue (1875–1941) gave another definition of what
is now known as the Lebesgue integral. This integral turns out to be
the correct one for almost all uses and is the one used almost exclusively
by professional mathematicians. In 1914 O. Perron proposed yet another definition, which had an additional advantage over the Lebesgue
definition: it included the Newton integral and all improper integrals
as well. All indications are that the Lebesgue (or Perron) theory is not
popular with non-mathematicians, the reason most likely being the level
of mathematical sophistication required for understanding it. In 1957
Kurzweil [20], in connection with research in differential equations, gave
an elementary definition of the integral equivalent to the Perron one.
For Kurzweil's own presentation of the theory see [20]. Henstock later
[14] independently rediscovered Kurzweil's approach and advanced it
further [15, 16, 18, 17]. The great advantages of the Kurzweil–Henstock
theory are that it preserves the intuitive geometrical background of the
Riemann theory, it is so simple that it can be presented in introductory courses, and it has the power of the Lebesgue theory. A further
essential contribution was made by McShane. He recaptured Lebesgue
integration in the Kurzweil–Henstock framework and by doing so made
it accessible to non-specialists (see [31], [32]). In the second chapter we
strive for the most elementary presentation of the Kurzweil–Henstock
theory, suitable as an introductory course replacing the usual one on
Riemann integration. The third chapter could serve as a first course

on the theory of the integral. The rest of the book is devoted to more advanced topics and surrounding ideas.

1.2 Notation and the Riemann definition

The sets of integers, positive integers, rationals, reals and positive reals are denoted by $\mathbb{Z}$, $\mathbb{N}$, $\mathbb{Q}$, $\mathbb{R}$ and $\mathbb{R}_+$, respectively. The positive or negative part of a real number a will be denoted by a^+ or a^-, respectively; i.e. $a^+ = (|a| + a)/2$, $a^- = (|a| - a)/2$. Unless something is specified to the contrary the word function means a real valued function. For a real valued function f then the meaning of f^+ and f^- is clear. Generally speaking, operations with functions are understood pointwise, for instance $f + g : x \mapsto f(x) + g(x)$, $\mathrm{Max}(f, g) : x \mapsto \mathrm{Max}(f(x), g(x))$ etc. Similarly with relations, $f \leq g$ means $f(x) \leq g(x)$ for every x from the common domain of definition of f and g. Likewise the inequality $f \leq K$ means $f(x) \leq K$ on the domain of f. The inverse function† to f is denoted by f_{-1}. If S is a set then $\mathbf{1}_S$ will denote the characteristic function of S, i.e. $\mathbf{1}_S(x) = 1$ for $x \in S$ and $\mathbf{1}_S(x) = 0$ for $x \notin S$. The sequence $n \mapsto c_n; n \in \mathbb{N}$ will be abbreviated to $\{c_n\}$, with a similar convention for sequences of functions. We use the term increasing (decreasing) in the wider sense, i.e. an increasing function might take the same value twice; an increasing (decreasing) function which is one-to-one will be called strictly increasing (decreasing). We shall use the symbol $\sup \{f; M\}$ for the supremum‡ of a function f over a set M and employ a similar notation for the infimum§. Sometimes we might write a defining relation for the set M instead of M itself, for instance $\sup \{a_n; n \geq N\}$ denotes $\sup\{a_N, a_{N+1}, \dots\}$. An interval $[a, b]$ will always be closed and (a, b) open. We shall use $|I|$ for the length of a bounded interval¶ I. Important for our further development are the concepts of a *division of an interval*, and that of a *partition of an interval*. By a division D of a compact interval $[a, b]$ we mean a set of intervals $[x_i, x_{i+1}]$ such that

$$a = x_0 < x_1 < x_2 < \cdots < x_n = b. \tag{1.1}$$

The points x_i are called the points of the division D. A function

$$\varphi : [a, b] \to \mathbb{R}$$

† Many authors use the notation f^{-1}; we reject that since f^{-1} could also legitimately denote $1/f$.
‡ The supremum of a set is its least upper bound.
§ greatest lower bound
¶ We accept as intervals also the sets $[a, a]$, consisting of one point, and $(a, a) = \emptyset$ the empty set. For these so-called degenerate intervals I the length is zero, $|I| = 0$.

is called *a step function* if there is a division (1.1) such that φ is constant on every interval (x_{i-1}, x_i). A partition of a compact interval $[a, b]$ is a set of couples (ξ_k, I_k) such that the points $\xi_k \in I_k$, the closed intervals I_k are non-overlapping† and

$$\bigcup_1^n I_k = [a, b].\tag{1.2}$$

We shall call the point ξ_k the *tag* of I_k. Often it will be convenient to have the intervals, $I_k = [u_k, v_k]$, ordered; hence for a partition

$$\pi \equiv \{(\xi_k, [u_k, v_k]); \ k = 1, 2, \ldots, n\}\tag{1.3}$$

we have

$$a = u_1 \le \xi_1 \le v_1 = u_2 \le \xi_2 \le v_2 \le \cdots \le v_n = b\tag{1.4}$$

The letters π and Π (possibly with subscripts) will denote partitions. A partition

$$\{(\xi_i, [u_i, v_i]); \ i = 1, 2, \ldots, n\}$$

can be abbreviated to $\{(\xi_i, [u_i, v_i])\}$ or even to $\{(\xi, [u, v])\}$ if the range of subscripts i is clear from the context or is not particularly important. If $\delta > 0$ then a partition π for which

$$\xi_i - \delta < u_i \le \xi_i \le v_i < \xi_i + \delta\tag{1.5}$$

for all i with $1 \le i \le n$ is called a δ-fine partition of $[a, b]$. It is obvious that a partition π is δ-fine if and only if the length of the largest interval of π, which we denote by $n(\pi)$, is less than 2δ. Similarly as with $n(\pi)$ we denote by $n(D)$ the length of the largest interval of the division D.

Given a function $f : [a, b] \to \mathbb{R}$ then a partition (1.4) has an associated Riemann sum‡

$$\sum_\pi f = \sum_{i=1}^n f(\xi_i)(v_i - u_i),\tag{1.6}$$

which we shall also abbreviate as $\sum_\pi f(\xi)(v - u)$. If the partition σ is given by (y_k, J_k) with $k = 1, 2, \ldots, m$, then, for the Riemann sum, we shall naturally use the notation

$$\sum_\sigma f = \sum_{k=1}^m f(y_k)|J_k| = \sum_\sigma f(y)|J|.\tag{1.7}$$

† i.e. they do not have any interior points in common.
‡ See Figure 1.1.

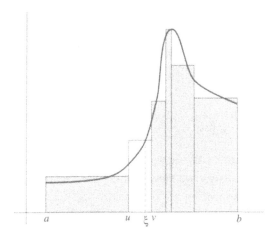

Fig. 1.1. Riemann sum

We extend our shorthand to similar sums, e.g. we shall denote by $\sum_\pi f(u, v)$ the sum $\sum_{i=1}^n [f(v_i) - f(u_i)]$.

The Riemann integral, $\int_a^b f$, is defined as the limit of Riemann sums. More precisely we define:

DEFINITION 1.2.1 *A number A is the Riemann integral of f from a to b (or on $[a, b]$) if for every positive ε there is a positive number δ such that for every δ-fine partition π*

$$\left| \sum_\pi f - A \right| < \varepsilon. \tag{1.8}$$

We denote the Riemann integral A as usual by $\int_a^b f$ or by $\int_a^b f(x)dx$. If we wish to distinguish the integral from another integral, e.g. the Newton integral, or if we wish to emphasize that the integral is to be understood in the sense of Definition 1.2.1, then we write $\mathcal{R} \int_a^b f$. We shall often abbreviate Riemann integral to R-integral and if there is no danger of confusion just to integral. The function f is called Riemann integrable, or briefly R-integrable, if the Riemann integral of f exists.

It is a consequence of Definition 1.2.1 that an R-integrable function f must be bounded. We choose $\varepsilon = 1$ and find a corresponding δ from Definition 1.2.1. We divide the interval into n equal intervals $[u_i, v_i]$ with $n^{-1}(b - a) < \delta$ and choose a number $C > \text{Max}(|f(u_j)|, j = 1, 2, \ldots, n)$. Let $x \in [a, b]$ be arbitrary; it lies in some $[u_\alpha, v_\alpha]$ and inequality (1.8) for the partition (1.3) with $\xi_j = u_j$ $(j \neq \alpha)$ and $\xi_\alpha = x$ yields

$$|f(x)(v_\alpha - u_\alpha)| < |A| + 1 + \sum_{i=1, i \neq \alpha}^{n} |f(u_i)|(v_i - u_i),$$

and consequently

$$|f(x)| < \frac{n[|A| + 1 + C(b-a)]}{b-a}.$$

Since x was arbitrary this shows f is bounded. Unless something is specified to the contrary we shall assume for the rest of this chapter that all functions appearing are bounded.†

1.3 Basic theorems, upper and lower integrals

For a division D as given in (1.1) we introduce the Darboux upper and lower sums $S(D)$ and $s(D)$ defined by

$$S(D) \;=\; \sum_{i=0}^{n-1} M_i(x_{i+1} - x_i), \qquad\qquad (1.9)$$

$$s(D) \;=\; \sum_{i=0}^{n-1} m_i(x_{i+1} - x_i), \qquad\qquad (1.10)$$

where $M_i = \sup\{f, [x_i, x_{i+1}]\}$ and $m_i = \inf\{f, [x_i, x_{i+1}]\}$. Obviously there is a division D_π naturally associated with a partition π; in the spirit of our shorthand writing we shall denote the the upper Darboux sum by $S(\pi)$ or by $S(D_\pi)$, in symbols

$$S(\pi) = S(D_\pi) = \sum_{\pi} M(v - u) = \sum_{i=1}^{n} M_i(v_i - u_i),$$

where now M_i stands for the supremum of f on $[u_i, v_i]$. Of course, we use the same convention for lower Darboux sums and other similar structured sums. A division $\tilde{D}$ is a *refinement* of D if all the points of D are also points of $\tilde{D}$. Adding points to a division increases the lower sums and decreases the upper sums,

$$S(D) \geq S(\tilde{D}) \text{ and } s(D) \leq s(\tilde{D}). \qquad\qquad (1.11)$$

We prove only the first of these inequalities and it is clearly sufficient to prove it if $\tilde{D}$ has only one additional point c. (In the general case we can move from D to $\tilde{D}$ by adding points one by one.) If $c \in (x_i, x_{i+1})$ then the contribution of the intervals $[x_i, c]$ and $[c, x_{i+1}]$ to $S(\tilde{D})$ is

$$\sup\{f, [x_i, c]\}(c - x_i) + \sup\{f, [c, x_{i+1}]\}(x_{i+1} - c) \leq M_i(x_{i+1} - x_i),$$

† The reader not interested in the Riemann theory can now start reading Chapter 2.

and the inequality $S(D) \geq S(\bar{D})$ follows. By comparing an upper sum $S(D_1)$ and a lower sum $s(D_2)$ with their common refinement we obtain

$$S(D_1) \geq s(D_2). \tag{1.12}$$

It is convenient to define the upper and lower Riemann integrals of f on $[a,b]$, in symbols $\overline{\int_a^b} f$ and $\underline{\int_a^b} f$, by

$$\overline{\int_a^b} f = \inf\{S(D) : D \text{ a division of } [a,b]\}, \tag{1.13}$$

$$\underline{\int_a^b} f = \sup\{s(D) : D \text{ a division of } [a,b]\}. \tag{1.14}$$

In view of relation (1.12) the upper and lower integrals always exist† and

$$\overline{\int_a^b} f \geq \underline{\int_a^b} f.$$

It follows easily that

$$\overline{\int_a^b} (-f) = -\underline{\int_a^b} f.$$

The connections between the upper, lower and Riemann integral are stated in the following theorem.

THEOREM 1.3.1 *The following statements are equivalent:*

(i) *The function f is Riemann integrable on $[a, b]$.*

(ii) *For every positive ε there is a positive δ such that*

$$S(D) - s(D) < \varepsilon, \tag{1.15}$$

whenever $n(D) < \delta$.

(iii) *For every positive ε there is a division D such that inequality (1.15) holds.*

(iv) *The upper and lower integrals of f are equal.*

Proof The implications (ii) $\Rightarrow$ (iii) $\Rightarrow$ (iv) are fairly obvious. Since every Riemann sum $\sum_\pi f$ lies between the corresponding Darboux sums $S(D_\pi)$ and $s(D_\pi)$ it is clear that

$$\sum_\pi f - \overline{\int_a^b} f > s(D_\pi) - S(D_\pi),$$

$$\overline{\int_a^b} f - \sum_\pi f < S(D_\pi) - s(D_\pi).$$

† According to the convention adopted f is bounded.

This shows (ii) $\Rightarrow$ (i) with $A = \overline{\int_a^b} f$. To prove (i)$\Rightarrow$ (ii) we find a positive δ such that for every partition π with $n(\pi) < \delta$

$$\left| \sum_\pi f(\xi)(v - u) - A \right| < \frac{\varepsilon}{4}.$$

On each interval $[u, v]$ we choose ξ so that

$$m + \frac{\varepsilon}{4(b - a)} > f(\xi).$$

It follows that $s(D_\pi) > A - \varepsilon/2$. Similarly $S(D_\pi) < A + \varepsilon/2$ and inequality (1.15) follows. To complete the proof it suffices to show (iv) $\Rightarrow$ (ii). This implication becomes obvious with the following lemma.   •

LEMMA 1.3.2 *For every positive ε there exists a positive δ such that*

$$S(D) - \overline{\int_a^b} f \quad < \varepsilon, \tag{1.16}$$

$$\underline{\int_a^b} f - s(D) \quad < \varepsilon, \tag{1.17}$$

whenever

$$n(D) < \delta. \tag{1.18}$$

Proof Only (1.16) needs proof since (1.17) follows from (1.16) applied to $-f$. For the proof of (1.16) we denote by A the value of the upper integral and find a division D_ε such that

$$S(D_\varepsilon) < A + \frac{\varepsilon}{2}. \tag{1.19}$$

Let N be the number of dividing points of D_ε and $|f| \leq K$. We show that the number $\varepsilon/4NK$ serves as the required δ. Let D be any division satisfying condition (1.18) and $\bar{D}$ a common refinement of D_ε and D. By inequalities (1.11) and (1.19)

$$S(\bar{D}) < A + \frac{\varepsilon}{2}. \tag{1.20}$$

We now estimate $S(D) - S(\bar{D})$. Every interval $[u, v]$ which is common to D and $\bar{D}$ makes the same contribution to both $S(D)$ and $S(\bar{D})$. An interval $[y, z]$ of D which is not an interval of $\bar{D}$ contains at least one point of D_ε. Hence there are at most N intervals $[y, z]$. The part of difference $S(D) - S(\bar{D})$ restricted to $[y, z]$ is at most $[M - (-M)](y - z)$ and that does not exceed $2M\delta$. Consequently $S(D) - S(\bar{D}) \leq N.2M\delta = \varepsilon/2$. This together with (1.20) establishes (1.16).   •

Every upper Darboux sum $S(D)$ defines a step function φ_D such that $S(D) = \int_a^b \varphi_D$ and $\varphi_D \geq f$. This step function can be modified into a continuous piecewise linear function H with a trapezoidal graph such that $\varphi_D \leq H$ and $\int_a^b H - S(D) < \varepsilon$. See Figure 1.2. The next lemma easily follows.

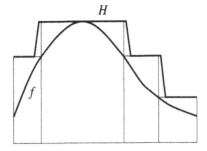

Fig. 1.2. Trapezoidal approximation from above

LEMMA 1.3.3 *For every positive ε there exist continuous functions h, H such that h ≤ f ≤ H and*

$$\int_{\underline{a}}^b f \quad < \quad \int_a^b h + \varepsilon, \tag{1.21}$$

$$\overline{\int_a^b} f \quad > \quad \int_a^b H - \varepsilon. \tag{1.22}$$

The basic properties of the Riemann integral follow easily from Definition 1.2.1 and Theorem 1.3.1. For ease of reference we state them here.

Homogeneity If $c > 0$ then

$$\overline{\int_a^b} cf = c\overline{\int_a^b} f. \tag{1.23}$$

The same equation holds for the lower integral and if f is integrable then it is valid for any c and for the R-integral.

Preservation of inequalities If $f \le g$ then

$$\overline{\int_a^b} f \le \overline{\int_a^b} g. \tag{1.24}$$

In particular, if $m \le f(x) \le M$ for all $x \in [a,b]$ then

$$m(b-a) \le \overline{\int_a^b} f \le M(b-a). \tag{1.25}$$

Consequently, if $|f| \le K$ then

$$\left| \overline{\int_a^b} f \right| \le K(b-a). \tag{1.26}$$

All these inequalities hold with the upper integral replaced by the lower integral and for an R-integrable f with $\int_a^b$ substituted for $\overline{\int_a^b}$.

Absolute value For any bounded f

$$\left| \int_{\underline{a}}^{b} f \right| \leq \int_{\underline{a}}^{b} |f|, \tag{1.27}$$

with a similar inequality holding for the upper integral. If f is integrable then so is $|f|$ and the above inequality holds with the lower integral replaced by the integral.

Integral as an additive function of intervals For any bounded f we have

$$\int_{a}^{\overline{b}} f = \int_{a}^{\overline{c}} f + \int_{c}^{\overline{b}} f, \tag{1.28}$$

with the same relation holding for the lower integrals. It follows that a function f is integrable on $[a, b]$ if and only if it is integrable on $[a, c]$ and $[c, b]$ with $a < c < b$ and then equation (1.28) holds with $\int_{a}^{\overline{b}}$ replaced by $\int_{a}^{b}$.

Additivity The upper integral is subadditive, the lower integral superadditive which means

$$\int_{a}^{\overline{b}} f + \int_{a}^{\overline{b}} g \ \geq \ \int_{a}^{\overline{b}} (f + g), \tag{1.29}$$

$$\int_{\underline{a}}^{b} f + \int_{\underline{a}}^{b} g \ \leq \ \int_{\underline{a}}^{b} (f + g). \tag{1.30}$$

It follows that if f and g are integrable then so is $f + g$ and

$$\int_{a}^{b} (f + g) = \int_{a}^{b} f + \int_{a}^{b} g. \tag{1.31}$$

1.4 Differentiability, continuity and integrability

We are assuming that f is bounded; let $|f| \leq K$. The functions

$$F(x) \ = \ \int_{a}^{x} f, \tag{1.32}$$

$$U(x) \ = \ \int_{a}^{\overline{x}} f,$$

$$L(x) \ = \ \int_{\underline{a}}^{x} f, \tag{1.33}$$

which we shall call the indefinite integral, the indefinite upper integral and the indefinite lower integral, will play an important rôle in this section. By inequality (1.26) and similar inequalities for the integral and for the lower integral, and by using equation (1.28) for the additivity of the (upper, lower) integral, we see that F, U, L are *Lipschitz continuous*† *with the constant K;*

---

† A function F is said to be Lipschitz continuous on the interval I with the constant L if $|F(x) - F(y)| \leq L|x - y|$ for x, y in I. Instead of Lipschitz continuous one often says just Lipschitz or merely L.

i.e. for $a \le x < y \le b$ we have

$$\left| \int_x^y f \right| = |F(y) - F(x)| \le K(y - x),$$

$$\left| \int_x^{\bar{}^y} f \right| = |U(y) - U(x)| \le K(y - x), \qquad (1.34)$$

$$\left| \int_{\underline{}x}^y f \right| = |L(y) - L(x)| \le K(y - x).$$

If f is continuous at x then it follows from the above inequalities that F, U, L are differentiable at x and $F'(x) = U'(x) = L'(x) = f(x)$. Indeed, combining the inequalities (1.27), (1.34) and $|f(x) - f(t)| < \varepsilon$ we obtain

$$\left| \int_x^{\bar{}^y} f(t)dt - f(x)(y - x) \right| = \left| \int_x^{\bar{}^y} [f(t) - f(x)]dt \right|$$

$$\int_x^{\bar{}^y} |f(t) - f(x)|dt \le \varepsilon(y - x).$$

This shows that the right-hand derivative $U'_+(x) = f(x)$. The proofs for the left-hand derivative and the function L are entirely similar. We have therefore established: For a function f continuous on all of $[a, b]$ we have $U'(x) = L'(x)$. Since also $\lim_{x \to a} U(x) = \lim_{x \to a} L(x) = 0$, it follows that $U = L$. Consequently, f is integrable and $F' = U' = f$. We have proved

THEOREM 1.4.1 *If f is continuous on $[a, b]$ then f is Riemann integrable there and*

$$F'(x) = f(x)$$

for all $x \in [a, b]$.†

A Riemann integrable function can clearly be discontinuous at some points. The nineteenth-century mathematicians were impressed by Riemann's example of an integrable function which was discontinuous on a dense set. An example of such a function can be found in Exercise 1.10. The next theorem shows that a function is Riemann integrable if and only if the set of its discontinuities is in some sense 'small'.

DEFINITION 1.4.2 *A set S is said to be of* **measure zero** *if for every positive ε there exists a countable system of open intervals $\{I_k : k = 1, 2, \dots\}$ such that*

$$S \subset \bigcup_{k=1}^{\infty} I_k \qquad (1.35)$$

and

$$\sum_{k=1}^{\infty} |I_k| < \varepsilon. \qquad (1.36)$$

† At a or b the derivative is one-sided.

A countable union of sets of measure zero is itself of measure zero. In particular every countable set is of measure zero. However, a set of measure zero need not be countable, see Appendix Section A.1. If something happens except on a set of measure zero, we say that it happens *almost everywhere*, or for almost all points. For instance, if f is continuous at points of $[a, b]$ not belonging to a set of measure zero then we would say that f is continuous almost everywhere on $[a, b]$. Now we can state

THEOREM 1.4.3 (Riemann integrability)   *A bounded function f is Riemann integrable on $[a, b]$ if and only if it is continuous almost everywhere on $[a, b]$.*

Proof We start with the only if part, which is easier to prove. For natural n we denote by E_n the set of points c such that

$$\limsup_{x \to c} f(x) - \liminf_{x \to c} f(x) \geq \frac{1}{n}.$$

Since the set of discontinuities of f is the union of the sets E_n, it suffices to show that E_n is of measure zero for every natural n. By (iii) of the theorem on alternative definitions of integrability (Theorem 1.3.1) there is a division D of $[a, b]$ such that

$$S(D) - s(D) < \frac{\varepsilon}{4n}. \tag{1.37}$$

If $[u, v]$ is an interval of D which contains a point of E_n in its *interior* then the contribution of $[u, v]$ to $S(D) - s(D)$ must exceed $(v - u)/2n$. Consequently, if l is the total length† of all intervals of D containing a point of E_n in their interiors then $l < \varepsilon/2$ by inequality (1.37). If we now cover the dividing points of D by a finite system of open intervals of total length less than $\varepsilon/2$ then this system together with all intervals $[u, v]$ covers E_n. Obviously the total length of this cover of E_n by finitely many intervals is less than ε.

Using the Heine–Borel‡ covering theorem we now give a very simple proof of sufficiency (due to Gee [10]). Given $\varepsilon > 0$ we find a countable system of open intervals J_n, $n = 1, 2, \ldots$ covering the set of discontinuities of f and such that $\bigcup_1^\infty |J_n| < \varepsilon$. For every point of continuity of f there is an open interval K_x containing x such§ that

$$\sup \{f; \bar{K}_x\} - \inf \{f; \bar{K}_x\} < \varepsilon.$$

The intervals J_n and K_x cover $[a, b]$. By the Borel theorem there exists a finite subcover of $[a, b]$. The endpoints of the intervals of the subcover, as long as they lie inside $[a, b]$, together with the points a,b define a division D of $[a, b]$. Each open subinterval of D is part of some J_n or some K_x. Hence

$$S(D) - s(D) < (\sup \{f; [a, b]\} - \inf \{f; [a, b]\}) \sum_1^\infty |J_n| + \varepsilon(b - a).$$

---

† By this we mean the sum of lengths of all individual intervals.
‡ Often referred to simply as the Borel theorem. See Exercise 2.2 and Appendix Section A.3. For a general version of the Borel theorem see [19] Chapter 5, especially Theorem 14.
§ $\bar{K}_x$ denotes K_x together with its endpoints.

Consequently f is Riemann integrable by (iii) of Theorem 1.3.1.   •
The fact that $F'(x) = f(x)$ for every x at which f is continuous combined
with the characterization of integrability Theorem 1.4.3 leads immediately to

THEOREM 1.4.4 *If f is integrable and F is defined by equation* (1.32)
then $F'(x) = f(x)$ almost everywhere.

Integrating f and then differentiating leads back to f (at least almost everywhere). It is a weakness of the Riemann integral that it is generally not
possible to recapture F from F' by R-integration. It was shown by Volterra
as far back as 1881 that F' need not be R-integrable even if F' is bounded and
exists everywhere on $[a, b]$, see [45]; his example is discussed in [7] on page
107 and in [13] page 56. We give our own version of an example by Goffman
[11].

EXAMPLE 1.4.5 (Not Riemann integrable derivative)   *There exists
a function L which has a derivative $L'(x) = f(x)$ for all $x \in [a, b]$ such that
the function f is bounded but not Riemann integrable on $[a, b]$.*

Proof Let $\{I_n : n \in \mathbb{N}\}$ be a system of open disjoint intervals such that
$I_n \subset [0, 1]$, the set $G = \bigcup_1^\infty I_n$ is dense, its complement Z is without isolated
points and†

$$\sum_1^\infty |I_n| = \frac{1}{2}. \tag{1.38}$$

Let J_n be a closed interval in the centre of I_n with $|J_n| = |I_n|^2$. We need the
following property of Z: If K, $K \subset [0, 1]$, is an interval containing a point of
Z then

$$\sum_{J_n \cap K \neq \emptyset} |J_n \cap K| \le 16|K|^2. \tag{1.39}$$

Indeed, first for $K \cap I_n \neq \emptyset$ we have

$$|K \cap I_n| \ge \frac{1}{2}[|I_n| - |J_n|] \ge \frac{1}{2}[|I_n| - \frac{1}{2}|I_n|] \ge \frac{1}{4}|I_n|$$

and then

$$|K \cap J_n| \le |J_n| = |I_n|^2 \le 16|K \cap I_n|^2.$$

Using

$$\sum_{J_n \cap K \neq \emptyset} |K \cap I_n|^2 \le \left(\sum_{J_n \cap K \neq \emptyset} |K \cap I_n| \right)^2 \le |K|^2$$

---

† Such a system can be constructed as follows: Take a series of positive terms with
$\sum a_n = 1/2$. Let I_1 be an open interval of length a_1 situated in the middle of
$[0, 1]$, then let I_2, I_3 be open intervals located in the middle of each remaining
interval and of total length a_2 and so on. The set Z is a *Cantor type set;* the
Cantor set is discussed in Section A.1 and is a set of measure zero which is not
countable.

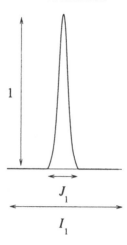

Fig. 1.3. Graph of f above I_1

gives inequality (1.39). First we define f and later we show that it is a derivative. Let f be continuous on I_n, 1 at the centre of I_n, 0 on $Z \cup \bigcup_1^\infty (I_n \setminus J_n)$ and always between 1 and 0. Since Z has no isolated points, any interval (α, β), with $x_1 \in Z \cap (\alpha, \beta)$, contains another point of Z, say x_2. The open interval with endpoints x_1, x_2 then contains I_n for some n and therefore a point where f has the value 1. This demonstrates the discontinuity of f on Z. It is easy to see from equation (1.38) (e.g. by use of the Borel covering theorem) that Z is not of measure zero, and consequently f is not R-integrable. Now we show that f is a derivative, i.e. there is a function L such that

$$L'(x) = f(x) \tag{1.40}$$

for every $x \in [0, 1]$. As our notation suggests, L is defined by equation (1.33). For $x \in G$ equation (1.40) follows from the continuity of f. Assume that $[u, v]$ meets Z and let $D \equiv \{[x_i, x_{i+1}]\}$ be a division of $[u, v]$. Let m_i be the infimum of f on $[x_i, x_{i+1}]$. If $m_i > 0$ then $[x_i, x_{i+1}]$ does not contain a point of Z or $\bigcup_1^\infty I_n \setminus J_n$. Therefore $[x_i, x_{i+1}] \subset J_n$ for some n. It follows from (1.39) that

$$s(D) \le \sum_{J_n \cap [u, v] \neq \emptyset} |J_n \cap [u, v]| \le 16(v - u)^2. \tag{1.41}$$

Consequently $0 \le L(v) - L(u) \le 16(v - u)^2$ and therefore $L'(x) = 0 = f(x)$ for $x \in Z$.   •

However, the problem of integrating the derivative within Riemann theory has a partially positive answer.

THEOREM 1.4.6 *If F is Lipschitz and has a Riemann integrable derivative on $[a, b]$ then*

$$F(x) = F(a) + \int_a^x F' \tag{1.42}$$

for $a < x \leq b$.

REMARK 1.4.7 If F' exists everywhere on $[a, b]$ then the assumption that F is Lipschitz is superfluous. The proof is easy and can be found in many calculus books (e.g. [40]). We prove equation (1.42) provided F' exists and is continuous almost everywhere on $[a, b]$. In this situation the assumption that F is Lipschitz is essential.

Proof of Theorem 1.4.6 The integral $\int_a^x F'$ exists; let us denote it by $H(x)$. The function H is Lipschitz and by Theorem 1.4.4 $H' = F'$ almost everywhere. The function $z : x \mapsto H(x) - F(x) - F(a)$ is also Lipschitz and $z' = 0$ almost everywhere. The proof will be complete if we establish the following lemma.•

LEMMA 1.4.8 *If z is a Lipschitz function on $[a, b]$ and $z' \leq 0$ almost everywhere then $z(b) \leq z(a)$.*

Proof Let $K > 0$ be a Lipschitz constant for z. Given $\varepsilon > 0$, there exists a countable system of open intervals J_n, $n = 1, 2, \ldots$ covering the set where z' either does not exist or is positive and such that

$$\sum_1^\infty |J_n| < \frac{\varepsilon}{2K}.$$

Let E be the set of all $x \in [a, b]$ such that

$$z(x) - z(a) \leq \frac{\varepsilon(x-a)}{2(b-a)} + \sum_1^\infty K\,|J_n \cap (a, x)| \tag{1.43}$$

and $S = \sup E$. It suffices to show that $S = b$, whence it would follow that $z(b) - z(a) \leq \varepsilon$. Assume contrary to what we want to prove that $S < b$. If $z'(S) \leq 0$ then there are $x \in E$ with† $x \leq S$ and v with $b \geq v > S$ such that

$$z(v) - z(x) \leq \frac{\varepsilon(v-x)}{2(b-a)}.$$

Simple calculation now shows that inequality (1.43) holds with x replaced by v. This is impossible and consequently $S \in J_k$ for some k. By the definition of S there is number $x \in E \cap J_k$ with† $x \leq S$ and for $v > S$ we have

$$z(v) - z(x) \leq K(v - x).$$

This again leads to inequality (1.43) with x replaced by v. This final contradiction completes the proof. •

REMARK 1.4.9 An alternative and simple proof of Lemma 1.4.8 is outlined in Exercise 2.1.

† The equality sign is to cover the possibility of $S = a$.

1.5 Limit and R-integration

In applications as well as in many problems in pure mathematics one needs to interchange limit and integration. However, theorems in the Riemann theory guaranteeing the formula

$$\lim_{n\to\infty} \int_a^b f_n = \int_a^b \lim_{n\to\infty} f_n \qquad (1.44)$$

are either not strong enough, like the next Theorem 1.5.1, or difficult to prove, like the Arzelà Theorem 1.5.6 below†. The most basic is

THEOREM 1.5.1 *If f_n are integrable and converge uniformly on $[a,\,b]$ then the limit function f is also integrable and equation (1.44) holds.*

Proof First we use Theorem 1.3.1 on the characterization of integrability to show that the limit function is integrable. Given $\varepsilon > 0$ there exists a natural N such that

$$|f_N - f| < \frac{\varepsilon}{2(b-a)}.$$

Let $S_N(D)$ or $s_N(D)$ be the upper or lower sum of f_N, respectively. By part (iii) of Theorem 1.3.1 on the characterization of integrability

$$S_N(D) - s_N(D) < \frac{\varepsilon}{2}.$$

Consequently‡

$$S(D) - s(D) < \frac{\varepsilon}{2} + \frac{\varepsilon}{2(b-a)}(b-a),$$

and this in turn implies the R-integrability of f. Equation (1.44) follows from the inequality

$$\left| \int_a^b f_n - \int_a^b f \right| \le (b-a)\sup\{|f_n - f|;\ [a,\,b]\}. \qquad \bullet$$

REMARK 1.5.2 The assertion of the integrability of f is an important part of Theorem 1.5.1. The limit of R-integrable functions need not be R-integrable, as the next example shows.

EXAMPLE 1.5.3 Let I_n be as in Example 1.4.5, h_n be a continuous function, always between 0 and 1, equal 1 at the centre of I_n and 0 outside I_n. Let $f_n = \sum_1^n h_k$. Obviously each f_n is continuous throughout $[0,\,1]$ and $f = \lim f_n$ is not Riemann integrable. Perhaps it is worth noting that§ no change of f on a set of measure zero could make it R-integrable.

---

† Another theorem on interchange of limit and integration is in Exercise 1.22.
‡ Naturally S and s are defined as in (1.9) and (1.10).
§ Since f is discontinuous on the complement of $\bigcup I_n$.

For the next lemma we need Dini's theorem† which asserts that a monotonic sequence of continuous functions with a continuous limit must converge uniformly.

LEMMA 1.5.4 *If f_1 is bounded, the sequence $\{f_n\}$ is decreasing and $f_n(x) \to 0$ for every $x \in [a, b]$ then*

$$\lim_{n \to \infty} \int_{\underline{a}}^{b} f_n = 0. \tag{1.45}$$

Proof By Lemma 1.3.3 for every $\varepsilon > 0$ there is a continuous function g_n such that $g_n \leq f_n$ and

$$\int_{\underline{a}}^{b} f_n < \int_{a}^{b} g_n + \frac{\varepsilon}{2^{n+2}}. \tag{1.46}$$

Since the above inequalities are not disturbed if g_n is replaced by $\mathrm{Max}(0, g_n)$ we can and shall assume that $g_n \geq 0$. We now set $h_1 = g_1$ and $h_n = \mathrm{Min}(g_n, h_{n-1})$ for $n \in \mathbb{N}$ and $n > 1$. The inequality

$$g_n \leq h_n + f_{n-1} - h_{n-1} \tag{1.47}$$

is easily checked, separately for the case when $h_n(x) = g_n(x)$ and the case when $h_n(x) = h_{n-1}(x)$. Using inequality (1.47) we prove by induction

$$\int_{\underline{a}}^{b} f_n \leq \int_{a}^{b} h_n + \sum_{k=1}^{n} \frac{\varepsilon}{2^{k+2}}. \tag{1.48}$$

This is clearly true for $n = 1$ and it follows from inequality (1.47) and the induction hypothesis that

$$\int_{a}^{b} g_n \leq \int_{a}^{b} h_n + \int_{\underline{a}}^{b} f_{n-1} - \int_{a}^{b} h_{n-1} < \int_{a}^{b} h_n + \sum_{1}^{n-1} \frac{\varepsilon}{2^{k+2}}. \tag{1.49}$$

Consequently

$$\int_{\underline{a}}^{b} f_n < \int_{a}^{b} g_n + \frac{\varepsilon}{2^{n+2}} \leq \int_{a}^{b} h_n + \sum_{1}^{n} \frac{\varepsilon}{2^{k+2}}. \tag{1.50}$$

By Dini's theorem $h_n \to 0$ uniformly and therefore by Theorem 1.5.1 we can find N such that for $n > N$ we have

$$\int_{a}^{b} h_n < \frac{\varepsilon}{2}.$$

Using this and inequality (1.48) we obtain

$$\int_{\underline{a}}^{b} f_n < \frac{\varepsilon}{2} + \sum_{1}^{n} \frac{\varepsilon}{2^{k+2}} < \varepsilon. \qquad \bullet$$

† The proof of which can be found in Section A.2.1.

This lemma leads to a theorem which allows the interchange of limit and integration even if the limit function is not integrable.

THEOREM 1.5.5 *If $\{f_n\}$ is an increasing sequence of uniformly bounded integrable functions with $f = \lim\limits_{n\to\infty} f_n$ then*

$$\lim_{n\to\infty} \int_a^b f_n = \int_{\underline{a}}^b f.$$

Proof Applying Lemma 1.5.4 to $f - f_n$, using inequality (1.30) with $g = -f_n$ and integrability of f_n gives

$$\int_{\underline{a}}^b f - \int_a^b f_n \le \int_{\underline{a}}^b (f - f_n) \to 0. \qquad \bullet$$

We close this section with Arzelà's dominated convergence theorem. A lot of effort has been devoted to finding an elementary proof of it; for the history and discussion see [27]. A more recent elementary proof is in [26]. For a short proof see Bullen and Výborný [5].

THEOREM 1.5.6 (Arzelà) *If a sequence of integrable functions $\{f_n\}$*

- *is uniformly bounded and*
- *converges on $[a, b]$ to a Riemann integrable function f*

then equation (1.44) holds.

Proof Since $\left| \int_a^b f - \int_a^b f_n \right| \le \int_a^b |f - f_n|$ there is no loss of generality in assuming $f = 0$ and $f_n \ge 0$. For each $n \in \mathbb{N}$ let $M_n = \sup\{f_n, f_{n+1}, \dots\}$. Then, for every $x \in [a, b]$, the sequence $\{M_n(x)\}$ is decreasing to zero, and consequently by Lemma 1.5.4

$$0 \le \int_a^b f_n \le \int_{\underline{a}}^b M_n \to 0. \qquad \bullet$$

1.6 Exercises

EXERCISE 1.1 *Recall that a function f is Newton integrable on $[a, b]$ if there is a differentiable function F such that $F'(x) = f(x)$ for all $x \in [a, b]$. Give examples of functions which are Newton integrable and which are not. [Hint: $f =$ a polynomial; $f(x) = x/|x|$.]*

EXERCISE 1.2 *Give an example of a function which is Riemann integrable but not Newton integrable. Give an example of a function which is Newton integrable but not Riemann integrable. Is there a discontinuous function which is both Riemann and Newton integrable? [Hint: $f(x) = \sin(1/x)$, $f(x) = 1/\sqrt{|x|}$, $f(x) = F'(x)$ where $F(x) = x^2 \sin(1/x)$.]*

EXERCISE 1.3 *Using the theorem on uniform continuity show (independently of Theorem 1.4.1) that a function continuous on $[a, b]$ is Riemann integrable there.*

EXERCISE 1.4 *Prove that a function continuous on $[a, b]$ is Newton integrable there. [Hint: Do not work hard, use Theorem 1.4.1.]*

EXERCISE 1.5 *Prove inequalities (1.24) to (1.30).*

EXERCISE 1.6 *Prove: If f and g are Riemann integrable on $[a, b]$ then so are $|f|$, fg, $\sqrt{f^2 + g^2}$, $\mathrm{Max}(f, g)$ and $\mathrm{Min}(f, g)$. [Hint: Use Theorem 1.4.3].*

EXERCISE 1.7 *Consider a function $c : \mathbb{R}^2 \mapsto \mathbb{R}$. Find a condition on c such that $c(f, g)$ is Riemann integrable when both f and g are. [Hint: Continuity, Theorem 1.4.3, use the Weierstrass Theorem to prove boundedness.]*

EXERCISE 1.8 *Prove: If f is Riemann integrable on $[a, b]$ and the function $1/f$ is well defined and bounded on $[a, b]$ then $1/f$ is Riemann integrable on $[a, b]$. [Hint: Use Theorem 1.4.3.]*

EXERCISE 1.9 ① *Do the previous three exercises without using Theorem 1.4.3.*

EXERCISE 1.10 ① *Let $f(x) = 0$ for irrational x and $f(x) = 1/q$ for a rational $x = p/q$ in the lowest terms. Prove that f is continuous at every irrational point and discontinuous at every rational point. Deduce that f is Riemann integrable on $[0, 1]$.*

EXERCISE 1.11 ① *Continuing with the previous exercise show that f is not differentiable at any point. If $g(x) = q^{-3}$ for rational $x = p/q$ in the lowest terms and $g(x) = 0$ otherwise, show that g is differentiable at every irrational point $c = \pm\sqrt{r/s}$ with r/s in the lowest terms. [Hint: for $x = p/q$ we have*

$$\frac{g(x) - g(c)}{x - c} = \frac{1}{q^3}\frac{1}{x - c} = \frac{1}{q^3}\frac{x + c}{x^2 - c^2} = \frac{1}{q}B$$

where

$$B = \frac{\pm\sqrt{r/s} + p/q}{sp^2 - rq^2}$$

is bounded as $x \to c$.]

EXERCISE 1.12 ①① *Prove: No function is continuous at rational and discontinuous at irrational points. [Hint: This is best done with the help of the so-called Baire category theorem. See reference [4] p. 56.]*

EXERCISE 1.13 ①① *A partition π_1 of $[a, b]$ is said to be a refinement of another partition π_2 of $[a, b]$ if every interval in π_1 is a subinterval of some interval in π_2. Prove: A function f is Riemann integrable with $\int_a^b f = A$ if* and only if for every positive ε there exists a partition π_0 such that inequality (1.8) holds whenever π is a refinement of π_0. [Hint: If f is R-integrable then there exists $\delta > 0$ such that (1.8) holds whenever $n(\pi) < \delta$. Fix π_1 with $n(\pi_1) < \delta$ and the condition is satisfied. Find π_1 according to the condition and a division D satisfying (1.16) and (1.17). Form a partition π whose intervals are intersections of intervals of π_1 and D. Show $S(\pi) < A + \varepsilon$, $s(\pi) < A - \varepsilon$, $\overline{\int_a^b} f - \underline{\int_a^b} f < 2\varepsilon$.]

EXERCISE 1.14 Prove some properties of the Riemann integral using the definition from the previous exercise. For example, prove relation (1.31) and the first inequality in (1.34).

EXERCISE 1.15 ⓘ① Let f, g be Riemann integrable on $[a, b]$ and c, k real numbers. Write $F(x) = \int_a^x f(x)\,dx + c$ and $G(x) = \int_a^x g(x)\,dx + k$. Prove that

$$\int_a^b Fg = F(b)G(b) - F(a)G(a) - \int_a^b Gf.$$

[Hint: For $H(x) = F(x)G(x) - F(a)G(a) - \int_a^x (Fg + Gf)$ show that $H'(x) = 0$ for all $x \in [a, b]$ or almost everywhere and then use Lemma 1.4.8.]

EXERCISE 1.16 Prove: If g is Riemann integrable on $[a, b]$ and $g(x) > 0$ for all $x \in [a, b]$ then $G(t) = \int_a^t g$ is strictly increasing on $[a, b]$. [Hint: It suffices to show $G(a) < G(b)$. Use Theorem 1.4.3 and continuity of g to show that on some subinterval of $[a, b]$ the function g is bounded away from zero.]

EXERCISE 1.17 Let D be the division (1.1) and ξ_i arbitrary points in $[x_{i-1}, x_i]$. Prove that

$$\sum_1^n |G(x_i) - G(x_{i-1}) - g(\xi_i)(x_i - x_{i-1})| \leq S(D) - s(D),$$

where $S(D)$ and $s(D)$ denote the upper sum and the lower sum of g, respectively.

EXERCISE 1.18 ⓘ① Prove: If $g \geq 0$ is Riemann integrable on $[a, b]$, the function G with $G(t) = \int_a^x g$ is strictly increasing and f is bounded on $[G(a), G(b)]$ then

$$\int_{G(a)}^{G(b)} f = \int_a^b (f \circ G).g,$$

provided that either f or $(f \circ G).g$ is integrable on $[G(a), G(b)]$ or $[a, b]$, respectively. [Hint: G provides a one-to-one correspondence between partitions of $[G(a), G(b)]$ and $[a, b]$. Use the previous exercise.]

EXERCISE 1.19 Let $\{f_n\}$ be a sequence of functions converging uniformly on $[a, b]$. If each f_n is continuous at $c \in [a, b]$ then $f = \lim_{n \to \infty} f_n$ is also continuous at c. Prove it.

EXERCISE 1.20 *Use the previous exercise and Theorem 1.4.3 to prove that a uniform limit of a sequence of R-integrable functions is R-integrable. (An alternative proof of Theorem 1.5.1.)*

EXERCISE 1.21 *Prove: A function f is Riemann integrable on $[a, b]$ if and only if there are two sequences of step functions $\{\phi_n\}$ and $\{\psi_n\}$ such that $\phi_n(x) \le f(x) \le \psi_n(x)$ for all x and all n, and $\int_a^b (\psi_n - \phi_n) \to 0$ as $n \to \infty$. Furthermore $\lim_{n \to \infty} \int_a^b \psi_n = \lim_{n \to \infty} \int_a^b \phi_n = \int_a^b f$.*

EXERCISE 1.22 *If the conditions in Exercise 1.21 hold with f and with f replaced by f_n where each f_n is Riemann integrable on $[a, b]$ then f is Riemann integrable on $[a, b]$ and $\lim_{n \to \infty} \int_a^b f_n = \int_a^b f$. Prove it.*

2

Basic Theory

2.1 Introduction

The aim of this chapter is to present the most elementary part of the Kurzweil–Henstock theory. Cousin's lemma in Section 2.3 is needed to show that Definition 2.4.1 is logically sound. Subsection 2.3.1 gives some examples on the use of Cousin's Lemma in elementary analysis but this subsection is not used anywhere in the rest of the book. The most important theorem in this chapter is the Fundamental Theorem 2.6.2, which will however be superseded in Section 3.9 in the next chapter. Section 2.7 on applications of the Fundamental Theorem aims to convince the reader that the Kurzweil–Henstock integral has simple but important applications in calculus. Sections 2.8 and 2.9 show that the so-called improper integrals are included in the Kurzweil–Henstock theory.

2.2 Motivation

The definition of the Riemann integral is based on approximation of the anticipated value of the integral by Riemann sums. Generally speaking a function behaves differently in different parts of an interval. It is natural to expect that better approximation is achieved if some of the intervals of a partition are *substantially smaller* than others.

EXAMPLE 2.2.1 Let $[a, b] = [0, 3]$, $f(1) = 2$ and $f(2) = 10000$ and $f(x) = 1$ otherwise. Since $f(x)$ is 1 except for $x = 1, 2$, geometric intuition tells us that $\int_0^3 f = 3$. The Riemann sum for a partition (1.4) differs from 3 only by contributions of intervals which have tags either

1 or 2. There can be at most four such intervals, say

$$[u_k, v_k], [u_{k+1}, v_{k+1}], [u_p, v_p], [u_{p+1}, v_{p+1}]$$

with $\xi_k = \xi_{k+1} = 1$ and $\xi_p = \xi_{p+1} = 2$. We can make

$$f(\xi_k)(v_k - u_k) + f(\xi_{k+1})(v_{k+1} - u_{k+1}) + f(\xi_p)(v_p - u_p)$$
$$+ f(\xi_{p+1})(v_{p+1} - u_{p+1})$$

small, say less than ε, by demanding

$$v_k - u_k \quad < \quad \frac{\varepsilon}{4}\frac{1}{2}, \tag{2.1}$$

$$v_{k+1} - u_{k+1} \quad < \quad \frac{\varepsilon}{4}\frac{1}{2}, \tag{2.2}$$

$$v_p - u_p \quad < \quad \frac{\varepsilon}{4}\frac{1}{10000}, \tag{2.3}$$

$$v_{p+1} - u_{p+1} \quad < \quad \frac{\varepsilon}{4}\frac{1}{10000}. \tag{2.4}$$

There is a common pattern in these inequalities and they can all be subsumed in one compact condition. We define a function $\delta : \{1, 2\} \mapsto \mathbb{R}_+$ by $\delta(\xi) = \varepsilon/4f(\xi)$ and require

$$v_j - u_j < \delta(\xi_j). \tag{2.5}$$

Then all inequalities (2.1)—(2.4) hold and

$$\left| \sum_\pi f - 3 \right| < \varepsilon. \tag{2.6}$$

Moreover, if only one tag is equal to 1 or 2 , condition (2.5) still ensures that inequality (2.6) is satisfied. We defined δ only at the troublesome points 1 and 2. Generally we can expect that the function δ might be needed at more points than two. The simplest way to deal with the domain of δ is to define it on all of $[a, b]$. In our present example it would not matter if we required inequality (2.5) for all ξ_j, $j = 1, 2, \ldots, n$ as long as $\delta(\xi) > 0$ on $[a, b]$.

This example suggests that it might be convenient to generalize the concept of a δ-fine partition to the case when δ is a *function*. See Figure 2.1.

DEFINITION 2.2.2 *Let* $\delta : [a, b] \mapsto \mathbb{R}_+$. *A partition* $\pi \equiv \{(\xi, [u, v])\}$ *of* $[a, b]$ *is said to be* δ-fine *if*

$$\xi - \delta(\xi) < u \leq \xi \leq v < \xi + \delta(\xi). \tag{2.7}$$

2 Basic Theory

We shall write $\pi \ll \delta$ to indicate that π is δ-fine.

Fig. 2.1. δ-fine

What was done in the last example can be criticized as unnecessarily complicated; after all inequality (2.6) can be guaranteed by simply demanding that $v_i - u_i < \varepsilon/40000$ for all $i = 1, 2, \ldots, n$. However, example 2.2.1 was only preparatory. For an unbounded f the condition that some intervals of the partition are *substantially smaller* than others becomes *essential*.

EXAMPLE 2.2.3 Let $f(x) = 0$ for $x \in [0,1]$, $x \neq k^{-1}$ and $f(k^{-1}) = k^2$ for $k \in \mathbb{N}$. The anticipated value of the integral is clearly zero. We show that

$$|\sum_{\pi} f| < \varepsilon \qquad (2.8)$$

for every δ-fine partition with a suitably defined *function δ*. For every positive ε choose $\delta(k^{-1}) = 2^{-k-2}k^{-2}\varepsilon$ for $k \in \mathbb{N}$, and $\delta(x) = 1$ otherwise. Let π be δ-fine. The number $\sum_{\pi} f$ equals to a sum of finitely many terms of the form $k^2(v_i - u_i)$, each is less than $\varepsilon 2^{-k-1}$, and there are at most two terms for the same k. Clearly we have that $\sum_{\pi} f < 2\varepsilon(1/2^2 + \cdots + 1/2^N)$ for some positive integer N, and inequality (2.8) follows. We could have taken any convergent series $\sum a_k$, where $a_k > 0$, in place of $\sum 2^{-k}$. Also we could have used another δ defined for $\xi > 0$ more compactly by

$$\delta(\xi) = \frac{\varepsilon}{2^{\frac{1}{\xi}+2}(f(\xi) + 1)}.$$

This formula shows rather clearly the dependence of δ on f and ξ.

With Example 2.2.3 we made some progress: we could feel comfortable defining an integral in exactly the same way as the Riemann integral but replacing the number δ with a positive function $\delta : [a, b] \mapsto \mathbb{R}_+$. This definition would include the Riemann integral as a special case and would have the advantage that many unbounded functions would become integrable. However, such a definition would make no sense if inequality (2.7) were satisfied only vacuously for some δ. In other words if there were no δ-fine partition of $[a, b]$. To illustrate this point we mention that there need not be a δ-fine partition if δ is zero at some point. For example there is no δ-fine partition of $[-1, 1]$ if $\delta(x) = |x|$. Indeed, if it existed then one interval, say $[u, v]$, would contain zero and the inequality $v - u < \delta(\xi)$ would imply $v - u < \text{Max}\{|u|, v\}$, which is impossible. The existence of a δ-fine partition becomes an important issue to which we devote the next section.

2.3 Cousin's lemma

We have seen in the last section that it is not immediately clear that given a function δ there is a δ-fine partition. It might look a little surprising that a δ-fine partition always exists no matter 'how badly' the function δ behaves. It is however important that $\delta > 0$ and the interval is compact. The discovery of the existence of a δ-fine partition for any positive δ has been traced back to the nineteenth century and to the Belgian mathematician Cousin.†

THEOREM 2.3.1 (Cousin's lemma) *If* $\delta : [a, b] \mapsto \mathbb{R}_+$ *and* $a \leq c < d \leq b$ *then there exists a δ-fine partition of* $[c, d]$.

COROLLARY 2.3.2 *If* $a < c < b$ *and* $\delta : [a, b] \mapsto \mathbb{R}_+$ *then there is a partition* π *of* $[a, b]$ *with* $\pi \ll \delta$ *and* c *as one of the endpoints of the intervals of* π.

Proof of Theorem 2.3.1 It is indirect and based on the method of nested intervals. Assume, contrary to what we want to prove, that there is no δ-fine partition of $[c, d]$. Then, either $[c, (c + d)/2]$ or $[(c + d)/2, d]$ has no δ-fine partition. Let us denote the half of $[c, d]$ without a δ-fine partition by $[c_1, d_1]$. Now continue this halving process indefinitely and obtain a sequence of nested intervals $[c_n, d_n]$ with $d_n - c_n = (d - c)2^{-n} \to 0$. There exists a point C which lies in all $[c_n, d_n]$. Since $\delta(C) > 0$ there exists a number N such that for $n > N$ we have

$$d_n - c_n < \delta(C).$$

This last inequality shows that if $\pi \equiv \{c_n = u_1 \leq \xi = C \leq v_1 = d_n\}$ then π is a δ-fine partition of $[c_n, d_n]$. This contradicts the definition of $[c_n, d_n]$. $\bullet$

† [6] p. 22

Theorem 2.3.6 in the next subsection states that Cousin's lemma is equivalent to the least upper bound axiom. The existence of δ-fine partitions can therefore be used as the fundamental principle in teaching elementary real analysis. For references as well as some proofs see [46]. The next subsection, containing applications of Cousin's lemma, is not needed in the rest of the book and can be skipped over.

2.3.1 Applications of Cousin's lemma

We begin with a proof of a classical theorem.

EXAMPLE 2.3.3 (Weierstrass theorem)   Let f be continuous on $[a, b]$. Then we want to prove that f attains its largest value on $[a, b]$. Assume the contrary. Then by continuity of f for every $y \in [a, b]$ there exist a number Y and a function $\delta : [a, b] \to \mathbb{R}_+$ such that

$$f(t) < f(Y) \quad \text{for} \quad y - \delta \leq t \leq y + \delta \tag{2.9}$$

and t in $[a, b]$. If $\{y_k, [u_k, v_k]\}$ is now a δ-fine partition of $[a, b]$ and $f(Y_p)$ the largest number among finitely many numbers $f(Y_k)$, $k = 1, 2, \ldots, n$ then Y_p lies in some interval $[u_i, v_i]$ and consequently by (2.9)

$$f(Y_p) < f(Y_i) \leq f(Y_p),$$

a contradiction.   ●

EXAMPLE 2.3.4   Let $S = \{c_1, c_2, \ldots\}$ with $c_i \neq c_j$ for $i \neq j$ and $F'(x) \geq 0$ for $x \in [a, b] \setminus S$. If F is continuous we wish to prove that F is increasing.† It is sufficient to prove $F(b) - F(a) \geq 0$. Given $\varepsilon > 0$ define $\delta(c_n) > 0$ by using continuity so that

$$|F(v) - F(u)| < \frac{\varepsilon}{2^n} \tag{2.10}$$

for $c_n - \delta(c_n) < u \leq c_n \leq v < c + \delta(c_n)$. The derivative can be defined as‡

$$F'(x) = \lim \frac{F(v) - F(u)}{v - u}$$

---

† Recall that we use 'increasing' and 'strictly increasing' rather than 'non-decreasing' and 'increasing' and similar terminology for 'decreasing.'

‡ For the proof see Lemma 2.6.1.

for $(u,v) \to (x,x)$ with $u \neq v$ and $u \leq x \leq v$. It follows that for $x \notin S$ there is a $\delta(x) > 0$ such that if

$$x - \delta(x) < u \leq x \leq v < x + \delta(x) \tag{2.11}$$

we have

$$F(v) - F(u) > -\varepsilon(v - u). \tag{2.12}$$

For a δ-fine partition π we obtain with the help of inequalities† (2.10) and (2.12)

$$F(b) - F(a) = \sum_\pi F(u,v) \;\; = \;\; \sum_{\xi \in S} F(u,v) + \sum_{\xi \notin S} F(u,v)$$

$$F(b) - F(a) \;\; \geq \;\; -\varepsilon \sum_1^\infty 2^n - \varepsilon \sum_{\xi \notin S}(v-u)$$

$$\geq \;\; -\varepsilon(1 + b - a).$$

Letting $\varepsilon \to 0$ completes the proof.   •

Our next example provides an alternative proof of sufficiency in Theorem 1.4.3 (Characterization of R-integrability) to the one given in Section 1.4.

EXAMPLE 2.3.5   We wish to prove that if $|f(x)| \leq M$ for all x in $[a,\,b]$ and the set $E = \{x; f$ discontinuous at $x\}$ is of measure zero‡ then f is R-integrable. For $x \notin E$ there is $\delta(x) > 0$ such that inequality (2.11) implies

$$\sup\{f;\, [u,\, v]\} - \inf\{f;\, [u,\, v]\} = M - m < \frac{\varepsilon}{2(b-a)}. \tag{2.13}$$

There exists a system of open disjoint intervals $\{I_n : n \in \mathbb{N}\}$ covering E with

$$\sum_1^\infty |I_n| < \frac{\varepsilon}{4M}. \tag{2.14}$$

For every $x \in E$ then there is a unique I_m which contains it and there is a $\delta > 0$ such that $[x - \delta(x),\, x + \delta(x)] \subset I_m$. We now have a positive $\delta(x)$ defined on all of $[a,\, b]$. Let $\pi \equiv \{(x, [u,\, v])\}$ be a δ-fine partition of

† Recall that $F(u,v) = F(v) - F(u)$.
‡ Sets of measure zero were introduced in Section 1.4.

$[a, b]$. Then

$$S(\pi) - s(\pi) = \sum_{x \notin E}(M - m)(v - u) + \sum_{x \in E}(M - m)(v - u).$$

Using inequality (2.13) on the first sum and inequality (2.14) on the second, we have

$$\begin{aligned} S(\pi) - s(\pi) \quad &< \quad \frac{\varepsilon}{2(b - a)}\sum_{x \notin E}(v - u) + 2M\sum_{x \in E}(v - u) \\ &< \quad \frac{\varepsilon}{2(b - a)}(b - a) + 2M\sum_{1}^{\infty}\sum_{x \in I_n}(v - u) \\ &= \quad \frac{\varepsilon}{2} + 2M\sum_{1}^{\infty}|I_n| < \varepsilon. \end{aligned}$$

This shows that f is Riemann integrable by (iii) of Theorem 1.3.1.   •

The next theorem states that Cousin's lemma is equivalent to the least upper bound axiom. Let us recall that an ordered field $\mathbf{F}$ is said to be complete if every set $X \subset \mathbf{F}$ which is non-empty and bounded above possesses a least upper bound.

THEOREM 2.3.6 (Cousin and l.u.b.)   *Let $\mathbf{F}$ be an ordered field. Then $\mathbf{F}$ is complete if and only if for every closed bounded interval $[a, b]$ and every function δ positive on $[a, b]$ there exists a δ-fine partition of $[a, b]$.*

Proof   The only if part is Cousin's lemma. For the if part we proceed indirectly. Let M be a non-empty set bounded from above which has no least upper bound; take $a \in M$ and b an upper bound for M. We now define δ as follows: If $\xi \in [a, b]$ and is not an upper bound for M then there exists $x > \xi$, $x \in M$. Let $\delta(\xi) = x - \xi$ in this case. Note that this defines δ at a. If $\xi \in (a, b]$ and is an upper bound for M then there exists $z < \xi$ which is also an upper bound (since by assumption ξ cannot be the least upper bound). Let $\delta(\xi) = \xi - z$. Let

$$\pi = \{(\xi_i, [u_i, v_i]); i = 1, \ldots, n\}$$

be a δ-fine partition of $[a, b]$. The partition π has the following properties:

(ν) if ξ_i is not an upper bound neither is v_i, i.e. if v_i is an upper bound so is ξ_i;

(μ) if ξ_i is an upper bound so is u_i.

The tag ξ_n is an upper bound by (ν) and ξ_1 is not an upper bound by (μ). There is a smallest i for which ξ_i is an upper bound; let us denote it by p. Clearly $p \geq 2$, consequently ξ_{p-1} exists and is not an upper bound. By (ν) the endpoint v_{p-1} is not an upper bound, on the other hand $v_{p-1} = u_p$ is an upper bound by (ν). •

Further examples on the use of Cousin's Lemma are in Exercises 2.1 and 2.2.

2.4 The definition

The Kurzweil–Henstock integral, $\int_a^b f$, is defined as the limit of Riemann sums, in very much the same way as the Riemann integral is, except that δ-fineness is measured by a function δ. More precisely we define:

DEFINITION 2.4.1 *A number I is the Kurzweil–Henstock integral (or just integral) of f from a to b (or on $[a,b]$) if for every positive ε there is a function $\delta : [a,b] \mapsto \mathbb{R}_+$ such that for every δ-fine partition π*

$$|\sum_\pi f - I| < \varepsilon. \tag{2.15}$$

We denote the Kurzweil–Henstock integral I as usual by $\int_a^b f$ or by $\int_a^b f(x)dx$ and refer to it as the KH-integral. If $K = [a,\, b]$ then we also write $\int_K f$ for $\int_a^b f$. If we wish to distinguish the Kurzweil–Henstock integral from another integral we may denote it by $\mathcal{KH} \int_a^b f$. We shall often refer to functions which have an integral according to Definition 2.4.1 as Kurzweil–Henstock integrable or as KH-integrable, and only when no confusion can arise as integrable.

The KH-integral is well defined; this is the content of the next theorem. First we look at some examples.

Examples 2.2.1 and 2.2.3 showed that $\mathcal{KH} \int_0^3 f = 3$ and $\mathcal{KH} \int_0^1 f = 0$, respectively. A definition of the integral must be judged on the power and usefulness of the theory which can be built on it and not on examples of its direct application. However, we still give three more examples of integral evaluation directly from the definition to illustrate now

- that the definition is easy to work with;
- that the class of KH-integrable functions is richer than the class of Riemann integrable functions.

EXAMPLE 2.4.2   The formula

$$\int_a^b f(x)dx = \int_{-b}^{-a} f(-x)dx$$

holds whenever one of the integrals exists. If

$$\pi \equiv \{\xi, [u, v]\}$$

is a δ-fine partition of $[a, b]$ then

$$\pi^- \equiv \{-\xi, [-v, -u]\}$$

is a δ-fine partition of of $[-b, -a]$ and vice versa. Since

$$\sum_\pi f(\xi)(v - u) = \sum_{\pi^-} f(-(-\xi))(-u - (-v)),$$

the assertion follows.   •

EXAMPLE 2.4.3 (An 'improper' integral)   We wish to prove
that

$$\int_0^A f(x)dx = 2\sqrt{A}, \qquad (2.16)$$

where $f(x) = \frac{1}{\sqrt{x}}$ for $x \neq 0$ and $f(0) = 0$. Define $\delta(x) = \varepsilon x$ for $x \neq 0$
and $\delta(0) = \varepsilon^2$. Let $\frac{1}{2} > \varepsilon > 0$ and $\pi \equiv \{(y_i, [u_i, v_i])\}$ be a δ-fine
partition of $[0, A]$. Note that this choice of δ implies $y_1 = 0$ (if $y_1 > 0$
then $y_1 \leq v_1 = v_1 - u_1 < 2\varepsilon y_1 < y_1$). Since

$$f(y_i)(v_i - u_i) \leq 2\sqrt{\frac{v_i}{y_i}}(\sqrt{v_i} - \sqrt{u_i})$$

and

$$\sqrt{\frac{v_i}{y_i}} \leq \sqrt{1 + \varepsilon}$$

for $y_i \neq 0$, we have

$$\sum_\pi f \leq 2\sqrt{1 + \varepsilon}\sum_{i=1}^n (\sqrt{v_i} - \sqrt{u_i}) = 2\sqrt{A}\sqrt{1 + \varepsilon} \qquad (2.17)$$

For an estimate from below note that

$$\sqrt{\frac{u_i}{y_i}} \geq \sqrt{1 - \varepsilon},$$

for $i \geq 2$. Hence

$$\sum_\pi f \;\geq\; \sum_{i=2}^n f(y_i)(v_i - u_i)$$

$$\geq\; 2\sqrt{1-\varepsilon}\sum_{i=2}^n (\sqrt{v_i} - \sqrt{u_i})$$

$$=\; 2\sqrt{1-\varepsilon}(\sqrt{A} - \sqrt{u_2})$$

$$=\; 2\sqrt{1-\varepsilon}(\sqrt{A} - \varepsilon). \qquad (2.18)$$

Since ε is arbitrary equation (2.16) follows from inequalities (2.17) and (2.18).   •

REMARK 2.4.4 In this example it was convenient to choose δ in such a way that for any δ-fine partition the tag of the first interval of the partition was 0. Generally any point c can be forced to become a tag of a δ-fine partition; if $a < c < b$ and $\delta(x) \leq |x - c|$ for $x \neq c$ and $\delta(c) > 0$ arbitrary then for any δ-fine partition c tags the interval in which it lies. We shall refer to choosing δ in such a way that c becomes a tag as *anchoring* the partition on c. It is easy to see that if S is a finite set in $[a,\, b]$ then there exists a positive function δ such that every δ-fine partition is anchored on S.

The next example will be superseded later but provides now an example of a function KH-integrable but not R-integrable.

EXAMPLE 2.4.5 If $S = \{c_1, c_2, c_3, \ldots\}$ with $c_i \neq c_j$ for $i \neq j$ and $f(x) = 0$ for $x \notin S$ then f is integrable and $\int_a^b f = 0$. The function f can be, for instance, the characteristic function of the rationals, which is often called *Dirichlet's function*. Define

$$\delta(c_j) = \frac{\varepsilon}{2^{j+2}(|f(c_j)| + 1)} \quad \text{and} \quad \delta(x) = 1 \text{ for } x \notin S.$$

Let $\pi \equiv \{(y_i, [u_i, v_i]); \ i = 1, 2, \ldots, n\}$ be a δ-fine partition of $[a, b]$. Clearly

$$\sum_\pi f = \sum_{y_i \in S} f(y_i)(v_i - u_i).$$

Since the same c_j can be equal to y_i for two distinct i and because of the choice of δ we have

$$\left| \sum_{y_i \in S} f(y_i)(v_i - u_i) \right| < 2\sum_{j=1}^\infty |f(c_j)| \frac{\varepsilon}{2^{j+1}(|f(c_j)| + 1)} < \varepsilon.$$

This proves that the function f is KH-integrable and $\int_a^b f = 0$. •

We still have to show that the definition is meaningful, namely that the value of the KH-integral is always uniquely determined. Note that the proof uses Cousin's lemma.

THEOREM 2.4.6 *There is at most one number I satisfying the condition from Definition 2.4.1.*

Proof Assume that for every positive ε there exists δ_1 such that

$$\left| \sum_\pi f - I \right| < \varepsilon \qquad (2.19)$$

whenever $\pi \ll \delta_1$ and also that there exists δ_2 such that

$$\left| \sum_\pi f - J \right| < \varepsilon \qquad (2.20)$$

whenever $\pi \ll \delta_2$. Let $\delta = \mathrm{Min}(\delta_1, \delta_2)$ and π be δ-fine. Then inequalities (2.19) and (2.20) hold and consequently

$$|I - J| < 2\varepsilon.$$

By letting $\varepsilon \to 0$ we obtain $|I - J| \leq 0$, that is $I = J$. •

2.5 Basic theorems

The next few theorems have almost the same proofs as in the Riemann theory. For the sake of completeness we prove them here.

THEOREM 2.5.1 (Homogeneity) *If f is integrable on $[a, b]$ and $c \in \mathbb{R}$, then cf is integrable and*

$$\int_a^b cf = c \int_a^b f. \qquad (2.21)$$

Proof If $c = 0$ the theorem is trivial. Let $c \neq 0$, $\varepsilon > 0$. There exists a positive function δ such that

$$\left| \sum_\pi f - \int_a^b f \right| < \frac{\varepsilon}{|c|}$$

whenever $\pi \ll \delta$. It follows that

$$\left| \sum_\pi cf - c \int_a^b f \right| = |c| \left| \sum_\pi f - \int_a^b f \right| < \varepsilon \qquad •$$

REMARK 2.5.2 It follows that if cf is integrable and $c \neq 0$ then f is also integrable and (2.21) holds.

THEOREM 2.5.3 (Additivity) *If f and g are integrable on $[a, b]$ then so is $f + g$ and*

$$\int_a^b (f + g) = \int_a^b f + \int_a^b g. \tag{2.22}$$

Proof For every positive ε there exist positive functions δ_1 and δ_2 such that

$$\left| \sum_\pi f - \int_a^b f \right| < \frac{\varepsilon}{2} \tag{2.23}$$

for $\pi \ll \delta_1$, and

$$\left| \sum_\pi g - \int_a^b g \right| < \frac{\varepsilon}{2} \tag{2.24}$$

for $\pi \ll \delta_2$. Define $\delta(x) = \text{Min}(\delta_1(x), \delta_2(x))$. If $\pi \ll \delta$ then inequalities (2.23) and (2.24) hold and since $\sum_\pi (f + g) = \sum_\pi f + \sum_\pi g$ we have

$$\left| \sum_\pi (f + g) - \int_a^b f - \int_a^b g \right| < \frac{\varepsilon}{2} + \frac{\varepsilon}{2} = \varepsilon. \qquad \bullet$$

REMARK 2.5.4 It is an easy exercise to extend formula (2.22) to a sum of n functions.

A set N is said to be *negligible* or a *null* set if every function h which is zero outside N is KH-integrable and $\int_a^b h = 0$ for every interval $[a, b]$. Every finite set is null and by Example 2.4.5 every countable set is null.† Section A.1 in the Appendix contains an example of a null set which is not countable. Obviously, the characteristic function of a null set is integrable and its integral is zero. It is interesting that in our theory the converse is also true. We have

THEOREM 2.5.5 *A set S is negligible if and only if the integral of its characteristic function 1_S over any interval $[a, b]$ is zero.*

---

† Section 2.11 contains the definition of sets of measure zero and the characterization of negligible sets as sets of measure zero.

Proof If S is negligible then by the very definition of a negligible set

$$\int_a^b \mathbf{1}_S = 0. \tag{2.25}$$

If equation (2.25) holds then for every positive ε there is $\delta_n : [a,b] \to \mathbb{R}_+$ such that if $\pi_n \ll \delta_n$ then

$$\sum_{\pi_n} \mathbf{1}_S(\xi_k)(v_k - u_k) < \frac{\varepsilon}{2^n n}. \tag{2.26}$$

Let f be an arbitrary function which is zero outside S and

$$E_n = \{x : n - 1 \le |f(x)| < n\}.$$

For $x \in E_n$ we define $\delta(x)$ to be $\delta_n(x)$. Let $\pi \equiv \{(\xi_k, [u_k, v_k]); k = 1, \dots, p\}$ be a δ-fine partition and M an integer with $|f(\xi_k)| \le M$ for $k = 1, \dots, p$. Then we have by (2.26)

$$\left| \sum_\pi f \right| \le \sum_{n=1}^M \sum_{\xi_k \in E_n} n(v_k - u_k)$$

$$= \sum_{n=1}^M n \sum_{\pi_n} \mathbf{1}_{E_n}(\xi_k)(v_k - u_k) < \sum_{n=1}^\infty \frac{\varepsilon}{2^n} < \varepsilon. \qquad \bullet$$

THEOREM 2.5.6 *Let f be an integrable function on $[a,b]$ and let g be a function which differs from f only at points of a null set. Then g is integrable and*

$$\int_a^b g = \int_a^b f.$$

Proof Since $g = f + (g - f)$ Theorem 2.5.6 is an immediate consequence of Theorem 2.5.3 and the definition of a null set. $\qquad \bullet$

REMARK 2.5.7 The last theorem can also be expressed as follows: Changing the definition of a function f at points of a null set affects neither the existence nor the value of $\int_a^b f$. This is used very often. It also allows us to assign a meaning to the integral of a function which is defined except on a null set. For example, the meaning of $\int_a^b \frac{1}{\sqrt{|x|}} dx$ is clear: it is equal to $\int_a^b f$ where $f(x) = \frac{1}{\sqrt{|x|}}$ for $x \ne 0$ and $f(0) = 0$ (or something else).

THEOREM 2.5.8 (Preservation of inequalities) *If f and g are integrable on $[a, b]$ and $f \leq g$ then*

$$\int_a^b f \leq \int_a^b g. \tag{2.27}$$

Proof For every positive ε there exist functions δ_1 and δ_2 such that

$$\int_a^b f - \varepsilon < \sum_\pi f < \int_a^b f + \varepsilon, \tag{2.28}$$

whenever $\pi \ll \delta_1$, and

$$\int_a^b g - \varepsilon < \sum_\pi g < \int_a^b g + \varepsilon \tag{2.29}$$

whenever $\pi \ll \delta_2$. Let $\delta = \text{Min}(\delta_1, \delta_2)$. If $\pi \ll \delta$ then inequalities (2.28) and (2.29) hold simultaneously. Since $\sum_\pi f \leq \sum_\pi g$, it follows that

$$\int_a^b f - \varepsilon < \int_a^b g + \varepsilon.$$

By letting $\varepsilon \to 0$ we obtain inequality (2.27). •

COROLLARY 2.5.9 *If f and g are integrable and $f(x) \leq g(x)$ for all $x \in [a, b]$ except on a null set then inequality (2.27) still holds.*

COROLLARY 2.5.10 *If f is integrable and $f(x) \leq M$ for all $x \in [a, b]$ except on a null set then*

$$\int_a^b f \leq M(b - a).$$

Similarly,

$$m(b - a) \leq \int_a^b f,$$

for an integrable f satisfying $f \geq m$ except on a null set.

COROLLARY 2.5.11 *If both the functions f and $|f|$ are integrable then*

$$\left| \int_a^b f \right| \leq \int_a^b |f|.$$

Proof We have $-|f| \leq f \leq |f|$ and therefore

$$\int_a^b -|f| = -\int_a^b |f| \leq \int_a^b f \leq \int_a^b |f|. \qquad \bullet$$

Warning. Even if f is integrable, the function $|f|$ need not be. This is shown by Example 2.6.7 in the next section.

THEOREM 2.5.12 *If f is integrable over $[a, c]$ and $[c, b]$ then f is integrable over $[a, b]$ and*

$$\int_a^c f + \int_c^b f = \int_a^b f. \qquad (2.30)$$

Proof For every positive ε there exist positive functions $\underline{\delta}$ and $\overline{\delta}$ such that if $\underline{\pi}$ is a $\underline{\delta}$-fine partition of $[a, c]$ and $\overline{\pi}$ is a $\overline{\delta}$-fine partition of $[c, b]$ then

$$\left| \sum_{\underline{\pi}} f - \int_a^c f \right| < \frac{\varepsilon}{2} \qquad (2.31)$$

and

$$\left| \sum_{\overline{\pi}} f - \int_c^b f \right| < \frac{\varepsilon}{2}. \qquad (2.32)$$

Let

$$\delta(x) \;=\; \text{Min}(\underline{\delta}(x), \tfrac{1}{2}(c - x)) \qquad \text{for} \qquad a \leq x < c,$$

$$\delta(x) \;=\; \text{Min}(\overline{\delta}(x), \tfrac{1}{2}(x - c)) \qquad \text{for} \qquad c < x \leq b,$$

$$\delta(c) \;=\; \text{Min}(\underline{\delta}(c), \overline{\delta}(c)).$$

If $\pi \equiv \{(\xi_i, [u_i, v_i]); \; i = 1, 2, \ldots, n\}$ and $\pi \ll \delta$ then π is anchored on c and hence $c = \xi_k$ for some $k \in \mathbb{N}$. We now have

$$\sum_{\pi} f \;=\; \sum_{i=1}^{k-1} f(\xi_i)(v_i - u_i) + f(\xi_k)(\xi_k - u_k)$$

$$+ \; f(\xi_k)(v_i - \xi_k) + \sum_{i=1}^{k-1} f(\xi_i)(v_i - u_i). \qquad (2.33)$$

Clearly

$$S_a^c = \sum_{i=1}^{k-1} f(\xi_i)(v_i - u_i) + f(\xi_k)(\xi_k - u_k) \qquad (2.34)$$

represents a Riemann sum for a $\underline{\delta}$-fine partition of $[a,\, c]$, hence by inequality (2.31)

$$\left| S_a^c - \int_a^c f \right| < \frac{\varepsilon}{2}. \qquad (2.35)$$

Denoting the last two terms in (2.33) by S_c^b we have for similar reasons

$$\left| S_c^b - \int_c^b f \right| < \frac{\varepsilon}{2}. \qquad (2.36)$$

Combining inequalities (2.35), (2.36) shows that for any δ-fine partition of $[a,\, b]$ we have

$$\left| \sum_\pi f - \int_a^c f - \int_c^b f \right| < \varepsilon.$$

This proves f integrable on $[a,\, b]$ and equation (2.30). $\qquad \bullet$

REMARK 2.5.13 In the proof of Theorem 2.5.12 we split the Riemann sum into two, one for the interval $[a, c]$ and another for $[c, b]$. This was possible because we anchored the partition on c. This trick is very useful and we will use it without further explanation in future.

On occasions like this when we split a Riemann sum into two, one corresponding to the interval $[a,\, c]$ and the other to $[c,\, b]$, we shall denote the first one as

$$\sum {}_a^c f \qquad \text{or} \qquad \sum_\pi {}_a^c f.$$

and the second similarly as

$$\sum {}_c^b f \qquad \text{or} \qquad \sum_\pi {}_c^b f.$$

We then have

$$\sum_\pi f = \sum_\pi {}_a^b f = \sum_\pi {}_a^c f + \sum_\pi {}_c^b f.$$

THEOREM 2.5.14 *If f is KH-integrable on $[a,\, b]$ then it is integrable on every $[\alpha,\, \beta] \subset [a,\, b]$.*

38 2 Basic Theory

REMARK 2.5.15 We shall give a two-line proof in the next chapter† using the Cauchy principle for Riemann sums. The proof here is elementary but somewhat lengthy.

Proof of Theorem 2.5.14 It is sufficient to consider the case when $[\alpha, \beta] = [a, c]$. The argument is entirely similar for $[\alpha, \beta] = [c, b]$ and the general case then follows. To every $n \in \mathbb{N}$ there exists a δ_n such that

$$\left| \sum_\Pi f - \int_a^b f \right| < \frac{1}{n} \tag{2.37}$$

whenever Π is δ_n-fine. Let π_1, π_2 be two δ_n-fine partitions of $[a, c]$. If we join these with a fixed δ_n-fine partition of $[c, b]$ we obtain two δ_n-fine partitions of $[a, b]$, say Π_1, Π_2, and

$$\left| \sum_{\pi_1} f - \sum_{\pi_2} f \right| = \left| \sum_{\Pi_1} f - \sum_{\Pi_2} f \right|. \tag{2.38}$$

The right-hand side of equation (2.38) is less than $2n^{-1}$ by inequality (2.37). Hence

$$\left| \sum_{\pi_1} f - \sum_{\pi_2} f \right| < \frac{2}{n}. \tag{2.39}$$

It follows that the sets

$$S_n = \{\sum_\pi f; \pi \text{ a } \delta_n\text{-fine partition of } [a, c]\}$$

are bounded and because we can assume $\delta_{n+1} \leq \delta_n$ the sets S_n are also nested. Let $[U_n, V_n]$ be the smallest closed interval containing S_n, i.e. $U_n = \sup S_n$, $V_n = \inf S_n$. By inequality (2.39) we have $V_n - U_n \leq 2n^{-1}$. Using the nested intervals theorem we find a number I which lies in all $[U_n, V_n]$. Let $\varepsilon > 0$ and $2n^{-1} < \varepsilon$. Let π be a δ_n-fine partition of $[a, c]$. Both numbers I and $\sum_\pi f$ lie in $[U_n, V_n]$, and consequently

$$\left| \sum_\pi f - I \right| \leq V_n - U_n \leq \frac{2}{n} < \varepsilon. \qquad \bullet$$

A *step function* is a linear combination of characteristic functions of bounded intervals‡. It follows from the last theorem and Theorems

† Remark 3.1.2.
‡ some of them possibly degenerating into one-point sets. We shall use this definition of a step function throughout this book. It is more general than the definition of a step function on page 4. Note that a step function is defined on all of $\mathbb{R}$.

2.5.1—2.5.3 that step functions are KH-integrable on any interval. This together with the next theorem makes it easy to identify some important classes of KH-integrable functions.

THEOREM 2.5.16 *If for every positive ε there exist KH-integrable functions h, H such that*

$$h(x) \leq f(x) \leq H(x), \qquad (2.40)$$

for all $x \in [a, b]$ except a null set, and

$$\int_a^b H - \int_a^b h \leq \varepsilon, \qquad (2.41)$$

then f is KH-integrable on $[a, b]$.

Proof We can and shall assume for the proof that inequalities (2.40) hold everwhere on $[a, b]$. Both

$$\sup\{\int_a^b h : h \leq f, h \text{ integrable}\}$$

and

$$\inf\{\int_a^b H : H \geq f, H \text{ integrable}\}$$

exist and are equal, say to I. We now choose h and H satisfying (2.41) and (2.40). Clearly

$$\int_a^b h \leq I \leq \int_a^b H.$$

There is a positive function δ such that for every δ-fine partition π

$$\left| \sum_\pi H - \int_a^b H \right| < \varepsilon,$$

$$\left| \sum_\pi h - \int_a^b h \right| < \varepsilon.$$

Since the Riemann sums for f are trapped between those for h and H we have for $\pi \ll \delta$

$$\sum_\pi f < \int_a^b H + \varepsilon < I + 2\varepsilon,$$

$$\sum_\pi f > \int_a^b h - \varepsilon > I - 2\varepsilon. \qquad \bullet$$

REMARK 2.5.17 Since step functions are KH-integrable Theorem 2.5.16 is often applied with h, H step functions.†

COROLLARY 2.5.18 *If f is monotonic and bounded on $[a, b]$ then it is KH-integrable.*

Proof For the sake of definiteness let f be increasing. Divide the interval $[a, b]$ into n intervals $[u_i, v_i]$ of equal length and let $H(x) = f(v_i)$, $h(x) = f(u_i)$ for $x \in [u_i, v_i)$. We have inequality (2.40) and

$$0 \leq \int_a^b H - \int_a^b h = \sum_{i=1}^n [f(v_i) - f(u_i)](v_i - u_i) = [f(b) - f(a)]\frac{b-a}{n}. \; \bullet$$

COROLLARY 2.5.19 *A function is KH-integrable on every closed and bounded interval on which it is continuous.*

Proof Let f be continuous on $[a, b]$ and $\varepsilon > 0$. For every $\xi \in [a, b]$ there is a positive $\delta(\xi)$ such that

$$f(\xi) - \varepsilon < f(x) < f(\xi) + \varepsilon \tag{2.42}$$

whenever $\xi - \delta(\xi) < x < \xi + \delta(\xi)$. Let $\pi \equiv \{(\xi, [u, v])\}$ be δ-fine and $h(x) = f(\xi) - \varepsilon$, $H(x) = f(\xi) + \varepsilon$ for x in $[u, v)$. The functions h and H satisfy inequalities (2.40) and

$$\int_a^b H - \int_a^b h = 2\varepsilon(b - a). \qquad\qquad \bullet$$

A function f is said to be *regulated* on $[a, b]$ if for every positive ε there is a step function φ such that

$$|\varphi(x) - f(x)| < \varepsilon \tag{2.43}$$

for all x in $[a, b]$. Obviously step functions are regulated. The proof of Corollary 2.5.19 shows that continuous functions are regulated.

THEOREM 2.5.20 *All regulated functions are KH-integrable.*

Proof We choose $H = \varphi + \varepsilon$ and $h = \varphi - \varepsilon$ and then apply Theorem 2.5.16. $\qquad\qquad\bullet$

---

† If h, H are step functions and inequalities (2.40) hold everywhere on $[a, b]$ then the function f becomes Riemann integrable. However, Theorem 2.5.16 can be applied e.g. if f is the Dirichlet function.

REMARK 2.5.21 What makes the last theorem more interesting is that functions having a limit both from the right and from the left at every point of $[a, b]$ are regulated. Let $\varepsilon > 0$ and denote by $f(x+)$ and $f(x-)$ the limits at x from right and left, respectively. There is a $\delta(x)$ such that

$$|f(t) - f(x-)| < \varepsilon$$

for $t \in (x - \delta(x), x) \cap [a, b]$ and

$$|f(t) - f(x+)| < \varepsilon$$

for $t \in (x, x + \delta(x)) \cap [a, b]$. Let $\{(x, [u, v])\}$ be a δ-fine partition of $[a, b]$ and

$$\varphi(t) = \begin{cases} f(x-) & \text{if } t \in (u, x) \neq \emptyset \\ f(x+) & \text{if } t \in (x, v) \neq \emptyset \\ f(t) & \text{otherwise.} \end{cases}$$

φ is clearly a step function and (2.43) holds on $[a, b]$.
It follows that monotonic functions are regulated.

In the Kurzweil–Henstock theory the question whether the product of two functions is integrable is not as easy as for instance in the Riemann or the Lebesgue theory. However, Theorem 2.5.16 leads to some practical criteria.

THEOREM 2.5.22 *If on $[a, b]$*

- *the function f is KH-integrable and bounded below*
- *g is regulated*

then fg is KH-integrable.

Proof It is sufficient to prove the theorem under the additional assumption that f is non-negative. The general case is obtained by considering $f - C$, where C is a lower bound for f. If f is integrable and φ a step function then $f\varphi$ is obviously integrable (by Theorems 2.5.1, 2.5.3, 2.5.12 and 2.5.14). Denote $\int_a^b f = A$. For a regulated g there is a step function φ such that

$$|g - \varphi| < \frac{\varepsilon}{2A}.$$

Consequently

$$h(x) < f(x)g(x) < H(x)$$

with

$$H(x) = f(x)\varphi(x) + \frac{\varepsilon}{2A}f(x),$$

$$h(x) = f(x)\varphi(x) - \frac{\varepsilon}{2A}f(x),$$

for $x \in [a, b]$. We can now apply Theorem 2.5.16.   •

COROLLARY 2.5.23   *If on* $[a, b]$

• *the function f is KH-integrable and bounded below*
• *g is either continuous or monotonic*

then fg is KH-integrable.

REMARK 2.5.24   The assumption that g is either continuous or monotonic on the the *closed* interval is essential, as the example of $f(x) = g(x) = x^{-\frac{1}{2}}$ and $[a, b] = [0, 1]$ shows. Also the assumption of f being bounded below (or above) is indispensable if g is continuous. Example 2.7.2 provides a KH-integrable f, a continuous G and a non-integrable Gf. On the other hand if f is KH-integrable and g monotonic and bounded then fg is KH-integrable. This will be proved later in this book (see Corollary 3.7.7). It follows, however, easily from Theorem 2.5.22 that if f and $|f|$ are KH-integrable and g is regulated then fg is KH-integrable. Indeed, f^+g and f^-g are KH-integrable by the theorem and $fg = f^+g - f^-g$.

THEOREM 2.5.25 (The first mean value theorem of integral calculus)   *If on* $[a, b]$

(ϕ) *f is non-negative and KH-integrable,*
(γ) *g is continuous*

then there is a $c \in [a, b]$ such that

$$\int_a^b fg = g(c) \int_a^b f. \tag{2.44}$$

REMARK 2.5.26   The special case of $f = 1$, when equation (2.44) reads

$$\int_a^b g = g(c)(a - b), \tag{2.45}$$

is also often called the mean value theorem. The geometrical meaning of (2.45) is evident from Figure 2.2.

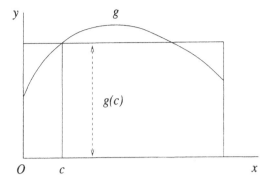

Fig. 2.2. Mean Value Theorem

Proof of Theorem 2.5.25 Let $m = \inf\{g; [a, b]\}$, $M = \sup\{g; [a, b]\}$. We know that fg is KH-integrable and hence

$$m \int_a^b f \leq \int_a^b fg \leq M \int_a^b f. \tag{2.46}$$

If $\int_a^b f = 0$ then by inequality (2.46) $\int_a^b fg = 0$ and the conclusion of the theorem obviously holds with any c in $[a, b]$. If $\int_a^b f > 0$ then the value

$$\mu = \frac{\int_a^b fg}{\int_a^b f} \tag{2.47}$$

lies between m and M and it must be taken by the continuous function g. •

The assumption of continuity of g can be weakened: one can assume the so-called intermediate value property instead. We say that a function g has the intermediate value property on $[a, b]$ if g takes on $[a, b]$ every value between $g(a)$ and $g(b)$. It is obvious that a function can have the intermediate value property on every subinterval of an interval I and be discontinuous at some point of I, e.g. the function f with $f(0) = 0$, $f(x) = \sin(1/x)$ and $I = [-1, 1]$. More interesting is an example of a function which has the intermediate property on every† $[\alpha, \beta] \subset [0, 1]$ and which is not continuous at any point of $[0, 1]$ (see [4] p. 71). On the other hand a function which has the intermediate value property on every subinterval of $[a, b]$ and takes each value exactly once is continuous and monotonic on $[a, b]$ (see [9]). The stronger version of the mean value theorem reads:

† No matter how small it is.

THEOREM 2.5.27   *The assertion of Theorem 2.5.25 remains valid if*

(ϕ) *f is non-negative and KH-integrable,*

(γ_1) *g has the intermediate value property on every interval $[\alpha, \beta]$ with $[\alpha, \beta] \subset [a, b]$, and*

($\phi\gamma$) *the product fg is KH-integrable.*

We postpone the proof of this theorem until Section 2.11.

DEFINITION 2.5.28 (Extension of the definition)   *We define*

$$\int_a^a f = 0$$

and if $a > b$

$$\int_a^b f = -\int_b^a f.$$

The convenience of this definition can be seen in the following examples. If f is integrable on every subinterval of $[a, b]$ then the function

$$F(x) = \int_c^x f,$$

with $c \in [a, b]$, is *now* defined for every $x \in (a, b)$. The formula

$$\int_a^b \alpha f(\alpha x) dx = \int_{\alpha a}^{\alpha b} f(t) dt,$$

which can be verified for $\alpha > 0$ directly from the definition, now holds for any real α.

Most of what we have said about the integral extends easily to the cases $a = b$ or $b < a$. However, there are differences, e.g. if $b < a$, inequality (2.27) in Theorem 2.5.8 is reversed and the inequality in Corollary 2.5.11 is replaced by

$$\left| \int_a^b f \right| \leq \left| \int_a^b |f| \right|.$$

Usually common sense suffices for modifications (if any are needed) of theorems stated for $a < b$ to include also the cases $a = b$ and $a > b$.

THEOREM 2.5.29 (Continuity of the indefinite integral) *If f is integrable on* $[a, b]$ *then the function*

$$F(x) = \int_a^x f \qquad (2.48)$$

is continuous on $[a, b]$.

Proof We prove only that F is continuous from the left at c with $a < c \le b$. Continuity from the right at c, with $a \le c < b$, follows by using the already proved part on the function $x \mapsto f(-x)$ and the interval $[-b, -a]$. Let $\varepsilon > 0$. There is a positive ω such that

$$|f(c)(c - t)| < \frac{\varepsilon}{3} \qquad (2.49)$$

whenever $c - \omega < t < c$ and a positive function δ_1 such that

$$\left| \sum_\pi f - [F(c) - F(a)] \right| < \frac{\varepsilon}{3} \qquad (2.50)$$

whenever π is a δ_1-fine partition of $[a, c]$. Define $\delta(c) = \text{Min}(\delta_1(c), \omega)$ and choose t with $c > t > c - \delta(c)$. There exists a positive function δ_2 such that if π_t is a δ_2-fine partition of $[a, t]$ then

$$\left| \sum_{\pi_t} f - [F(t) - F(a)] \right| < \frac{\varepsilon}{3}. \qquad (2.51)$$

Let

$$\delta(\xi) = \text{Min}[\delta_1(\xi), \delta_2(\xi)] \quad \text{for} \quad a \le \xi \le t,$$
$$\delta(\xi) = \delta_1(\xi) \quad \text{for} \quad t < \xi < c.$$

We now choose a partition π consisting of a δ-fine partition π_t of $[a, t]$ and a pair $(c, [t, c])$. The following identity is obvious:

$$\sum_\pi f - [F(c) - F(a)] = \sum_{\pi_t} f - [F(t) - F(a)]$$
$$+ f(c)(c - t) - [F(c) - F(t)]. \qquad (2.52)$$

Consequently

$$|F(c) - F(t)| \le \left| \sum_{\pi_t} f - [F(t) - F(a)] \right|$$
$$+ \left| \sum_\pi f - [F(c) - F(a)] \right| + |f(c)(c - t)|. \qquad (2.53)$$

Clearly $\pi \ll \delta$ and therefore by inequalities (2.53), (2.49), (2.50), (2.51)

$$|F(c) - F(t)| < \varepsilon,$$

for any t with $c - \delta(c) < t < c$.   •

2.6 The Fundamental Theorem of calculus

The Fundamental Theorem of calculus asserts that

$$F(b) - F(a) = \int_a^b F'. \tag{2.54}$$

Formula (2.54) is often called the Newton–Leibniz formula. It is perhaps not as much realized as it ought to be that the validity of this formula depends on the concept of the integral used. In in Section 1.4 (Differentiability, continuity and integrability) we discussed it in the framework of Riemann's integration. The Kurzweil–Henstock theory has the best theorem in this direction and we shall see in the next chapter. The theorem which we prove now is still better than anything the Riemann theory can offer. First we need

LEMMA 2.6.1 (Alternative definition of derivative) *F has a derivative $f(x)$ at x if and only if for every $\varepsilon > 0$ there is a $\delta > 0$ such that if $u \neq v$ and*

$$x - \delta < u \leq x \leq v < x + \delta \tag{2.55}$$

then

$$|F(v) - F(u) - f(x)(v - u)| < \varepsilon(v - u). \tag{2.56}$$

Proof In inequality (2.56) we choose u or v equal to x and have

$$|F(v) - F(x) - f(x)(v - x)| < \varepsilon(v - x), \tag{2.57}$$

$$|F(x) - F(u) - f(x)(x - u)| < \varepsilon(x - u). \tag{2.58}$$

The if part is now evident. For the only if part we note that given $\varepsilon > 0$ there is a $\delta > 0$ such that inequalities (2.57) and (2.58) hold if u and v satisfy (2.55). It follows that

$$|F(v) - F(u) - f(x)(v - u)| < \varepsilon(v - x) + \varepsilon(x - u). \qquad •$$

THEOREM 2.6.2 (Fundamental Theorem. First version) *If $F'(x) = f(x)$ for all $x \in (a, b)$ and F is continuous on $[a, b]$ then f is KH-integrable and*

$$\int_a^b f = F(b) - F(a).$$

REMARK 2.6.3 We prove the theorem under the additional assumption that $F'(b)$ exists. The more general case then follows by the already proved part of the theorem to the interval $[a, c]$† and to the function $x \mapsto f(-x)$ and the interval $[-b, -c]$. Alternatively, one can give a similar argument for the point b as we give for the point a.

REMARK 2.6.4 The function F is called the primitive of f. More precisely F is said to be a primitive of f on $[a, b]$ if F is continuous on $[a, b]$ and $F'(x) = f(x)$ for all x in (a, b). The Fundamental Theorem can be now rephrased thus: *If f has a primitive F on $[a, b]$ then f is KH-integrable and equation (2.54) holds.*

Proof of Theorem 2.6.2 Let $\varepsilon > 0$. By the continuity of F at a there exists a positive ω_1 such that $a \leq v < a + \omega_1$ implies that

$$|F(v) - F(a)| < \frac{\varepsilon}{3}. \qquad (2.59)$$

Since $\lim_{v \downarrow a} f(a)(v - a) = 0$ there exists a positive ω_2 such that

$$|f(a)(v - a)| < \frac{\varepsilon}{3} \qquad (2.60)$$

whenever $a \leq v < a + \omega_2$. Since $F'(x) = f(x)$ for $x \in (a, b]$ we can find a positive $\eta = \eta(x)$ such that

$$|F(v) - F(u) - f(x)(v - u)| < \frac{\varepsilon}{3(b - a)}(v - u), \qquad (2.61)$$

whenever $x - \eta(x) < u \leq x \leq v < x + \eta(x)$, by Lemma 2.6.1. Now define δ by

$$\begin{aligned}
\delta(a) &= \text{Min}(\omega_1, \omega_2), \\
\delta(x) &= \text{Min}(x - a, \eta(x)).
\end{aligned} \qquad (2.62)$$

† $a < c < b$

For the rest of the proof, let $\pi \equiv \{(\xi_i, [u_i, v_i])\}$ be δ-fine. Note that $(x - a)$ appears in equation (2.62) to anchor x_1 at $a = u_1$. Now we write

$$F(b) - F(a) = \sum_{i=1}^{n} F(u_n, v_n)$$

and have by inequalities (2.59)—(2.61)

$$\left| F(b) - F(a) - \sum_{i=1}^{n} f(\xi_i)(v_i - u_i) \right| \leq |F(v_1) - F(a)|$$

$$|f(a)(v_1 - a)| + \sum_{i=2}^{n} |F(v_i) - F(u_i) - f(\xi_i)(v_i - u_i)|$$

$$\frac{\varepsilon}{3} + \frac{\varepsilon}{3} + \frac{\varepsilon}{3(b - a)} \sum_{i=2}^{n} (v_i - u_i) < \varepsilon. \qquad \bullet$$

REMARK 2.6.5 It is obvious that F may fail to have a derivative on a finite set of points and the Fundamental Theorem still hold, provided F remains continuous on $[a, b]$. Moreover, provided F is continuous, the above proof can be extended by allowing a countable exceptional set $S = \{c_n; n = 1, 2, \dots\}$ where F' does not exist or is not equal to f. Since we shall prove a more general theorem in the next chapter we shall not prove this generalization here. We only indicate how to define the function δ. We can assume without loss of generality that the endpoints a, b are in S. For $x \neq c_n$ we define δ as before: $\delta(x) = \eta(x)$. The function F is continuous on S. Hence there exists $\delta_1 : S \to \mathbb{R}_+$ such that $|F(v) - F(u)| < \varepsilon 2^{-n-1}$ for $c_n - \delta_1 < u \leq c_n \leq v < c_n + \delta_1$ and there exists $\delta_2 : S \to \mathbb{R}_+$ such that $|(v - u)f(c_n)| < \varepsilon 2^{-n-1}$ for $c_n - \delta_2 < u \leq c_n \leq v < c_n + \delta_2$; and one can take $\delta = \text{Min}(\delta_1, \delta_2)$.

EXAMPLE 2.6.6 We have

$$\int_0^A \frac{dx}{\sqrt{x}} = 2\sqrt{A}.$$

Just apply the Newton–Leibniz formula. Note that $\sqrt{x}$ is not differentiable at zero. In this example we can see how convenient it was to assume in the theorem the existence of the derivative only in the open interval (a, b). However, for the validity of the theorem it is essential that F is continuous on the closed interval $[a, b]$.

EXAMPLE 2.6.7 Let F be defined by $F(x) = x^2 \cos(\pi/x^2)$ for $x \neq 0$ and $F(0) = 0$. By Theorem 2.6.2 the function F' is integrable on $[0, 1]$ and $\int_0^1 F' = -1$. However, $|F'|$ is not integrable. For an indirect proof assume it is and define f_i to be $|F'|$ on $(1/\sqrt{i+1}, 1/\sqrt{i}) \equiv (a_i, b_i)$ and $f_i = 0$ otherwise. Then

$$\int_0^1 f_i = \int_{a_i}^{b_i} f_i = \int_{a_i}^{b_i} |F'| \geq \left| \int_{a_i}^{b_i} F' \right|$$

$$= |F(b_i) - F(a_i)| = \frac{1}{i} + \frac{1}{i+1} \geq \frac{2}{i+1}.$$

For every $n \in \mathbb{N}$ and every $x \in [0, 1]$ we have

$$|F'(x)| \geq \sum_{i=1}^{n} f_i(x).$$

Consequently

$$\int_0^1 |F'| \geq \sum_{i=1}^{n} \frac{2}{i+1}.$$

Since the right-hand side of this inequality diverges as $n \to \infty$ we have a contradiction. •

Let f be integrable on $[a, b]$ and continuous from the left at c, with $a < c \leq b$, and let

$$F(x) = \int_a^x f. \tag{2.63}$$

Given $\varepsilon > 0$ there is a positive η such that

$$f(c) - \varepsilon < f(t) < f(c) + \varepsilon$$

for $c - \eta < t \leq c$. Hence $|F(x) - F(c) - f(c)(c-x)| < \varepsilon(c-x)$ for $c - \eta < x \leq c$. This shows that the left-hand derivative $F'_-(c)$ exists and equals $f(c)$. A similar argument shows that $F'_+(c) = f(c)$ if f is continuous from the right at c and $a \leq c < b$. We have proved

THEOREM 2.6.8 (Fundamental Theorem. Second version) *If f is integrable on $[a, b]$ and F is defined by equation (2.63) then*

- *F has a right-hand derivative at every c with $a \leq c < b$ at which f is continuous from the right and then $F'_+(c) = f(c)$;*
- *F has a left-hand derivative at every c with $a < c \leq b$ at which f is continuous from the left and then $F'_-(c) = f(c)$.*

COROLLARY 2.6.9 (Existence of a primitive) *If f is continuous on $[a, b]$ then there exists a function G such that† $G' = f$ on $[a, b]$.*

The first version of the fundamental theorem can be schematically stated

$$\text{if} \quad F'(x) = f(x) \quad \text{then} \quad F(x) = F(a) + \int_a^x f; \qquad (2.64)$$

and the second version as

$$\text{if} \quad F(x) = F(a) + \int_a^x f \quad \text{then} \quad F'(x) = f(x). \qquad (2.65)$$

Implications (2.64) and (2.65) show that differentiation and integration can be regarded as inverse processes. However, in the Kurzweil–Henstock theory the validity of equations (2.64) and (2.65) is ensured under different assumptions. For (2.64) we need the mere existence of F' whereas (2.65) requires continuity of f as well. Continuity is really needed‡ as the following example shows. If $f(x) = x/|x|$ for $x \neq 0$ and $f(0) = $ anything then F fails to have a derivative at 0. A function f can be KH-integrable and discontinuous at every point of $[a, b]$, e.g. the Dirichlet function from Example 2.4.5. This leaves the possibility open that $F'(x)$ may fail to exist or to be equal to $f(x)$ at every x in $[a, b]$. Fortunately this is not so; in the next chapter we prove that $F'(x) = f(x)$ except for a null set.

2.7 Consequences of the Fundamental Theorem

The Kurzweil–Henstock theory provides powerful theorems on substitution and integration by parts as we shall see later in this book. Here we show that the Fundamental Theorem leads naturally and easily to reasonably good theorems, far better than the ones obtained for the Riemann integral.

THEOREM 2.7.1 (Integration by parts) *If F and G are continuous on $[a, b]$, $F' = f$ and $G' = g$ on $[a, b]$ except on a finite set§ then*

$$\int_a^b (Fg + Gf) = F(b)G(b) - F(a)G(a). \qquad (2.66)$$

---

† At a or b the derivative is one-sided.
‡ Not only in KH-theory but in any reasonable theory of integration.
§ If one is prepared to use Remark 2.6.5 to its full extent, this set can be countable.

Proof We have $(FG)' = Fg + Gf$ except on a finite set, the function FG is continuous on $[a, b]$ and by the Fundamental Theorem equation (2.66) follows.   •

Provided that one of the integrals in equation (2.67) below exists we obtain the usual formula

$$\int_a^b Fg = F(b)G(b) - F(a)G(a) - \int_a^b fG. \tag{2.67}$$

EXAMPLE 2.7.2   The assumptions of the theorem alone do not guarantee the existence of either integral in equation (2.67) as the following example shows. For $F(x) = x^2 \sin x^{-4}, G(x) = x^2 \cos x^{-4}$ for $x \neq 0$ and $F(0) = G(0) = 0$ neither $\int_0^1 Fg$ nor $\int_0^1 Gf$ exists because $F(x)g(x) - G(x)f(x) = 4x^{-1}$. For F and G chosen as above, equation (2.66) holds but formula (2.67) does not make sense. Moreover this example also shows that the product of a KH-integrable function f and a continuous function G need not be integrable.

REMARK 2.7.3 (The Taylor theorem)   It is sometimes stated even in very good texts that the Lagrange and Cauchy forms of remainder are more general than the integral form. This is not so much a statement about the nature of the theorem as about the concept of the integral used in these formulae. When using the KH-integral the opposite is true as the following theorem shows.

THEOREM 2.7.4 (Integral remainder) *If $f, f', \ldots, f^{(n)}$ are continuous on $[a, b]$ and $f^{(n+1)}$ exists except possibly on a finite set†, then the function $t \mapsto f^{(n+1)}(t)(b - t)^n/n!$ is KH-integrable and*

$$f(b) = f(a) + f'(a)(b - a) + \cdots + f^n(a)\frac{(b - a)^n}{n!} + R_n, \tag{2.68}$$

where

$$R_n = \int_a^b f^{(n+1)}(t)\frac{(b - t)^n}{n!}\, dt. \tag{2.69}$$

Proof By induction. For $n = 0$ the result is just the Fundamental Theorem. By equation (2.67) with $F = f^{(n+1)}$ and $G(t) = (b-t)^{(n+1)}/(n+1)!$ we have (the existence of the first integral is guaranteed by the induction

---

† If one is prepared to use Remark 2.6.5 to its full extent, this set can be countable.

hypothesis)

$$\int_a^b f^{(n+1)}(t)\frac{(b-t)^n}{n!}dt \;=\; f^{(n+1)}(a)\frac{(b-a)^{n+1}}{(n+1)!}$$
$$+ \int_a^b f^{(n+2)}(t)\frac{(b-t)^{n+1}}{(n+1)!}\,dt.$$

This means

$$R_n = f^{(n+1)}(a)\frac{(b-a)^{n+1}}{(n+1)!} + R_{n+1}. \qquad \bullet$$

Since the derivative need not exist everywhere on $[a,\,b]$ the Lagrange
(or Cauchy) remainder is not valid. Moreover the Lagrange (or Cauchy)
remainder follows from the above theorem under the additional assumption that $f^{(n+1)}$ exists everywhere on $[a,b]$. Indeed, $f^{(n+1)}$ has the
intermediate value property† and the function

$$f^{(n+1)}(t)\frac{(b-t)^n}{n!}$$

has a primitive. This follows by induction from

$$\frac{d}{dt}f^{(n+1)}(t)\frac{(b-t)^{n+1}}{(n+1)!} = f^{(n+2)}(t)\frac{(b-t)^{n+1}}{(n+1)!} - f^{(n+1)}(t)\frac{(b-t)^n}{n!}.$$

By Theorem 2.5.27

$$R_n = f^{(n+1)}(c)\int_a^b \frac{(b-t)^n}{n!}dt = f^{(n+1)}(c)\frac{(b-a)^{(n+1)}}{(n+1)!}.$$

Using equation (2.45) gives

$$R_n = f^{(n+1)}(c)\frac{(b-c)^n}{n!}(b-a),$$

i.e. the Cauchy remainder. Other remainder forms can also be derived;
for more details we refer to [43]. The point worth making is that with
the use of the Kurzweil–Henstock integral the increased validity of the
integral form of the remainder makes this form preferable to the other
ones.

THEOREM 2.7.5 (Substitution I) *If*

(i) *the function* $\varphi : [a,\,b] \mapsto [A,\,B]$,
(ii) φ *is continuous on* $[a,\,b]$,
(iii) φ *has a derivative on* $(a,\,b)$;

† see e.g. [4] p. 122.

(iv) *there is a function F, continuous on [A, B] with F'(x) = f(x) for x ∈ (A, B),*

then

$$\int_{\varphi(a)}^{\varphi(b)} f(x)dx = \int_a^b f(\varphi(t))\varphi'(t)dt \qquad (2.70)$$

REMARK 2.7.6 The theorem as stated suggests that $a < b$. However, a moment's reflection shows that, with obvious modifications of the assumptions, formula (2.70) is also valid for $a \geq b$. In the proof we assume $a < b$.

Proof of Theorem 2.7.5 By the Fundamental Theorem the left-hand side of (2.70) is $F(\varphi(b)) - F(\varphi(a))$. Noting that $[F(\varphi(t))]' = f(\varphi(t))\varphi'(t)$ for every $t \in (a, b)$ and applying the Fundamental Theorem to the integral on the right-hand side of (2.70) proves the theorem. •

REMARK 2.7.7 The strength of this theorem lies in the fact that no assumption is made concerning the monotonicity of φ. It is applicable for instance with $\varphi(t) = t \sin t^{-1}$, and we have

$$\int_0^{\frac{1}{\pi}} [\varphi(t)]^2 \varphi'(t)dt = 0.$$

Assumption (iv) is a little inconvenient. The most likely reason for using the formula is that we do not know F but then it might be difficult to justify the use of (2.70).† The next theorem does not have this weakness.

THEOREM 2.7.8 (Substitution II) *If we assume* (i)–(iii) *of Theorem 2.7.5 and if φ is strictly monotonic then formula (2.70) holds in the following sense: If one of the integrals appearing in (2.70) exists then so does the other and (2.70) is valid.*

Proof We shall consider the case of strictly increasing φ. The function φ associates with a partition π of $[a, b]$ a partition $\varphi \circ \pi$ of $[\varphi(a), \varphi(b)]$: if $\pi \equiv \{(t, [u, v])\}$ then $\varphi \circ \pi \equiv \{(\varphi(t), [\varphi(u), \varphi(v)])\} \equiv \{(x, [y, z])\}$. Also every partition $\omega \equiv \{(x, [y, z])\}$ of $[\varphi(a), \varphi(b)]$ is of the form $\varphi \circ \pi$ for some partition π of $[a, b]$, namely‡

$$\pi \equiv \{(\varphi_{-1}(x), [\varphi_{-1}(y), \varphi_{-1}(z)])\}.$$

---

† If f is continuous on $[a, b]$ then (iv) holds because of Corollary 2.6.9.
‡ φ_{-1} denotes the inverse of φ.

Given

$$\eta : [\varphi(a),\, \varphi(b)] \to \mathbb{R}_+$$

it is possible using continuity of φ to find

$$\delta_\eta : [a,\, b] \to \mathbb{R}_+$$

such that $\pi \ll \delta_\eta$ implies $\varphi \circ \pi \ll \eta$. Conversely, given $\delta : [a,\, b] \to \mathbb{R}_+$ there is a positive η_δ such that $\varphi \circ \pi \ll \eta_\delta$ implies $\pi \ll \delta$, where η_δ is given by

$$\eta_\delta(t) = \mathrm{Min}[\varphi(t) - \varphi(t - \delta(t)),\, \varphi(t + \delta(t)) - \varphi(t)].$$

The following lemma will facilitate the proof.

LEMMA 2.7.9 *For every $\varepsilon > 0$ there is a $\Delta : [a,\, b] \to \mathbb{R}_+$ such that for every Δ-fine partition π*

$$\left| \sum_\omega f(x)(z - y) - \sum_\pi f(\varphi(t))\varphi'(t)(v - u) \right| < \varepsilon, \qquad (2.71)$$

where $\omega = \varphi \circ \pi$.

Proof of the lemma From the alternative definition of derivative (Lemma 2.6.1) there is a positive $\Delta(t)$ such that

$$|\varphi(v) - \varphi(u) - \varphi'(t)(v - u)| < \frac{\varepsilon(v - u)}{[1 + |f(\varphi(t))|](b - a)}, \qquad (2.72)$$

whenever $t - \Delta(t) < u \le t \le v < t + \Delta(t)$. If $\pi \ll \Delta$ then inequality (2.71) follows.   ●

We now finish **proof of Theorem 2.7.8.** If the integral on the right-hand side of equation (2.70) exists then given $\varepsilon > 0$ there is a $\delta_1 : [a,\, b] \to \mathbb{R}_+$ such that if $\pi \ll \delta_1$ then

$$\left| \sum_\pi f(\varphi(t))\varphi'(t)(v - u) - \int_a^b f(\varphi(t))\varphi'(t)dt \right| < \varepsilon. \qquad (2.73)$$

Set $\delta = \mathrm{Min}(\delta_1, \Delta)$ and find the corresponding η_δ. If now $\omega \ll \eta_\delta$ then π is both δ-fine and Δ-fine and by inequalities (2.71) and (2.73)

$$\left| \sum_\omega f(x)(z - y) - \int_a^b f(\varphi(t))\varphi'(t)dt \right| < 2\varepsilon. \qquad (2.74)$$

This proves the existence of $\int_{\varphi(a)}^{\varphi(b)} f$ and equation (2.70). If the integral on the left-hand side of equation (2.70) exists then there is η_1 such that

$$\left| \sum_\omega f(x)(z-y) - \int_a^b f \right| < \varepsilon, \qquad (2.75)$$

whenever $\omega \ll \eta_1$. Let $\eta = \text{Min}(\eta_1, \eta_\Delta)$. If $\pi \ll \delta_\eta$ then by inequalities (2.71) and (2.73)

$$\left| \sum_\pi f(\varphi(t))\varphi'(t)(v-u) - \int_a^b f \right|. \qquad \bullet$$

2.8 Improper integrals

The function f, $f(x) = \log x$, has a primitive $F(x) = x \log x - x$ on every interval $[c, 1]$ with $0 < c$, and by the Fundamental Theorem

$$\int_c^1 \log x \, dx = -1 - c \log c + c.$$

It is not possible to apply the Fundamental Theorem directly to the interval $[0, 1]$, i.e. to set $c = 0$ in the above formula, because F is not defined at 0. This difficulty is however easily circumvented. The limit $\lim_{c \downarrow 0} F(c)$ exists and if we extend the definition of F by defining $F(0) = \lim_{c \downarrow 0} F(c) = 0$ then F becomes continuous on the whole of $[0, 1]$ and we obtain $\int_0^1 \log x \, dx = -1$. We calculated $\int_0^1 f$ as $\lim_{c \downarrow 0} \int_c^1 f$. It is easily seen that this can always be done if the primitive F has a limit at 0. Naturally, one asks: is it always true that

$$\int_a^b f = \lim_{c \downarrow a} \int_c^b f \ ? \qquad (2.76)$$

Equation (2.76) can be interpreted in two different ways:

- if the integral exists then so does the limit and equation (2.76) holds;
- if the limit exists then so does the integral and equation (2.76) holds.

The first interpretation is, indeed, true, by the the continuity of the in-definite integral (Theorem 2.5.29). This, however, is not of great help if we want to use equation (2.76) to evaluate the integral $\int_a^b f$ because the existence of this integral may not be known. It is an advantage of the KH-theory that the second interpretation is also correct: as long as we can evaluate the right-hand side of (2.76) we also have the integral. In other theories of integration the second interpretation of equation

(2.76) could be false. For example $\mathcal{R}\int_0^1 \log x\, dx$ does not exist because the the logarithm is not bounded on $(0, 1]$, but we have just seen that $\lim_{c\downarrow 0}\int_c^1 \log x\, dx = -1$. In some theories of integration, e.g. in the Riemann theory, equation (2.76) is used to *define* the integral on the left-hand side. This is often referred to as the Cauchy extension of the integral; also the integral so defined is often referred to as an improper integral. Our statement above concerning the validity of the second interpretation of equation (2.76) can be rephrased by saying: There are no improper integrals in the KH-theory. Before we prove this we need

LEMMA 2.8.1 *Let ε be positive, f integrable on $[a, b]$ and δ such that*

$$\left| \sum_\pi f - \int_a^b f \right| < \varepsilon \qquad (2.77)$$

whenever $\pi \ll \delta$. Then for every c, $a < c < b$,

$$\left| \sum_\pi {}_c^b f - \int_c^b f \right| < 2\varepsilon. \qquad (2.78)$$

$$\left| \sum_\pi {}_a^c f - \int_a^c f \right| < 2\varepsilon. \qquad (2.79)$$

REMARK 2.8.2 This is a special case of Henstock's lemma proved in the next chapter; this simpler version will suffice here. Henstock's lemma marks the departure point from the most elementary part of the KH-theory and most of the important advanced theorems depend directly or indirectly on it.

Proof of Lemma 2.8.1 If π is any partition with c as one of the dividing points then

$$\left| \sum_\pi {}_c^b f - \int_c^b f \right| \leq \left| \sum_\pi {}_a^b f - \int_a^b f \right| + \left| \sum_\pi {}_a^c f - \int_a^c f \right|. \qquad (2.80)$$

Since f is integrable on $[a, c]$ there is a positive δ_c such that

$$\left| \sum_{\pi_c} {}_a^c f - \int_a^c f \right| < \varepsilon \qquad (2.81)$$

whenever the partition π_c of $[a, c]$ is δ_c-fine. We now choose a partition of $[a, b]$ consisting of a δ-fine partition of $[c, b]$ and a partition π_c of $[a, c]$

such that both $\pi_c \ll \delta_c$ and $\pi_c \ll \delta$. Then by inequalities (2.80), (2.77) and (2.81)

$$\left| \sum_{\pi} {}_c^b f - \int_c^b f \right| < 2\varepsilon. \tag{2.82}$$

The inequality (2.79) can be proved similarly. •

THEOREM 2.8.3 *If f is integrable on $[c, b]$ for every c with $a < c < b$ and*

$$\lim_{c \downarrow a} \int_c^b f \text{ exists and equals } A \tag{2.83}$$

then f is integrable on $[a, b]$ and $\int_a^b f = A$.

REMARK 2.8.4 A similar theorem holds for $\lim_{c \uparrow b} \int_a^c f$.

The above theorem and this remark are sometimes expressed by saying that the KH-integral is closed under the Cauchy extension. We have seen that the theorem is false if integrable is taken to mean Riemann integrable. The Riemann integral is not closed under the Cauchy extension.

Proof of Theorem 2.8.3 We can and shall assume without loss of generality that $f(a) = 0$. Take a strictly decreasing sequence $\{c_n\}$ with $c_n \downarrow a$ and $c_0 = b$. For every positive ε there exists δ_n such that if π_n is a δ_n-fine partition of $[c_n, c_{n-1}]$ then

$$\left| \sum_{\pi_n} f - \int_{c_n}^{c_{n-1}} f \right| < \frac{\varepsilon}{2^{n+1}}. \tag{2.84}$$

According to the hypothesis there exists r such that

$$\left| \int_c^b f - A \right| < \frac{\varepsilon}{2} \tag{2.85}$$

whenever $a < c < r$. Now define $\delta : [a, b] \mapsto \mathbb{R}_+$ which has, for all $n \in \mathbb{N}$, the following properties.†

$$\delta(x) \leq \delta_n(x) \text{ for } x \in [c_n, c_{n-1}],$$

$$[x - \delta(x), x + \delta(x)] \subset (c_n, c_{n-1}) \text{ for } x \in (c_n, c_{n-1}), \tag{2.86}$$

† The first inequality implies that $\delta(c_n) \leq \text{Min}(\delta_n(c_n), \delta_{n+1}(c_n))$.

$$(c_n - \delta(c_n),\, c_n + \delta(c_n)) \subset (c_{n+1},\, c_{n-1}). \tag{2.87}$$

In addition we also require that

$$\delta(a) \;=\; r - a, \tag{2.88}$$

$$\delta(b) \;=\; \frac{1}{2}(b - c_1). \tag{2.89}$$

Let now π be a δ-fine partition of $[a, b]$,

$$\pi \equiv \{(x_i, [u_i, v_i]); i = 1, 2, \dots, p\}.$$

Relations (2.86) and (2.87) imply that $x_1 = a$ and hence $f(x_1) = 0$. The Riemann sum $\sum_\pi f$ starts with the term $f(x_2)(v_2 - u_2)$. Let N be the first integer such that $c_N < v_1 = u_2$. As the next step we show that the Riemann sum $\sum_\pi f$ over $[a, b]$ divides† into smaller sums over $[c_n, c_{n-1}]$, i.e.

$$\sum_\pi f = \sum_\pi {}_{u_2}^{c_{N-1}} f + \sum_{i=1}^{N-1} \sum_\pi {}_{c_i}^{c_{i-1}} f. \tag{2.90}$$

Any c_n with $n < N$ must lie in some $[u_k, v_k]$ with $k \geq 2$. Inclusions (2.86), (2.87) imply that neither $c_n < x_k \leq v_k$ nor $u_k \leq x_k < c_n$. Consequently $x_k = c_n$ and there is at most one c_j in any $[u_i, v_i]$. The partition π anchors on all c_n with $n < N$ and we have established formula (2.90). Now we estimate $\sum_\pi f - A$ by

$$\left| \sum_\pi f - A \right| \leq \left| \sum_\pi f - \int_{u_2}^b f \right| + \left| \int_{u_2}^b f - A \right|. \tag{2.91}$$

By inequality (2.88) we have $u_2 = v_1 < r$. Consequently the second term on the right-hand side of inequality (2.91) is less than $\varepsilon/2$ by (2.85). For the estimate of the first term we write

$$\left| \sum_\pi f - \int_{u_2}^b f \right| \leq \left| \sum_\pi {}_{u_2}^{c_{N-1}} f - \int_{u_2}^{c_{N-1}} f \right| + \sum_{i=1}^{N-1} \left| \sum_\pi {}_{c_i}^{c_{i-1}} f - \int_{c_i}^{c_{i-1}} f \right|. \tag{2.92}$$

Now we apply Lemma 2.8.1 with the rôles of $[a, b]$ and c played by $[c_N, c_{N-1}]$ and $u_2 = v_1$, respectively.

$$\left| \sum_\pi {}_{u_2}^{c_{N-1}} f - \int_{u_2}^{c_{N-1}} f \right| \leq 2\frac{\varepsilon}{2^{N+1}}. \tag{2.93}$$

---

† If $u_2 = c_{N-1}$ then the first sum on the right-hand side of the next equation is interpreted as 0.

For the remaining estimate we use inequality (2.84).

$$\sum_{i=1}^{N-1} \left| \sum_{\pi} {}^{c_{i-1}}_{c_i} f - \int_{c_i}^{c_{i-1}} f \right| < \sum_{i=1}^{N-1} \frac{\varepsilon}{2^{i+1}}. \tag{2.94}$$

Combining inequalities (2.91)—(2.94) we have

$$\left| \sum_{\pi} f - A \right| < \varepsilon. \qquad \bullet$$

Theorem 2.8.3 leads naturally to the following test of integrability.

Integrability test *If, for every c, $a < c < b$, f is KH-integrable on $[c, b]$ and there are KH-integrable functions G, g such that*

$$g(x) \le f(x) \le G(x)$$

for all $x \in [a, b]$ then f is KH-integrable on $[a, b]$.

For every positive ε there is a d such that

$$\left| \int_{a'}^{a''} g \right| < \varepsilon \quad \text{and} \quad \left| \int_{a'}^{a''} G \right| < \varepsilon$$

whenever $a < a'$, $a'' < d$. Consequently

$$\left| \int_{a'}^{a''} f \right| < \varepsilon$$

and the limit in (2.83) exists.

An analogous test holds for $c \uparrow b$ if f is KH-integrable on $[a, c]$. For an application of the integrability test see Exercise 2.12.

2.9 Integrals over unbounded intervals

In applications one encounters the integral

$$\int_0^\infty \frac{\sin x}{x} \, dx.$$

In calculus or in the Riemann† theory of integration this integral is defined as a limit, i.e.

$$\int_0^\infty \frac{\sin x}{x} \, dx \overset{\text{def}}{=} \lim_{b \to \infty} \int_0^b \frac{\sin x}{x} \, dx.$$

In the Kurzweil–Henstock theory it is possible to recast Definition 2.4.1 in such a way as to include in it also integrals over unbounded intervals

† Even in the Lebesgue theory.

like the integral above. We have defined the KH-integral over closed intervals. Since the values of the function at two points do not influence the existence or the value of the integral we could have used open intervals instead but we did not. For consistency we define the integral over unbounded intervals also over closed intervals but we have to introduce these first. Let us make the following definitions. We add two elements ∞ and $-\infty$ to the reals and denote the extended set by $\overline{\mathbb{R}}$, i.e. $\overline{\mathbb{R}} = \mathbb{R} \cup \{\infty\} \cup \{-\infty\}$. Here ∞ and $-\infty$ are not real numbers and are sometimes referred to as infinite numbers. Noting that intervals (A, B) are already well defined even if A or B is ∞ or $-\infty$ we declare the sets $[A, B] = (A, B) \cup \{A\} \cup \{B\}$ as closed intervals.† The use of capital letters is a matter of emphasis, and should alert the reader to the possibility that A or B could be ∞ or $-\infty$. However, sometimes a closed interval in $\overline{\mathbb{R}}$ will be denoted by $[a, b]$ or $[u, v]$.

The geometric meaning behind the definition below is that for a non-negative function f on an unbounded interval $[A, B]$ the area under the graph of f is approximated by Riemann sums for δ-fine partitions of sufficiently large bounded intervals.

DEFINITION 2.9.1 *A number I is the Kurzweil–Henstock integral of f over $[A, B]$ if for every positive ε there are a function $\delta : [A, B] \to \mathbb{R}_+$ and a positive number Δ such that*

$$\left| \sum_\pi f - I \right| < \varepsilon \tag{2.95}$$

whenever π is a δ-fine partition of a bounded interval $[a, b]$ with $[a, b] \supset [A, B] \cap [-\Delta, \Delta]$.

The integral will be denoted as usual by $\int_A^B f$ or $\int_A^B f(x)dx$ or by other symbols used for the KH-integral. There is at most one number I satisfying the requirements of the definition, in other words the number I is well defined. The proof is rather similar to the proof of Theorem 2.4.6 and is omitted. To see that for a closed bounded interval $[A, B]$ the above definition agrees with the original one (Definition 2.4.1) it is sufficient to choose $\Delta > \text{Max}(|A|, |B|)$.

EXAMPLE 2.9.2 We wish to show that $\int_1^\infty x^{-2}dx = 1$. We choose $\delta(x) = \varepsilon x/3(1+\varepsilon)$ and $\Delta > 3/\varepsilon$. Let $\pi \equiv (\xi, [u, v])$ be a δ-fine partition

† This is, of course, consistent in case A or B is in $\mathbb{R}$.

of $[1, v_n]$ with $v_n > \Delta$. Motivated by the Fundamental Theorem and by the fact that $-\frac{d}{dx}x^{-1} = x^{-2}$ we shall approximate

$$\frac{1}{\xi^2}(v - u) \quad \text{by} \quad \frac{1}{u} - \frac{1}{v}.$$

Firstly

$$1 \geq \sum_{\pi}\left(\frac{1}{u} - \frac{1}{v}\right) = 1 - \frac{1}{v_n} > 1 - \frac{\varepsilon}{3}, \qquad (2.96)$$

secondly

$$1 \leq \frac{v}{\xi} < \frac{\xi + \delta(\xi)}{\xi} < 1 + \frac{\varepsilon}{3},$$

$$1 \leq \frac{\xi}{u} < \frac{\xi}{\xi - \delta(\xi)} = \frac{3(1 + \varepsilon)}{3 + 2\varepsilon} < 1 + \frac{\varepsilon}{3}.$$

This leads to

$$\left|\frac{1}{\xi^2}(v - u) - \left(\frac{1}{u} - \frac{1}{v}\right)\right| = \left|\frac{uv}{\xi^2} - 1\right|\left(\frac{1}{u} - \frac{1}{v}\right)$$

$$\leq \left|\frac{v}{\xi} - \frac{\xi}{u}\right|\left(\frac{1}{u} - \frac{1}{v}\right) \leq \frac{\varepsilon}{3}\left(\frac{1}{u} - \frac{1}{v}\right).$$

Using these estimates on the Riemann sum yields

$$\sum_{\pi}\left[\frac{1}{\xi^2}(v - u) - \left(\frac{1}{u} - \frac{1}{v}\right)\right] \leq \frac{\varepsilon}{3}\sum_{\pi}\left(\frac{1}{u} - \frac{1}{v}\right).$$

Combining this inequality with (2.96) gives

$$\left|\sum_{\pi}\frac{1}{\xi^2}(v - u) - 1\right| < \frac{\varepsilon}{3} + \frac{\varepsilon}{3}. \qquad \bullet$$

THEOREM 2.9.3 *f is KH-integrable on* $[a, \infty]$ *if and only if*

$$\lim_{b \to \infty}\int_a^b f$$

exists and then

$$\lim_{b \to \infty}\int_a^b f = \int_a^\infty f. \qquad (2.97)$$

REMARK 2.9.4 Similar theorems hold for $\int_{-\infty}^b f$ and for $\int_{-\infty}^\infty f = \int_{-\infty}^0 f + \int_0^\infty f$.

Proof of Theorem 2.9.3 I. Assume that the integral exists and for $\varepsilon > 0$ let δ and Δ be as in Definition 2.9.1. Given $b > \Delta$, we find δ_1 such that for every δ_1-fine partition π of $[a, b]$

$$\left| \sum_\pi f - \int_a^b f \right| < \varepsilon. \tag{2.98}$$

Define $\delta_2 = \text{Min}(\delta, \delta_1)$ and let π be δ_2-fine. Then by (2.95) and (2.98)

$$\left| \int_a^b f - I \right| < 2\varepsilon.$$

This proves (2.97).

II. Assume the limit in (2.97) exists and equals l. For $\varepsilon > 0$ there is $\Delta > 0$ such that

$$\left| \int_a^b f - l \right| < \frac{\varepsilon}{2} \tag{2.99}$$

for $b > \Delta$. Take a strictly increasing sequence $n \mapsto b_n$ with $b_1 = a$ and $\lim_{n \to \infty} b_n = \infty$. For $\varepsilon > 0$ there is $\delta_n : [b_n, b_{n+1}] \to \mathbb{R}_+$ such that

$$\left| \sum_{\pi_n} f - \int_{b_n}^{b_{n+1}} f \right| < \frac{\varepsilon}{2^{n+2}} \tag{2.100}$$

whenever π_n is a δ_n-fine partition of $[b_n, b_{n+1}]$. Let $\delta : [a, \infty] \to \mathbb{R}_+$ be such that for all $n \in \mathbb{N}$

$$\delta(\xi) \leq \delta_n(\xi) \quad \text{if} \quad \xi \in [b_n, b_{n+1}], \tag{2.101}$$
$$[\xi - \delta(\xi), \xi + \delta(\xi)] \subset (b_n, b_{n+1}) \quad \text{if} \quad \xi \in (b_n, b_{n+1}).$$

If

$$b_N < b \leq b_{N+1} \tag{2.102}$$

and $\pi \ll \delta$ then, for reasons similar to those given in the proof of Theorem 2.8.3, the partition π anchors on all b_n with $n \leq N$ and

$$\sum_\pi {}_a^b f = \sum_1^{N-1} \sum_\pi {}_{b_n}^{b_{n+1}} f + \sum_\pi {}_{b_N}^b f. \tag{2.103}$$

According to (2.100) and (2.101)

$$\left| \sum_\pi {}_{b_n}^{b_{n+1}} f - \int_{b_n}^{b_{n+1}} f \right| < \frac{\varepsilon}{2^{n+2}}. \tag{2.104}$$

Let $b > \Delta$ and N be as in (2.102). If π is a δ-fine partition of $[a, b]$ then by (2.99), (2.103) and (2.104)

$$\left| \sum_\pi f - l \right| < \frac{\varepsilon}{2} + \frac{\varepsilon}{4} \sum_1^{N-1} \frac{1}{2^n} + \left| \sum_\pi {}^b_{b_N} f - \int_{b_N}^b f \right|.$$

The last term in this inequality is less than $\varepsilon/2^{N+2}$ by inequality (2.79) in Lemma 2.8.1 where the rôles of $[a, b]$ and c are played by $[b_N, b_{N+1}]$ and b, respectively. $\bullet$

EXAMPLE 2.9.5 (The Frullani integral) As an example we evaluate the integral

$$\int_0^\infty \frac{f(bx) - f(ax)}{x} dx, \qquad (2.105)$$

by using Theorem 2.9.3. We make the following assumptions:

(a) $0 < a < b$,
(b) f is continuous on $[0, \infty)$,
(c) $\lim_{x\to\infty} f(x)$ exists and we denote it by $f(\infty)$.

We write

$$\int_0^\infty \frac{f(bx) - f(ax)}{x} dx = \lim_{c\to 0} \int_c^1 \frac{f(bx) - f(ax)}{x} dx$$
$$+ \lim_{C\to\infty} \int_1^C \frac{f(bx) - f(ax)}{x} dx \quad (2.106)$$

and evaluate the first limit; the second can be obtained similarly. Using substitution and the mean value theorem 2.5.25 we have

$$\int_c^1 \frac{f(bx) - f(ax)}{x} dx = \int_{cb}^b \frac{f(t)}{t} dt - \int_{ca}^a \frac{f(t)}{t} dt$$
$$= \int_a^b \frac{f(t)}{t} dt - \int_{ca}^{cb} \frac{f(t)}{t} dt = -f(\xi) \int_{ca}^{cb} \frac{dt}{t} + \int_a^b \frac{dt}{t},$$

with ξ between ca and cb. Sending $c \downarrow 0$ gives $f(\xi) \to f(0)$. Applying similar arguments to the second limit in (2.106) leads to

$$\int_0^\infty \frac{f(bx) - f(ax)}{x} dx = f(0) \log \frac{a}{b} - f(\infty) \log \frac{a}{b}. \qquad \bullet$$

Theorem 2.9.3 leads naturally to the following test of integrability.

Integrability test *If, for every* c, $a < c < \infty$, f *is KH-integrable on* $[a, c]$ *and there are KH-integrable functions* G, g *such that*

$$g(x) \le f(x) \le G(x)$$

for all $x \in [a, \infty)$ *then* f *is KH-integrable on* $[a, \infty)$.
The proof is very similar to the proof of the integrability test in Section 2.8. For an application of this integrability test see Exercise 2.13. An analogous test holds for $c \to -\infty$ if f is KH-integrable on $[c, b]$.

2.10 Alternative approach to integration over unbounded intervals

In the previous section we introduced the set $\overline{\mathbb{R}}$ and closed intervals in it. In this alternative approach open intervals in $\overline{\mathbb{R}}$ are needed. First we systematically state the basic arithmetic rules† in $\overline{\mathbb{R}}$ and extend the order from $\mathbb{R}$ to $\overline{\mathbb{R}}$. For $c \in \mathbb{R}$ we define:

$$\infty + c = c + \infty = \infty \text{ for any } -\infty < c. \tag{2.107}$$

$$(-\infty) + c = c + (-\infty) = -\infty \text{ for any } c < \infty. \tag{2.108}$$

$$\text{If } c > 0 \text{ then } c.\infty = \infty.c = \infty \text{ and } c.(-\infty) = (-\infty).c = -\infty. \tag{2.109}$$

$$\text{If } c < 0 \text{ then } c.\infty = \infty.c = -\infty \text{ and } c.(-\infty) = (-\infty).c = \infty. \tag{2.110}$$

Note that $\infty + (-\infty)$ and similar expressions are not defined. The following definition‡ is a little surprising, but it is convenient and we shall adhere to it.

$$0.\infty = \infty.0 = 0 \quad \text{and also} \quad 0.(-\infty) = (-\infty).0 = 0. \tag{2.111}$$

We declare

$$-\infty < c < \infty \tag{2.112}$$

for any real number c. For the closed intervals in $\overline{\mathbb{R}}$ we now have $[A, B] = \{x; A \leq x \leq B\}$, exactly as in $\mathbb{R}$. With A,B in $\overline{\mathbb{R}}$ we declare as open intervals in $\overline{\mathbb{R}}$ all intervals $(A, B) \subset \mathbb{R}$ as well as $\overline{\mathbb{R}}$ itself and the sets

$$[-\infty, B) = \{x; x \in \overline{\mathbb{R}}, x < B\},$$
$$(A, \infty] = \{x; x \in \overline{\mathbb{R}}, x > A\}.$$

The supremum of a set which is either not bounded above or contains ∞ is ∞, similarly the infimum of a set which is either not bounded below or contains $-\infty$ is $-\infty$.

---

† Not all these rules are needed now but for sake of reference we keep them together.
‡ One has to regard such definitions purely formally as a matter of convenience and devoid of any deep meaning. In particular one should divorce such definitions from rules of calculating limits of products. However, the definition $\infty.0 = 0$ is also often used in Lebesgue integration theory.

The concept of being δ-fine can be rephrased in a way which is suitable to generalization to infinite intervals. With

$$\delta : [a, b] \mapsto \mathbb{R}_+$$

we can associate a map δ from $[a, b]$ into the set of open intervals in $\mathbb{R}$ by setting

$$\delta(\xi) = (\xi - \delta(\xi), \xi - \delta(\xi)). \qquad (2.113)$$

A partition $\pi \equiv \{(\xi, [u, v])\}$ is then δ-fine if and only if

$$[u, v] \subset \delta(\xi). \qquad (2.114)$$

We use (2.114) for defining δ-fine for unbounded intervals. If now δ is defined as in (2.113) and moreover $\delta(\infty) = (\Delta, \infty]$ and $\delta(-\infty) = [-\infty, -\Delta)$ then a partition of any closed bounded interval $[a, b]$ containing $[A, B] \cap [-\Delta, \Delta]$ is δ-fine if and only if inclusion (2.114) holds. In future we shall call a map $\hat{\gamma}$ from a closed interval $[A, B]$ into the set of open intervals in $\overline{\mathbb{R}}$ a gauge if $\xi \in \hat{\gamma}(\xi)$ for every $\xi \in [A, B]$ and $\hat{\gamma}(\xi)$ is a *bounded* open interval for every $\xi \in \mathbb{R} \cap [A, B]$. A partition of a closed interval in $\overline{\mathbb{R}}$ is, similarly as in Section 1.2, a set of pairs (ξ_k, I_k), $k = 1, \ldots, n$, such that the points $\xi_k \in I_k$, and the closed intervals I_k are non-overlapping with $\bigcup_1^n I_k = [A, B]$. We shall say that a partition $\pi \equiv \{(\xi_k, J_k)\}$ of $[A, B]$ is δ-fine if δ is a gauge defined on $[A, B]$ and $J_k \subset \delta(\xi_k)$. It is worth noting that at most two intervals of a partition are unbounded and if δ is a gauge the unbounded intervals of a δ-fine partition are tagged by either ∞ or $-\infty$. We need to extend Riemann sums to partitions of possibly unbounded intervals. If $\{(x_k, J_k)\}$ is such a partition then we shall include in the sum $\sum_\pi f = \sum_\pi f(x_k)|J_k|$ only those terms for which J_k is a bounded interval. This can be achieved by simply postulating it or, as some authors prefer, by defining the length of an unbounded interval as ∞, demanding that $f(\pm\infty) = 0\dagger$. If we adopt the definition (2.111) then $f(x_k)|J_k| = 0.\infty = 0$ for unbounded J_k and $f(x_k)|J_k|$ can be included into the Riemann sum without affecting it. We are now in a position to formulate the 'all embracing' definition of the KH-integral.

DEFINITION 2.10.1 *A number I is the Kurzweil–Henstock integral*

$\dagger$ The function f to be integrated usually comes undefined at ∞ and $-\infty$ but there is no harm in making a convention that all integrands automatically satisfy $f(\pm\infty) = 0$.

*(or just integral) of f from A to B (or on [A, B]) if for every positive ε
there is a gauge δ such that for every δ-fine partition π*

$$\left| \sum_\pi f - I \right| < \varepsilon. \tag{2.115}$$

The gauge definition of the KH-integral has the advantage that it
reads word by word the same for a bounded or unbounded interval.
Since mathematics is a deductive science it would have been preferable
if we had started with one of the more general definitions 2.9.1 or 2.10.1.
This would be also more economical because after the introduction of a
more general definition one is obliged to prove all the theorems afresh.
This we shall not do because the new proofs would be very similar to
those already supplied. The reason why we integrated over bounded
intervals first was methodological, we did not want to obscure the main
ideas by technicalities involving the infinite numbers ∞ and $-\infty$. In
Chapter 4 (The SL-integral) we explain yet another approach to KH-
integration and the definition there covers integrals over both bounded
and unbounded intervals.

2.11 Negligible sets

The next two theorems provide a full characterization of null sets. Recall
that in Section 1.4 we defined a set S to be of *measure zero* if for every
positive ε there was a countable system of intervals $\{I_j; j = 1, 2, \dots\}$
covering S with

$$\sum_{j=1}^\infty |I_j| < \varepsilon. \tag{2.116}$$

A set of measure zero can always be covered by a *disjoint* system of
intervals $\{I_j\}$ such that inequality (2.116) holds. However we shall not
use this fact here. For the proof see Exercises 2.25 and 2.26. If, for every
natural j, the set S_j is of measure zero then so is $\bigcup_1^\infty S_j$. This can be
proved by covering each S_j by a system I_k^j with $\sum_{k=1}^\infty |I_k^j| < \varepsilon 2^{-j-1}$. In
particular every countable set is of measure zero. If something happens
except on a set of measure zero we say that it happens *almost everywhere*. For example, if $\{x; f(x) \le 0\}$ is of measure zero we say that f is
positive almost everywhere. The phrase almost everywhere will be often
abbreviated to a.e.

THEOREM 2.11.1 *If S is of measure zero then it is null.*

Proof Let $\varepsilon > 0$. Cover S by a system $\{I_j; j = 1, 2, \dots\}$ such that inequality (2.116) holds. Set $E_j = I_j \setminus (I_1 \cup \cdots \cup I_{j-1})$. The sets E_j are disjoint and $S \subset \bigcup_1^\infty E_j$. If $x \in S$ there is a uniquely determined k for which $x \in E_k$. Let $\delta(x)$ be positive and such that $(x - \delta(x), x + \delta(x)) \subset I_k$. If $x \notin S$ let $\delta(x) = 1$. For a δ-fine partition π of an arbitrary interval $[a, b]$ the total length of intervals of π in I_j is less than $|I_j|$ and therefore

$$\sum_\pi 1_S \leq \sum_1^\infty |I_j| < \varepsilon.$$

Consequently $\int_a^b 1_S = 0$. ●

THEOREM 2.11.2 *If S is null then it is of measure zero.*

It is convenient for later purposes to prove without additional effort a more general result:

THEOREM 2.11.3 *If $S \subset \mathbb{R}$ and 1_S is integrable then for every positive ε there is a system of open intervals J_n, $n = 1, 2, \dots$ such that*

$$S \ \subset \ \bigcup_1^\infty J_n, \tag{2.117}$$

$$\sum_1^\infty |J_n| \ \leq \ \int_{-\infty}^\infty 1_S + \varepsilon. \tag{2.118}$$

For the proof we need a lemma.

LEMMA 2.11.4 (Covering lemma†) *If S is an arbitrary set, $S \subset \mathbb{R}$,*

$$\delta : S \to \mathbb{R}_+$$

then there exists a countable system of closed non-overlapping intervals $\{K_i; i = 1, 2, \dots\}$ such that for every K_i there exists $x_i \in S$ with

$$K_i \subset (x_i - \delta(x_i), x_i + \delta(x_i)) \tag{2.119}$$

and

$$S \subset \bigcup_1^\infty K_i. \tag{2.120}$$

† [30] page 143.

Proof of the lemma We construct the sequence $\{K_i\}$ in steps. In the i-th step we divide the interval $[-i, i]$ into $2i2^i$ closed non-overlapping intervals J_α of equal length. If some J_α contains $x \in S$ with $J_\alpha \subset (x - \delta(x), x + \delta(x))$ then we denote it by K, add it to the intervals previously selected, and number it and the point x accordingly. Now it remains to prove inclusion (2.120). For $x \in S$ we find $i \in \mathbb{N}$ so large that

$$-i \le x \le i \qquad \text{and} \qquad 2^{-i} < \delta(x). \tag{2.121}$$

If x lies in some interval selected before the i-th step there is nothing to prove, hence assume the contrary. There is an interval of the form $[k2^{-i}, (k+1)2^{-i}]$ which contains x and by (2.121) we have $[k2^{-i}, (k+1)2^{-i}] \subset (x - \delta(x), x + \delta(x))$. The interval $[k2^{-i}, (k+1)2^{-i}]$ becomes selected as one of the K_j in the i-th step.   •

Proof of Theorem 2.11.3 It is sufficient to prove the theorem under the additional assumption† that $S \subset [a, b] \subset \mathbb{R}$. Denote $A = \int_a^b 1_S = \int_{-\infty}^\infty 1_S$ and take $\eta > 0$. By the definition of the KH-integral there exists a $\delta : [a, b] \mapsto \mathbb{R}_+$ such that for every δ-fine partition π of $[a, b]$ we have

$$0 \le \sum_\pi 1_S < A + \eta. \tag{2.122}$$

Let $\{K_n\}$ and x_n be as in Lemma 2.11.4 and $N \in \mathbb{N}$. The set $[a, b] \setminus \bigcup_1^N K_i$ consists of finitely many disjoint intervals. Therefore it is possible to find a δ-fine partition π_1 of $[a, b]$ such that for all $i \le N$ the intervals K_i are intervals of π_1. Consequently

$$\sum_1^N |K_n| \le \sum_{\pi_1} 1_S < A + \eta,$$

the last inequality holding because inequality (2.122) applies with π replaced by π_1. Sending $N \to \infty$ gives $\sum_1^\infty |K_n| \le A + \eta$. Let J_n be an open interval of length $(1 + \eta)|K_n|$ and containing K_n. The intervals J_n are open, they cover S and

$$\sum_1^\infty |J_n| \le A + \eta(A + 1 + \eta).$$

The choice $\eta = \text{Min}(1, \dfrac{\varepsilon}{A + 2})$ gives (2.118).   •

THEOREM 2.11.5   *If $f \ge 0$ and $\int_a^b f = 0$ then $f = 0$ almost everywhere.*

† See Exercise 2.15.

Proof Let $S_n = \{x : f(x) \geq 1/n\}$ and let $\mathbf{1}_n$ be the characteristic function of S_n. Then $\mathbf{1}_n \leq nf$ and consequently

$$\int_a^b \mathbf{1}_n = 0.$$

For every natural n the set S_n is of measure zero, and consequently so is the union $\bigcup_1^\infty S_n = \{x : f(x) \neq 0\}$. •

Proof of Theorem 2.5.27 If $f = 0$ almost everywhere then both sides of equation (2.44) are zero and this equation holds with any $c \in (a, b)$. If f is not zero a.e. then $\int_a^b f > 0$ and equation (2.47) defines μ and then

$$\int_a^b f(g - \mu) = 0.$$

If

$$g(x) \neq \mu \quad \text{for every} \quad x \in (a, b) \tag{2.123}$$

then because of the intermediate value property either $g(x) > \mu$ or $g(x) < \mu$ for all $x \in (a, b)$. In either case $f(x)(g(x) - \mu) = 0$ almost everywhere, which is impossible. Consequently relation (2.123) is untenable. •

2.12 Complex valued function

We presume that the reader is familiar with complex numbers, but we review notation and some definitions for complex valued functions in this section. A complex number z is a number of the form $z = x + \jmath y$ with $x, y \in \mathbb{R}$ and $\jmath^2 = -1$. The real part of z is x, the imaginary part of z is y, in symbols $x = \Re z$ and $y = \Im z$. The absolute value of z is denoted by $|z|$, $|z| = \sqrt{x^2 + y^2}$. For every complex number $z \neq 0$ there exists a unique φ with $0 \leq \varphi < 2\pi$ such that $z = |z|(\cos \varphi + \jmath \sin \varphi)$. The set of all complex numbers is denoted by $\mathbb{C}$. A complex valued function of a real variable is a rule which associates with every x from some set E, the domain of definition of f, a complex number $f(x)$. Hence we have $f(x) = f_1(x) + \jmath f_2(x)$, where f_1 and f_2 are real valued functions, $f_1(x) = \Re f(x)$ and $f_2(x) = \Im f(x)$. Most concepts from real analysis are easily transferred to complex valued functions of a real variable either by mimicking the definition of the real valued situation or by applying the relevant definition to real and imaginary parts. We illustrate this duality in case of a limit.

Definition *A function $f : (a, b) \mapsto \mathbb{C}$ has a limit L at $c \in (a, b)$ if and only if for every positive ε there is a positive δ such that $|f(x) - L| < \varepsilon$ whenever $|x - c| < \delta$.*

Characterization *A function $f : (a, b) \mapsto \mathbb{C}$ has a limit L at $c \in (a, b)$ if and only if $\lim\limits_{x \to c} \Re f(x) = \Re L$ and $\lim\limits_{x \to c} \Im f(x) = \Im L$.*

A similar situation exists with limit from the right, limit from the left, continuity and derivative. For the KH-integral we have

Definition *A complex number I is the Kurzweil–Henstock integral (KH-integral) of a function*

$$f : [A, B] \mapsto \mathbb{C}$$

if and only if for every positive ε there is a gauge δ such that for every δ-fine partition π the inequality (2.115) is satisfied.

Characterization *A complex number I is the Kurzweil–Henstock integral (KH-integral) of a function $f : [A, B] \mapsto \mathbb{C}$ if and only if*

$$\mathcal{KH} \int_A^B \Re f = \Re I \quad and \quad \mathcal{KH} \int_A^B \Im f = \Im I.$$

Nearly all the theorems in this chapter† can be extended to complex valued functions by using either the definition or the characterization. For instance, the Fundamental Theorem holds for complex valued functions because Theorem 2.6.2 can be applied to both the real part and the imaginary part of the function. Corollary 2.5.11 needs a different proof but its validity is in no doubt since

$$\left| \sum_\pi f \right| \le \sum_\pi |f|.$$

Sometimes care is needed: Theorem 2.5.16 makes no sense for complex valued functions and, as the following example shows, the mean value theorem (Theorem 2.5.25) is false for complex valued functions.

$$\int_0^1 (x + \jmath x^3)\, dx = \frac{1}{2} + \jmath \frac{1}{4}.$$

There is no c with $c + \jmath c^3 = 1/2 + \jmath 1/4$. We trust readers will be able to decide for themselves whether or not a particular theorem can be extended from real valued to the complex valued functions.

---

† And in the rest of the book.

2.13 Exercises

EXERCISE 2.1 ⓘ *Use Cousin's lemma to prove Lemma 1.4.8. [Hint: Define J_k as in the proof of Lemma 1.4.8. For ξ in some J_k let $(\xi - \delta(\xi), \xi + \delta(\xi)) \subset J_k$. For $\xi \notin \cup J_k$ use the same idea as in the proof of Example 2.3.4.]*

EXERCISE 2.2 ⓘ *Use Cousin's lemma to prove the Heine–Borel theorem in $\mathbb{R}$. If F is a closed and bounded set $F \subset \mathbb{R}$ and for every $x \in F$ there is an open interval I_x with $x \in I_x$ then there exist finitely many intervals I_{x_k}, $k = 1, \dots m$ such that $F \subset \bigcup_{k=1}^{m} I_{x_k}$. [Hint: $F \subset [a, b]$, let $\delta(x)$ be such that the interval $(x - \delta(x), x + \delta(x))$ does not intersect F if $x \notin F$ and is included in I_x if $x \in F$.]*

EXERCISE 2.3 *Let $f(x) = 1$ when x is rational, and 0 when x is irrational. This is known as the Dirichlet function. It follows from Example 2.4.5 that the Dirichlet function is KH-integrable on any interval $[a, b]$. Prove directly that the Dirichlet function is KH-integrable on $[0, 1]$. [Hint: For $x = p/q$ in the lowest terms define $\delta(x) = 2^{-q-1}q^{-1}\varepsilon$.]*

EXERCISE 2.4 *Prove that f, $f(x) = 1/x$ for $x \neq 0$ is not KH-integrable on $[0, 1]$. Give your own example of a function f which is not KH-integrable on $[a, b]$.*

EXERCISE 2.5 *Let $F(x) = x^2 \sin x^{-2}$ when $x \neq 0$ and $F(0) = 0$. Prove by definition that the derivative F' of F is KH-integrable on $[0, 1]$.*

EXERCISE 2.6 *Recall from Section 1.1 that a function f is said to be Newton integrable on $[a, b]$ if there is another function F such that the derivative $F'(x) = f(x)$ for every $x \in [a, b]$. Then $F(a, b) = F(b) - F(a)$ is called the Newton integral of f on $[a, b]$. Prove that if f is Newton integrable on $[a, b]$ then it is KH-integrable there. Show by an example that the converse is not true.*

EXERCISE 2.7 ⓘ *Show that a function f is regulated on $[a, b]$ if and only if f has one-sided limits at every point in $[a, b]$.*

EXERCISE 2.8 *Determine whether the following functions are KH-integrable on $[0, 1]$: (a) $x^{-1} \sin x^{-1}$ for $x \neq 0$; (b) $x^{-2} \sin x^{-2}$ for $x \neq 0$. [Hint: Differentiate $x \cos x^{-1}$ and $x \cos x^{-2}$.]*

EXERCISE 2.9 *Continuing with the previous exercise, consider* $f(x) = x^{-p} \sin x^{-q}$ *for* $x \neq 0$ *and for positive integers* p, q. *Give values for* p *and* q *other than those above such that the function* f *is KH-integrable on* $[0,1]$ *and further values of* p *and* q *such that* f *is not.*

EXERCISE 2.10 ⓘ*Let* J_n, $n = 1, 2, \ldots$ *be a system of non-overlapping intervals with* $\bigcup J_k \subset [a, b]$. *Show that the function* $f = \sum_1^\infty c_k 1_{J_k}$ *is KH-integrable if and only if the series* $\sum_1^\infty c_k |J_k|$ *converges and then* $\int_a^b f = \sum_1^\infty c_k |J_k|$.

EXERCISE 2.11 *A function* f *is absolutely integrable if both* f *and* $|f|$ *are integrable (see Section 3.4). Example 2.6.7 exhibited a function* f *which is KH-integrable but not absolutely integrable. Give your own example of such a function (preferably without the use of the Fundamental Theorem). Show further that if* f *and* g *are KH-integrable then* $\text{Max}(f, g)$ *may or may not be KH-integrable.*

EXERCISE 2.12 ⓘ*If* f *is continuous on the interval* $[c, b]$ *with* $c > 0$ *and* $\lim_{x \downarrow 0} x^\alpha f(x) = L$ *for* $\alpha < 1$ *then* f *is absolutely KH-integrable on* $[0, b]$. *Prove this and show also that if* $L \neq 0$ *and* $\alpha \geq 1$ *then* f *is not absolutely KH-integrable on* $[0, b]$. *Give an example of a KH-integrable* f *for which* $L \neq 0$ *and* $\alpha \geq 1$.

EXERCISE 2.13 ⓘ*Prove results analogous to those of the previous exercise but for* $[1, \infty]$ *and* $c \uparrow \infty$. *[Hint:* $\alpha > 1$ *for convergence.]*

EXERCISE 2.14 *Theorem 2.5.14 says that if* f *is KH-integrable on* $[a, b]$ *then the same holds on any subinterval* $[c, d]$ *of* $[a, b]$. *A function* f *is said to be KH-integrable on a set* $S \subset [a, b]$ *if* $f 1_S$ *is KH-integrable on* $[a, b]$. *Is it true that if* f *is KH-integrable on* $[a, b]$ *then the same holds on any subset of* $[a, b]$?

EXERCISE 2.15 ⓘ*Prove: If* f *is KH-integrable on* $[a, \infty)$, $c_n \uparrow \infty$ *and* $c_1 = a$ *then* $\int_a^\infty f = \sum_1^\infty \int_{c_n}^{c_{n+1}} f$. *[Hint: Do not work hard, use Theorem 2.9.3.]*

EXERCISE 2.16 *In order that* f *is KH-integrable on* $[a, \infty]$, *it is necessary and sufficient that the following conditions hold: there is a number* I, *and for every positive* ε *there is* $\delta : [a, \infty] \to \mathbb{R}_+$, *such that for any division* D *of* $[a, \infty]$ *given by* $a = x_0 < x_1 < \ldots < x_n < \ldots$

with $x_n \uparrow \infty$ and such that $\xi_n \in [x_{n-1}, x_n] \subset (\xi_n - \delta(\xi_n), \xi_n + \delta(\xi_n))$ the series $\sum_{n=1}^{\infty} f(\xi_n)(x_n - x_{n-1})$ converges and

$$|\sum_{n=1}^{\infty} f(\xi_n)(x_n - x_{n-1}) - I| < \varepsilon.$$

[Hints: Necessity. Take $\delta_n : [a + n - 1, a + n] \to \mathbb{R}_+$ such that for any δ_n-fine partition π of $[a+n-1, a+n]$ we have $|\sum_\pi f - \int_{a+n-1}^{a+n} f| < \varepsilon/2^n$. Sufficiency. Take Δ such that for every $x_m \geq \Delta$ we have

$$|\sum_{n=m+1}^{\infty} f(\xi_n)(x_n - x_{n-1})| < \varepsilon$$

and apply Definition 2.9.1.]

EXERCISE 2.17 *We define the Henstock variation of a point-interval function $h(\xi, [u, v])$ as follows:*

$$V_H(h; [a,b]) = \inf \{V(\delta); \delta : [a, b] \mapsto \mathbb{R}_+\}$$

where

$$V(\delta) = \sup \{\sum_\pi |h(\xi, [u, v])|; \pi \ll \delta\}.$$

(The infimum is over all positive functions δ, the supremum over all δ-fine partitions π of $[a, b]$.) Show that if $V_H(h; [a, b]) = 0$ where $h(\xi, [u, v]) = F(v) - F(u) - f(\xi)(v - u)$, then f is KH-integrable on $[a, b]$ with primitive F. Show that the converse also holds. [Hint: $V_H(h; [a, b]) = 0$ means for every positive ε there is $\delta : [a, b] \to \mathbb{R}_+$ such that for every δ-fine partition π we have $\sum_\pi |h(\xi, [u, v])| < \varepsilon$.]

EXERCISE 2.18 *Given a function F defined on $[a, b]$, we differentiate and then integrate. Given another function f defined on $[a, b]$, we integrate first then differentiate. Find condition on F or f such that we get back to the same function each time.*

EXERCISE 2.19 ⓒⓘ *We define the Kurzweil–Henstock–Stieltjes integral as follows. A number I is the Stieltjes integral of f with respect to g on $[a, b]$ if for every positive ε there is a function $\delta : [a, b] \to \mathbb{R}_+$ such that for every δ-fine partition $\pi \equiv \{(\xi, [u, v])\}$*

$$|\sum_\pi f(\xi)[g(v) - g(u)] - I| < \varepsilon.$$

We write $I = \int_a^b f dg$. Suppose either $\int_a^b f dg$ or $\int_a^b g df$ exists. Then both integrals exist and

$$\int_a^b f dg + \int_a^b g df = f(b)g(b) - f(a)g(a)$$

is true if and only if $V_H(h; [a, b]) = 0$ where $h(u, v) = [f(v) - f(u)][g(v) - g(u)]$ for all $[u, v] \subset [a, b]$.

EXERCISE 2.20 ☺☺ Suppose $\int_a^b f dG$ exists where $G(x) = \int_a^x g$ for all $x \in [a, b]$. Show that if $|f| \leq K$ then the integral $\int_a^b f g$ exists and $\int_a^b f dG = \int_a^b f g$.

Exercises 2.21 to 2.24 come from a paper [25].

EXERCISE 2.21 ☺☺ Given a function $\delta : [0, 1] \rightarrow \mathbb{R}_+$, define a function ν mapping a closed interval I into a subset of I as follows:

$$\nu(I) = \{x \in I : I \subset (x - \delta(x), x + \delta(x))\}.$$

Show that the function ν satisfies the following conditions:
(1) $\nu(I) \subset I$ for each closed interval $I \subset [0, 1]$,
(2) $I_1 \subset I_2$ implies $\nu(I_2) \cap I_1 \subset \nu(I_1)$,
(3) $\bigcup_{I \subset [0,1]}(\nu(I) \cap I^\circ) = [0, 1]$, where I° denotes the interior of I, except that if $I = [0, a]$ or $[b, 1]$, then I° means $[0, a)$ or $(b, 1]$ respectively.
A function which satisfies the above three conditions is called a tag function.

EXERCISE 2.22 ☺☺ Show that if f is KH-integrable on $[0, 1]$ then for every positive ε there exists a tag function ν as defined in Exercise 2.21 such that f is bounded on $\nu(I)$ for each closed interval $I \subset [0, 1]$.

EXERCISE 2.23 ☺☺ Let ν be a tag function defined on the collection of closed subintervals of $[0, 1]$. A partition D of $[0, 1]$ is said to be ν-fine if $\nu([u, v])$ is non-empty for every $[u, v]$ in D. Show that the following condition is necessary and sufficient for f to be KH-integrable to a number I on $[0, 1]$: for every positive ε there is a tag function ν such that for any ν-fine partition D of $[0, 1]$ we have

$$|\sum_\pi f(\xi)(v - u) - I| < \varepsilon,$$

where $\xi \in \nu([u, v])$ for all $[u, v] \in \pi$.

EXERCISE 2.24 ⓢⓛ *Given a tag function* ν, *we define an upper sum as follows:*

$$S(\nu, f) = \sup_{\pi} \sum_{\pi} f(\xi)(v - u)$$

where the supremum is taken over all ν-*fine partitions* π *of* $[0, 1]$ *with* $\xi \in \nu([u, v])$ *for all* $[u, v] \in \pi$. *Then the upper KH-integral of* f *is the infimum of all upper sums over* ν. *Similarly, we define lower sums and the lower KH-integral. Show that* f *is KH-integrable on* $[0, 1]$ *if and only if the upper and lower integrals are equal.*

EXERCISE 2.25 *Prove: If* $\mathbb{R} \supset G = \bigcup_{k=1}^{\infty} I_k$ *with* I_k *open intervals then there exists a countable system of disjoint intervals* J_k *such that* $G = \bigcup_{k=1}^{\infty} J_k$. *[Hint: For each* $x \in G$ *there is a maximal* J_x *containing* x *and contained in* G. *Two* J_x *are either identical or disjoint. By selecting a rational in each* J_x *show that the system* J_x *is countable.]*

EXERCISE 2.26 *Prove: If* $\mathbb{R} \supset G = \bigcup_{k=1}^{\infty} I_k$ *with* I_k *open intervals and* $\sum_{k=1}^{\infty} |I_k| \leq M$ *then there exists a system of disjoint open intervals* J_k *such that* $G = \bigcup_{k=1}^{\infty} J_k$ *and* $\sum_{k=1}^{\infty} |J_k| \leq M$. *[Hint: Use the previous exercise. Replace each* J_k *by a closed interval* $\bar{J}_k$ *with the same endpoints and use the Heine–Borel theorem to show that* $\sum_{k=1}^{N} |\bar{J}_k| \leq M$ *for* $N \in \mathbb{N}$.]*

3
Development of the Theory

3.1 Equivalent forms of the definition

Our basic Definition 2.4.1 can be rephrased in several ways. Before we can state the appropriate theorems we need a few new concepts. We shall say that *the Riemann sums are Cauchy*† for f on $[a, b]$ if for every positive ε there is a $\delta : [a, b] \mapsto \mathbb{R}_+$ such that for two δ-fine partitions π_1 and π_2 we have

$$\left| \sum_{\pi_2} f - \sum_{\pi_1} f \right| < \varepsilon.$$

A function M is said to be a major function to f on $[a, b]$, or simply a major function, if there exists a $\delta_M : [a, b] \to \mathbb{R}_+$ such that

$$f(x) \leq \frac{M(v) - M(u)}{v - u}$$

for

$$x - \delta_M(x) < u \leq x \leq v < x + \delta_M(x)$$

and $[u, v] \subset [a, b]$. A function m is a minor function to f on $[a, b]$, or simply a minor function, if $-m$ is a major function to $-f$. If now $\pi \ll \delta_M$ then $f(x_i)(v_i - u_i) \leq M(v_i) - M(u_i)$ and consequently

$$\sum_\pi f \leq \sum_\pi (M(v_i) - M(u_i)) = M(b) - M(a). \tag{3.1}$$

---

† The necessary and sufficient condition for the existence of a limit of a sequence, function etc. is known as the Cauchy convergence principle. It was known before Cauchy to a Czech Jesuit priest B. Bolzano living in Prague but writing in German. Some authors, as we do, prefer the term Bolzano–Cauchy condition, but the term *Cauchy sequence* is used exclusively and we have followed this custom for Riemann sums.

76

Similarly if $\pi \ll \delta_m$ then

$$\sum_\pi f \geq m(b) - m(a). \tag{3.2}$$

Using (3.1) and (3.2) for a subinterval $[x, y]$ we obtain

$$m(y) - m(x) \leq \sum_\pi {}_x^y f \leq M(y) - M(x), \tag{3.3}$$

i.e. $M - m$ is always increasing (not necessarily strictly increasing). We can now state

THEOREM 3.1.1 (Characterization of integrability) *The following four conditions are equivalent*

(A) *f is KH-integrable on $[a, b] \subset \mathbb{R}$;*

(B) *the Riemann sums are Cauchy;*

(C) *for every positive ε there are major and minor functions M and m, respectively, such that*

$$M(b) - M(a) - [m(b) - m(a)] < \varepsilon;$$

(D) *for the following supremum and infimum*

$$\underline{I}(f) \quad = \sup\{m(b) - m(a) : m \text{ a minor function}\} \tag{3.4}$$
$$\overline{I}(f) \quad = \inf\{M(b) - M(a) : M \text{ a major function}\} \tag{3.5}$$

the relation

$$-\infty < \overline{I}(f) = \underline{I}(f) < \infty \tag{3.6}$$

holds.

Proof The implications (A) $\Rightarrow$ (B) and (C) $\Rightarrow$ (D) are obvious. To prove (D) $\Rightarrow$ (A) we find for $\varepsilon > 0$ a minor m and a major M such that

$$M(b) - M(a) < \overline{I}(f) + \varepsilon \tag{3.7}$$

and

$$m(b) - m(a) > \underline{I}(f) - \varepsilon = \overline{I}(f) - \varepsilon. \tag{3.8}$$

If $\delta = \text{Min}(\delta_M, \delta_m)$ and π is δ-fine then in view of (3.1), (3.2), (3.7) and (3.8)

$$|\overline{I}(f) - \sum_\pi f| < \varepsilon. \tag{3.9}$$

This shows (D) $\Rightarrow$ (A). We have also obtained: If relation (3.6) holds then

$$\overline{I}(f) = \underline{I}(f) = \int_a^b f. \tag{3.10}$$

It remains to prove (B) $\Rightarrow$ (C). If π_1 and π_2 are two partitions containing c as a dividing point† and having the same dividing points and tags in $[c, b]$ then

$$\sum_{\pi_2} f - \sum_{\pi_1} f = \sum_{\pi_1} {}_a^c f - \sum_{\pi_2} {}_a^c f. \tag{3.11}$$

According to (B) for $\varepsilon > 0$ there is a positive δ such that if π_1 and π_2 are δ-fine then

$$\left| \sum_{\pi_2} f - \sum_{\pi_1} f \right| < \varepsilon \tag{3.12}$$

and consequently

$$\sum_{\pi_1} {}_a^c f - \varepsilon \le \sum_{\pi_2} {}_a^c f \le \sum_{\pi_1} {}_a^c f + \varepsilon. \tag{3.13}$$

These inequalities also hold with c replaced by x, $a < x \le b$. We define

$$M_\delta(x) = \sup \{ \sum_\pi {}_a^x f : \pi \ll \delta \} \tag{3.14}$$

and

$$m_\delta(x) = \inf \{ \sum_\pi {}_a^x f : \pi \ll \delta \}, \tag{3.15}$$

for $x > a$ and $m_\delta(a) = M_\delta(a) = 0$. It follows from (3.13) that M_δ and m_δ are well defined and finite, and

$$M_\delta(b) - \varepsilon \le m_\delta(b). \tag{3.16}$$

If we prove that m_δ and M_δ are minor and major functions, respectively, then inequality (3.16) establishes (C). Clearly

$$\sum_\pi {}_a^u f + f(x)(u - v) \le M_\delta(v), \tag{3.17}$$

if $\pi \ll \delta$, $[u, v] \subset (a, b]$ and $x - \delta(x) < u \le x \le v \le x + \delta(x)$. By taking the supremum in (3.17) we obtain

$$f(x)(v - u) \le M_\delta(v) - M_\delta(u). \tag{3.18}$$

† $a < c < b$

The proof that

$$f(x)(u-v) \geq m_\delta(v) - m_\delta(u) \tag{3.19}$$

is similar. Inequalities (3.18) and (3.19) hold also with $u = a$ by the definition of M_δ and m_δ. •

REMARK 3.1.2 In the course of the proof we showed in (3.11)— (3.13) that if Riemann sums are Cauchy on $[a,\ b]$ and $a < c < b$ then they are also Cauchy on $[a,\ c]$. This combined with the theorem proves that a function integrable on $[a,\ b]$ is also integrable on $[a,\ c]$ and consequently on any subinterval of $[a,\ b]$. We have obtained an easy proof of Theorem 2.5.12.

The fundamental theorem of calculus is an easy consequence of the Characterization of Integrability. If F is a primitive of f, i.e. $F'(x) = f(x)$ for $x \in [a,\ b]$†, then $x \to F(x) + \varepsilon x$ is a major function and $x \to F(x) - \varepsilon x$ is a minor function, hence

$$\overline{I}(f) = \underline{I}(f) = F(b) - F(a),$$

and consequently by (3.10)

$$\int_a^b f = F(b) - F(a).$$

The characterization of integrability opens a way to relate the KH-integral to other known integrals. If $f : [a,\ b] \to \mathbb{R}$ then M_p is called a P-major function of f on $[a,\ b]$ if

$$f(x) \leq \underline{D}M_p(x) \text{ for } x \in [a,\ b].$$

The function m_p is called a P-minor function of f on $[a,\ b]$ if

$$f(x) \geq \overline{D}m_p(x) \text{ for } x \in [a,\ b].$$

As usual, $\overline{D}$ and $\underline{D}$ denote the upper and the lower derivatives, i.e.

$$\overline{D}h(x) = \limsup_{t \to 0} \frac{h(x+t) - h(x)}{t},$$

$$\underline{D}h(x) = \liminf_{t \to 0} \frac{h(x+t) - h(x)}{t}.$$

† At a and b the derivatives are one-sided.

Once a function has a P-major function we can define the upper Perron integral,

$$\overline{P} \int_a^b f = \inf\{M_p(b) - M_p(a); M_p \text{ a P-major function}\},$$

and, provided there is a P-minor function, a lower Perron integral,

$$\underline{P} \int_a^b f = \sup\{m_p(b) - m_p(a) : m_p \text{ a P-minor function}\}.$$

If both exist it can be shown that

$$-\infty < \underline{P} \int_a^b f \leq \overline{P} \int_a^b f < \infty. \tag{3.20}$$

If the middle inequality turns into an equality then f is Perron integrable, or briefly P-integrable, and the common value of the upper Perron integral and the lower Perron integral is by definition the Perron integral (of f over $[a, b]$), denoted by $\mathcal{P} \int_a^b f$.

If M is a major function then it is also a P-major function, and if M_p is a P-major function then $x \to M_p(x) + \varepsilon x$ is a major function. A similar relation holds for minor and P-minor functions (with $m_p(x) - \varepsilon x$ replacing $M_p(x) + \varepsilon x$). Hence

$$\overline{I}(f) = \overline{P} \int_a^b f \text{ and } \underline{I}(f) = \underline{P} \int_a^b f.$$

We have proved

THEOREM 3.1.3 *A function f is KH-integrable on $[a, b]$ if and only if it is Perron integrable on $[a, b]$ and then*

$$\mathcal{P} \int_a^b f = \mathcal{KH} \int_a^b f.$$

A theorem in the theory of Perron integration states that f is Lebesgue integrable if and only if both f and $|f|$ are P-integrable. Thus, in view of the equivalence of P-integration and KH-integration, this last statement is also true when P-integrable is replaced by KH-integrable. However, we shall look at the relation between Lebesgue and KH-integration independently of the Perron theory later.

3.2 Henstock's lemma

For the more advanced theory of KH-integration the following lemma is critical, since the proofs of important theorems depend in various ways on it. Recall that a system of couples

$$\{(x_k, [u_k, v_k]); \; k = 1, 2, \ldots, n\} \qquad (3.21)$$

such that $[u_k, v_k]$ were not overlapping for $k = 1, 2, \ldots, n$ and

$$\bigcup_{k=1}^{n} [u_k, v_k] = [A, B] \qquad (3.22)$$

was said to be a partition of $[A, B]$. We shall use the term tagged division for a partition. If condition (3.22) is not necessarily satisfied we shall call (3.21) a *subpartition* or *tagged partial division* (of $[A, B]$). Some authors use also the term *system*. If $[A, B]$ is not bounded then *by definition* only bounded intervals are allowed in a subpartition. If S is a set and π is a tagged partial division with all its tags in S we say that π is a *tagged partial division on S* or that π is a *partial division (subpartition) tagged in S*.

The concepts of Riemann sum and δ-fineness obviously extend to subpartitions. Also we shall use the notation π, possibly with subscripts, for subpartitions and $\pi \ll \delta$ to indicate that a subpartition π is δ-fine.

THEOREM 3.2.1 (Henstock's lemma) *If f is integrable on $K = [A, B]$ then for every positive ε there is a $\delta : [A, B] \mapsto \mathbb{R}_+$ such that*

$$\sum_{\pi} \left| f(x_i) |I_i| - \int_{I_i} f \right| < \varepsilon \qquad (3.23)$$

whenever the tagged partial division $\pi \equiv \{(x_i, I_i); i = 1, \ldots, n\}$ is δ-fine.

There are many applications of this lemma, for instance:

If $\int_a^c f = 0$ for every $c \in [a, b]$ then by (3.23) we have $\sum_{\pi} |f| < \varepsilon$ and this implies $\int_a^b |f| = 0$. Consequently $|f| = 0$ almost everywhere by Theorem 2.11.5.

Using this lemma it is easy to prove the continuity of the integral, namely Theorem 2.5.29. Let

$$F(x) = \int_a^x f.$$

We wish to prove that F is continuous. Let $a \leq c < b$, $\varepsilon > 0$ and let

$\delta : [a, b] \to \mathbb{R}_+$ be associated with ε by the definition of integrability of f. For

$$a \leq c < x < c + \mathrm{Min}\left(\frac{1}{2}\delta(c), \frac{\varepsilon}{|f(c)| + 1}\right)$$

the single pair $(c, [c, x])$ is a δ-fine subpartition and consequently

$$\left|\int_c^x f\right| \leq \left|\int_c^x f - f(c)(x - c)\right| + |f(c)(x - c)| < 2\varepsilon.$$

Continuity from the left can be proved similarly.

Inequality (3.23) may look a little surprising. On the other hand the proof is not difficult. It is obvious that the function δ can be replaced by a gauge δ without changing the meaning of Henstock's lemma. It is in this version that we prove it, using the notation and convention of Section 2.10.

Proof of Henstock's lemma For a given positive ε there is a gauge δ such that the inequality

$$\left|\sum_\Pi f - \int_K f\right| < \frac{\varepsilon}{2} \tag{3.24}$$

holds for every δ-fine partition Π. Let

$$\pi \equiv \{(x_k, I_k); \, k = 1, 2, \ldots, n\}$$

be a δ-fine tagged partial division on $[A, B]$. We first prove the inequality

$$\left|\sum_{i=1}^n \left(f(x_i)|I_i| - \int_{I_i} f\right)\right| \leq \frac{\varepsilon}{2}. \tag{3.25}$$

This is true if π is a partition of K. If the set $K \setminus \bigcup_1^n I_k$ is non-empty then it is a union of finitely many disjoint intervals; consequently there are closed intervals I_k, $k = n + 1, n + 2, \ldots, N$ such that $\bigcup_1^N I_k = K$ and all intervals I_k, $k = 1, 2, \ldots, N$ are non-overlapping. For a positive η and $n < k \leq N$ we find $\delta_k(x) \subset \delta(x)$ such that if π_k is a δ_k-fine partition of I_k then

$$\left|\sum_{\pi_k} f - \int_{I_k} f\right| < \frac{\eta}{N - n}. \tag{3.26}$$

Now $\Pi = \pi \cup \bigcup_{k+1}^N \pi_k$ is a partition of K and it is δ-fine; therefore we have (3.24). Since

$$\sum_{\Pi} f = \sum_{\pi} f + \sum_{k=n+1}^{N} \sum_{\pi_k} f$$

we have

$$\left| \sum_{i=1}^{n} \left(f(x_i)|I_i| - \int_{I_i} f \right) \right|$$

$$\leq \frac{\varepsilon}{2} + \sum_{k=n+1}^{N} \left| \sum_{\pi_k} f - \int_{I_k} f \right|$$

$$\leq \frac{\varepsilon}{2} + (N - n) \frac{\eta}{N - n} = \frac{\varepsilon}{2} + \eta.$$

By letting $\eta \to 0$ we obtain (3.25). Let π_1 and π_2 be the partial divisions tagged by those x_i for which

$$f(x_i)|I_i| - \int_{I_i} f \geq 0$$

and

$$f(x_i)|I_i| - \int_{I_i} f \leq 0,$$

respectively. By what we have already proved

$$\sum_{\pi_1} \left| f(x_i)|I_i| - \int_{I_i} f \right| \leq \frac{\varepsilon}{2}$$

and

$$\sum_{\pi_2} \left| f(x_i)|I_i| - \int_{I_i} f \right| \leq \frac{\varepsilon}{2}.$$

Hence

$$\sum_{\pi} \left| f(x_i)|I_i| - \int_{I_i} f \right|$$

$$= \sum_{\pi_1} \left| f(x_i)|I_i| - \int_{I_i} f \right| + \sum_{\pi_2} \left| f(x_i)|I_i| - \int_{I_i} f \right| \leq \varepsilon. \qquad \bullet$$

3.3 Functions of bounded variation

In this section we shall study the smallest class of functions which contains all monotonic functions and which contains with F_1 and F_2 also

$F_1 - F_2\dagger$. If D is a division (1.1) of $[a, b]$ and F_1, F_2 are monotonic on $[a, b]$ and $F = F_1 - F_2$ then

$$\sum_1^n |F(x_i) - F(x_{i-1})| \leq |F_1(b) - F_1(a)| + |F_2(b) - F_2(a)|. \quad (3.27)$$

In other words the sum on the left-hand side of (3.27) is bounded by a number independent of the division D. We denote by $\text{Var}_a^b F$ the least upper bound of the numbers

$$\sum_D |F(x_{i-1}, x_i)| = \sum_1^n |F(x_i) - F(x_{i-1})|$$

for all possible divisions D of $[a, b]$ and call it the *variation of F on $[a, b]$*. Since the above sum can only increase when adding another point to the division, we can assume when seeking the variation of F that a given point c, $a < c < b$, is a point of the division, say $c = x_p$. Then we have$\ddagger$

$$\sum_D {}_a^b |F(x_{i-1}, x_i)| = \sum_D {}_a^c |F(x_{i-1}, x_i)| + \sum_D {}_c^b |F(x_{i-1}, x_i)|.$$

By passing to the least upper bounds we obtain successively

$$\text{Var}_a^b F \geq \text{Var}_a^c F + \text{Var}_c^b F,$$
$$\text{Var}_a^b F \leq \text{Var}_a^c F + \text{Var}_c^b F.$$

Consequently

$$\text{Var}_a^b F = \text{Var}_a^c F + \text{Var}_c^b F. \quad (3.28)$$

If $\text{Var}_a^b F < \infty$ then we say that F *is of bounded variation*. By (3.27) a difference of two increasing functions is of bounded variation. Conversely we have

THEOREM 3.3.1 *If F is of bounded variation on $[a, b]$ then there are two functions F_1 and F_2 increasing on $[a, b]$ such that $F = F_1 - F_2$.*

Proof We set $F_1(x) = \text{Var}_a^x F$ for $a < x \leq b$, $F_1(a) = 0$ and $F_2(x) = F_1(x) - F(x)$. Clearly F_1 is increasing. To complete the proof it is sufficient to show that F_2 is increasing. Let $a \leq x < y \leq b$. We have by (3.28)

$$F_2(y) = F_1(y) - F(y) = F_1(x) + \text{Var}_x^y F - F(y)$$

† Or $F_1 + F_2$.
‡ For the notation see p. 5 and p. 37.

and therefore

$$F_2(y) - F_2(x) = \text{Var}_x^y F - [F(y) - F(x)] \geq 0.$$ •

If F is of bounded variation on $[a, b]$ then F is bounded on $[a, b]$. Indeed $|F(x)| \leq |F(a)| + \text{Var}_a^x F \leq |F(a)| + \text{Var}_a^b F$.

THEOREM 3.3.2 *If F and G are of bounded variation on $[a, b]$ then so are the functions $|F|$, $F \pm G$, FG, $\text{Max}(F, G)$ and $\text{Min}(F, G)$. If moreover $1/G$ is bounded then F/G is also of bounded variation.*

Proof Let K be such that $|F| \leq K$ and $|G| \leq K$ on $[a, b]$. It is easy to check the following inequalities†:

$$\big| |F(x_i)| - |F(x_{i-1})| \big| \leq |F(x_{i-1}, x_i)|,$$
$$|(F(x_i) \pm G(x_i)) - (F(x_{i-1}) \pm G(x_{i-1}))|$$
$$\leq |F(x_{i-1}, x_i)| + |G(x_{i-1}, x_i)|,$$

$$\left. \begin{array}{c} |F(x_i)G(x_i) - F(x_{i-1})G(x_{i-1})| \\ = |F(x_i)G(x_{i-1}, x_i) + G(x_{i-1})F(x_{i-1}, x_i)| \\ \leq K\,|G(x_{i-1}, x_i)| + K\,|F(x_{i-1}, x_i)|, \end{array} \right\}$$

$$|\text{Max}\,(F(x_i),\, G(x_i)) - \text{Max}\,(F(x_{i-1}),\, G(x_{i-1}))|$$
$$\leq |F(x_{i-1}, x_i)| + |G(x_{i-1}, x_i)|,$$
$$|\text{Min}\,(F(x_i),\, G(x_i)) - \text{Min}\,(F(x_{i-1}),\, G(x_{i-1}))|$$
$$\leq |F(x_{i-1}, x_i)| + |G(x_{i-1}, x_i)|.$$

If, in addition, $|1/G| \leq L$ then

$$\left| \frac{1}{G(x_i)} - \frac{1}{G(x_{i-1})} \right| = \left| \frac{G(x_{i-1}, x_i)}{G(x_i)G(x_{i-1})} \right| \leq L^2\,|G(x_{i-1}, x_i)|.$$

It follows from (3.3) that

$$\sum_1^n |F(x_i)G(x_i) - F(x_{i-1})G(x_{i-1})| \leq K\text{Var}_a^b F + K\text{Var}_a^b G,$$

which shows that FG is of bounded variation. The rest of the theorem follows similarly. •

THEOREM 3.3.3 *Let f be of bounded variation on $[a, b]$ and $c \in [a, b)$. Denote $v(x) = \text{Var}_a^x f$, where $x \in (a, b]$ and $v(a) = 0$. Then v is continuous from the right at c if and only if f is. Similarly for the continuity from the left if $c \in (a, b]$.*

† $F(u, v)$ stands for $F(v) - F(u)$.

Proof Since $|f(x) - f(c)| \leq v(x) - v(c)$ for $c < x < b$ continuity of v at c from the right implies the same continuity for f. Going in the other direction we find, for every positive ε, a partition

$$c = x_0 < x_1 < x_2 < \cdots < x_n = b$$

such that

$$\sum_{k=1}^{n} |f(x_{k-1}) - f(x_k)| > \mathrm{Var}_c^b f - \frac{\varepsilon}{2}$$

and a positive δ such that $|f(x) - f(c)| < \varepsilon/2$ whenever $c < x < c + \delta$. Then we have that, for $c < x < \mathrm{Min}(x_1, c + \delta)$,

$$\mathrm{Var}_c^b f \quad < \quad |f(c) - f(x)| + |f(x) - f(x_1)| + \sum_{k=2}^{n} |f(x_{k+1}) - f(x_k)| + \frac{\varepsilon}{2}$$

$$< \quad \mathrm{Var}_x^b f + \varepsilon.$$

Consequently

$$0 \leq v(x) - v(c) < \varepsilon. \qquad \bullet$$

3.4 Absolute integrability

Recall that a function f is called absolutely integrable on $[A, B]$ if both f and $|f|$ are integrable there. The next theorem characterizes absolute integrability in terms of the primitive.

THEOREM 3.4.1 (Criterion of absolute integrability) *A function f is absolutely integrable on $[a, b] \subset \mathbb{R}$ if and only if the function F, $F(x) = \mathcal{KH} \int_a^x f$, is of bounded variation and then*

$$\mathrm{Var}_a^b F = \int_a^b |f|. \tag{3.29}$$

Proof If f is absolutely integrable on $[a, b]$, then for any division

$$D \equiv a = x_0 < x_1 < x_2 \cdots < x_n = b \tag{3.30}$$

we have

$$\sum_D |F(x_{i-1}, x_i)| = \sum_1^n \left| \int_{x_{i-1}}^{x_i} f \right| \leq \sum_1^n \int_{x_{i-1}}^{x_i} |f| = \int_a^b |f|.$$

This shows F is of bounded variation. For the proof of the converse let $\mathrm{Var}_a^b F = V$. For $\varepsilon > 0$ there exists a division D as in (3.30) such that

$$V - \varepsilon < \sum_D |F(x_{i-1}, x_i)| \leq V. \tag{3.31}$$

Now we find a δ_1 such that any partition which is δ_1-fine anchors on the points x_i, i.e. any Riemann sum $\sum_\pi f$ for a partition $\pi \ll \delta_1$ can be expressed as

$$\sum_\pi f = \sum_{i=1}^n \sum_\pi {}_{x_{i-1}}^{x_i} f. \tag{3.32}$$

This equation shows that if π is δ_1-fine we can include all points x_i as dividing points of π without altering the Riemann sum and we do this for the rest of the proof. Let δ_2 correspond to f and ε by Henstock's lemma and set $\delta = \mathrm{Min}(\delta_1, \delta_2)$. Then we have

$$\sum_1^n \sum_\pi {}_{x_{i-1}}^{x_i} |f(\xi)(v - u) - F(v, u)| \leq \varepsilon,$$

for any partition π which is δ-fine. Consequently

$$\sum_\pi |F(u, v)| - \varepsilon \leq \sum_\pi |f(\xi)(v - u)| \leq \sum_\pi |F(u, v)| + \varepsilon. \tag{3.33}$$

Since adding more dividing points to D can only increase the sum in (3.31) we obtain

$$V - \varepsilon < \sum_\pi |F(u, v)| \leq V.$$

Combining this with (3.33) gives

$$V - 2\varepsilon < \sum_1^m |f(\xi)|(v - u) \leq V + \varepsilon.$$

This means that

$$\int_a^b |f| = V. \qquad \bullet$$

COROLLARY 3.4.2 *If f and g are integrable and $|f| \leq g$ on $[A, B] \subset \mathbb{R}$ then f is absolutely integrable. In particular, a KH-integrable function is absolutely integrable over $[a, b] \subset \mathbb{R}$ if it is bounded there.*

Proof Indeed, if $F(x) = \int_a^x f$ then

$$\sum_{i=1}^{n} |F(x_{i-1}, x_i)| \le \sum_{i=1}^{n} \int_{x_{i-1}}^{x_i} g \le \int_a^b g < \infty.$$

Consequently f is absolutely integrable on any finite interval. The function $x \mapsto \int_c^x |f|$ is increasing and bounded, hence finite limits at A and B exist. By Theorem 2.9.3 and/or Theorem 2.8.3 the integral $\int_A^B |f|$ exists. •

It is an immediate consequence of the corollary that $f_1 \pm f_2$ are absolutely integrable if f_1 and f_2 are. The integrability of $\text{Min}(f_1, f_2)$ and of $\text{Max}(f_1, f_2)$ now follows immediately from the absolute integrability of f_1 and f_2 because

$$\text{Max}(f_1, f_2) = \frac{1}{2}[f_1 + f_2 + |f_1 - f_2|],$$

$$\text{Min}(f_1, f_2) = \frac{1}{2}[f_1 + f_2 - |f_1 - f_2|].$$

3.5 Limit and KH-integration

The advantage of KH-theory is that it has powerful and easily proved theorems on the interchange of limit and integration. However, one cannot expect that the formula

$$\int_a^b \lim_{n \to \infty} f_n = \lim_{n \to \infty} \int_a^b f_n \qquad (3.34)$$

holds without restriction, as the following example shows.

EXAMPLE 3.5.1 If $f_n = n\mathbf{1}_{(0,n^{-1})}$ and $a = 0$, $b = 1$ then the left-hand side of (3.34) is 0 whereas the right-hand side is 1.

As in the Riemann theory, uniform convergence of f_n to f guarantees the validity of (3.34) for the KH-integral. However, on many occasions, the need for the uniform convergence is too restrictive. Moreover on an infinite interval the uniform limit need not be integrable and even if it is, equation (3.34) may fail. For instance if $f_n = \frac{1}{n}\mathbf{1}_{(n,2n)}$ then

$$\int_0^\infty f_n = 1 \ne \int_0^\infty \lim_{n \to \infty} f_n = 0.$$

The next theorem guaranteeing (3.34) is valid for infinite intervals and verification of the assumptions is often easier than proving uniform convergence.

THEOREM 3.5.2 (Monotone convergence theorem.) *If*

(i) *the sequence $\{f_n(x)\}$ is monotone for almost all $x \in [A, B] \subset \overline{\mathbb{R}}$,*

(ii) *the functions f_n are KH-integrable and the sequence $\left\{\int_A^B f_n\right\}$ is bounded, i.e. $\left|\int_A^B f_n\right| < K$ for some K all $n \in \mathbb{N}$,*

(iii) $\lim_{n \to \infty} f_n = f$ *is finite a.e.*

then f is KH-integrable on $[A, B]$ and

$$\int_A^B f = \lim_{n \to \infty} \int_A^B f_n. \tag{3.35}$$

The theorem is restated and an alternative proof is also given in Subsection 6.5.4 on page 224.

Proof of the monotone convergence theorem Since the change of a function on a set of measure zero influences neither the existence nor the value of the integral we can and shall assume that the functions f_n and f are defined and finite everywhere on $[A, B]$. By considering $-f_n$ or $f_n - f_1$ instead of f_n, if need be, we can achieve that the sequence $\{f_n\}$ is increasing and $f_n \geq 0$. Since the sequence $\left\{\int_A^B f_n\right\}$ is monotonic and bounded the limit on the right-hand side of (3.35) exists; let us denote it by L. Given ε we can find N such that

$$\int_A^B f_N > L - \frac{\varepsilon}{3}.$$

By the easy part of Theorem 2.9.3 or Remark 2.9.4 there exists a closed bounded interval $[a_N, b_N]$ such that

$$\int_{a_N}^{b_N} f_N > L - \frac{\varepsilon}{3}. \tag{3.36}$$

Next we find $n(x) \geq N$ such that, for $n \geq n(x)$,

$$\frac{3L + 3\varepsilon}{3L + \varepsilon} f_n(x) \geq f(x). \tag{3.37}$$

If $f(x) > 0$ this is possible because the left-hand side of (3.37) has a limit strictly larger than the right-hand side; if $f(x) = 0$ we can take $n(x) = N$ since equality holds in (3.37) for any n. By Henstock's lemma there is

$$\delta_n : [A, B] \mapsto \mathbb{R}_+$$

such that

$$\sum_{\pi} \left| f_n(x)(v-u) - \int_u^v f_n \right| < \frac{\varepsilon}{3.2^n} \qquad (3.38)$$

whenever the tagged partial division $\pi \ll \delta_n$. We define

$$\delta(x) = \delta_{n(x)}(x).$$

Let π be a partition of an interval containing $[a_N, b_N]$, $\pi \ll \delta$. The proof will be accomplished if we show that

$$\left| \sum_{\pi} f(x_i)(v_i - u_i) - L \right| < \varepsilon. \qquad (3.39)$$

The way from the Riemann sum of f to L goes through the sums

$$\sum_{\pi} f_{n(x_i)}(x_i)(v_i - u_i), \qquad (3.40)$$

$$\sum_{\pi} \int_{u_i}^{v_i} f_{n(x_i)}. \qquad (3.41)$$

It is easy to check that the first sum is close to the Riemann sum of f and the second to L. Indeed

$$\sum_{\pi} f_{n(x_i)}(x_i)(v_i - u_i) \le \sum_{\pi} f(x_i)(v_i - u_i)$$

$$\le \frac{3L + 3\varepsilon}{3L + \varepsilon} \sum_{\pi} f_{n(x_i)}(x_i)(v_i - u_i), \qquad (3.42)$$

and

$$\sum_{\pi} \int_{u_i}^{v_i} f_{n(x_i)} \ge \sum_{\pi} \int_{u_i}^{v_i} f_N \ge \int_{a_N}^{b_N} f_N > L - \frac{\varepsilon}{3}. \qquad (3.43)$$

Denoting by $\bar{N}$ the largest $n(x_i)$ we also have

$$\sum_{\pi} \int_{u_i}^{v_i} f_{n(x_i)} \le \sum_{\pi} \int_{u_i}^{v_i} f_{\bar{N}} \le \int_A^B f_{\bar{N}} \le L. \qquad (3.44)$$

It remains to estimate the difference between (3.40) and (3.41), i.e.

$$\left| \sum_{\pi} \left[f_{n(x_i)}(x_i)(v_i - u_i) - \int_{u_i}^{v_i} f_{n(x_i)} \right] \right|. \qquad (3.45)$$

The $n(x_i)$ are not necessarily distinct; let $i_1, i_2, \ldots, i_\kappa$ be the distinct i such that $n(x_i) = \kappa$. In (3.45) we group together the terms with the same $n(x_i) = \kappa$ and estimate that sum using Henstock's lemma (3.38):

$$\left| \sum_{j=1}^{\kappa} \left[f_\kappa(v_{i_j} - u_{i_j}) - \int_{u_{i_j}}^{v_{i_j}} f_\kappa \right] \right| < \frac{\varepsilon}{3.2^\kappa}.$$

Consequently

$$\left| \sum_{\pi} \left[f_{n(x_i)}(x_i)(v_i - u_i) - \int_{u_i}^{v_i} f_{n(x_i)} \right] \right| < \sum_{1}^{\infty} \frac{\varepsilon}{3.2^k} = \frac{\varepsilon}{3}. \qquad (3.46)$$

Collecting (3.42), (3.43), (3.44) and (3.46) gives on one hand

$$\sum_{\pi} f(x_i)(v_i - u_i) \geq \sum_{\pi} f_{n(x_i)}(x_i)(v_i - u_i)$$

$$\geq \sum_{\pi} \int_{u_i}^{v_i} f_{n(x_i)} - \frac{\varepsilon}{3} > L - \frac{2\varepsilon}{3},$$

and on the other

$$\frac{3L + \varepsilon}{3(L + \varepsilon)} \sum_{\pi} f(x_i)(v_i - u_i) \leq \sum_{\pi} f_{n(x_i)}(x_i)(v_i - u_i)$$

$$\leq \sum_{\pi} \int_{u_i}^{v_i} f_{n(x_i)} + \frac{\varepsilon}{3} < \frac{3L + \varepsilon}{3}.$$

The last four inequalities imply (3.39).   •

REMARK 3.5.3 Since a countable union of sets of measure zero is of measure zero and since changing a function on a set of measure zero does not influence either the existence or the value of the integral it is irrelevant whether the limit in (iii) in Theorem 3.5.2 is required to exist everywhere or a.e. A similar situation will occur often in this book but we shall not comment on it in future.

EXAMPLE 3.5.4 If G is an open set† $G = \bigcup_1^\infty J_n$ with pairwise disjoint J_n then

$$\int_{-\infty}^{\infty} \mathbf{1}_G = \sum_{1}^{\infty} |J_n|, \qquad (3.47)$$

---

† The reader who is not familiar with open and closed sets is advised to read Appendix Section A.3.

provided the series on the right-hand side converges. The proof is immediate by the monotone convergence theorem with $f_n = 1_{G_n}$, $G_n = \bigcup_1^n J_n$.

It is convenient to define $\int_A^B f = \infty$ if there exists an increasing sequence of integrable functions f_n converging to f such that $\lim_{n\to\infty} \int_A^B f_n = \infty$. It is then true, by the monotone convergence theorem, that

$$\lim_{n\to\infty} \int_A^B g_n = \infty$$

for any increasing sequence $\{g_n\}$ which converges to f almost everywhere. The advantage of allowing the integral to have an infinite value is that many formulae hold with less restriction or without any restriction. e.g. the formula (3.47) always holds even if the series diverges. For the sake of reference let us state formally

DEFINITION 3.5.5 *If there exists an increasing sequence of integrable functions f_n converging to f a.e. and such that*

$$\lim_{n\to\infty} \mathcal{KH} \int_A^B f_n = \infty \quad \text{then} \quad \mathcal{KH} \int_A^B f = \infty.$$

We define also

$$\mathcal{KH} \int_A^B f = -\infty \quad \text{if} \quad \mathcal{KH} \int_A^B (-f) = \infty.$$

By accepting infinite values of integrals the phrases 'f is integrable' and 'the integral of f exists' cease to be equivalent: if $\int_A^B f = \infty$ or $-\infty$ then f is not integrable. If $\int_A^B f$ is finite we say that the integral converges.

EXAMPLE 3.5.6 If $f_n \geq f_{n+1} \geq 0$ and $\int_a^b f_n \to 0$ then $f_n \to 0$ a.e. For $f = \lim_{n\to\infty} f_n$ we have $f \geq 0$ and by the monotone convergence Theorem $\int_a^b f = 0$.   •

EXAMPLE 3.5.7 One often associates with a function f a function f^N which is obtained from f by truncating it from above by N and from below by $-N$. More precisely

$$f^N(x) = \begin{cases} N & \text{if } f(x) > N \\ f(x) & \text{if } |f(x)| \leq N \\ -N & \text{if } f(x) < -N. \end{cases} \tag{3.48}$$

If f is absolutely integrable then so is the function f^N since $f^N = \text{Max}(\text{Min}(f, N), -N)$. If f is absolutely KH-integrable then

$$\lim_{N \to \infty} \int_A^B f^N = \int_A^B f. \tag{3.49}$$

If f is non-negative (3.49) follows directly from the monotone convergence theorem; for general f one can apply the result for non-negative f to f^+ and f^-.

It is interesting to observe that (3.49) may fail if f is not absolutely integrable. The idea when constructing a counterexample is to define f asymmetrically with respect to zero so that truncation from above has a bigger effect than truncation from below. Let us divide the interval $[(i-1)/i,\, i/(i+1)]$ by a point x_i in the ratio $1 : i$, i.e. let

$$i\left(x_i - \frac{i-1}{i}\right) = \frac{i}{i+1} - x_i$$

which leads to

$$x_i = \frac{i-1}{i+1} + \frac{1}{(i+1)^2}.$$

Now we define

$$f(x) = \begin{cases} i^2 & \text{if } x \in \left[\dfrac{i-1}{i}, x_i\right) \\[2ex] -i & \text{if } x \in \left[x_i, \dfrac{i}{i+1}\right). \end{cases}$$

Then, for $x \in \left[\dfrac{i-1}{i}, \dfrac{i}{i+1}\right]$,

$$0 \leq \int_0^x f \leq \int_{\frac{i-1}{i}}^{x_i} = i^2 \frac{1}{i+1} \frac{1}{i(i+1)} = \frac{i}{(i+1)^2}.$$

Consequently

$$\lim_{x \to 1} \int_0^x f = 0 = \int_0^1 f.$$

On the other hand

$$\int_0^1 f^{N^2} = \int_{\frac{N-1}{N}}^{\frac{N^2}{N^2+1}} f^{N^2}$$

$$= \sum_{N+1}^{N^2}\left[\frac{1}{(i+1)^2} - \frac{i}{(i+1)^2}\right] \le -\frac{1}{2}\sum_{N+1}^{N^2}\frac{1}{i+1}.$$

The right-hand side clearly diverges. This example is due to Lu Shipan.

The next theorem shows that assumption (iii) in the monotone convergence theorem is superfluous, but we use the theorem itself to prove it.

THEOREM 3.5.8 (Beppo Levi) *If*

(i) *the sequence $\{f_n(x)\}$ is monotone for almost all $x \in [A, B] \subset \overline{\mathbb{R}}$,*

(ii) *the functions f_n are KH-integrable and the sequence $\left\{\int_A^B f_n\right\}$ is bounded, $\left|\int_A^B f_n\right| < K$ for all $n \in \mathbb{N}$,*

then $\lim_{n\to\infty} f_n(x) = f(x)$

(a) *is finite a.e.*

(b) *is KH-integrable on $[A, B]$ and*

$$\int_A^B f = \lim_{n\to\infty} \int_A^B f_n. \tag{3.50}$$

Proof It is sufficient to prove (a). As in the previous proof we assume that f_n are defined everywhere and $0 \le f_n \le f_{n+1}$. For a given x the limit of $f_n(x)$ can be infinite; denote $Z = \left\{x : \lim_{n\to\infty} f_n(x) = \infty\right\}$. Let $g_n = \text{Min}(\frac{1}{i}f_n, 1)$. By the monotone convergence theorem

$$h_i = \lim_{n\to\infty} g_n \text{ is integrable and } \int_A^B h_i \le \frac{1}{i}K.$$

If $x \in Z$ then $h_i(x) = 1$; if $x \notin Z$ then $i > f(x)$ for some i and consequently $\lim_{i\to\infty} h_i(x) = 0$. This means that $\lim_{i\to\infty} h_i(x) = 1_Z$. Another application of the monotone convergence theorem gives

$$\int_A^B \lim_{i\to\infty} h_i = \int_A^B 1_Z = \lim_{i\to\infty}\int_A^B h_i \le \lim_{i\to\infty}\frac{K}{i} = 0. \qquad \bullet$$

COROLLARY 3.5.9 *If w_n are KH-integrable and non-negative and the series $\sum_1^\infty \int_A^B w_n$ converges then $\sum_1^\infty w_n$ converges a.e. and*

$$\sum_1^\infty \int_A^B w_n = \int_A^B \sum_1^\infty w_n.$$

Proof Apply the Beppo Levi theorem to $f_n = \sum_1^n w_i$.

EXAMPLE 3.5.10

$$\int_0^1 \frac{\log(1-x)}{x}\,dx = \sum_1^\infty \int_0^1 \frac{x^{n-1}}{n}\,dx = \sum_1^\infty \frac{1}{n^2}. \qquad (3.51)$$

This sum is known† to be $\pi^2/6$.

COROLLARY 3.5.11 *If w_n are absolutely KH-integrable and the series $\sum_1^\infty \int_A^B |w_n|$ converges then $\sum_1^\infty w_n$ converges a.e. and*

$$\sum_1^\infty \int_A^B w_n = \int_A^B \sum_1^\infty w_n.$$

Proof Apply the previous corollary to w_n^+ and w_n^-. •

EXAMPLE 3.5.12

$$\int_0^1 \frac{\log(1+x)}{x}\,dx = \sum_1^\infty \int_0^1 \frac{(-1)^{n-1}x^{n-1}}{n}\,dx = \sum_1^\infty \frac{(-1)^{n-1}}{n^2}. \qquad (3.52)$$

THEOREM 3.5.13 (Fatou's lemma) *If f_n are KH-integrable and non-negative on $[A, B] \subset \overline{\mathbb{R}}$ then*

$$\int_A^B \liminf_{n\to\infty} f_n \le \liminf_{n\to\infty} \int_A^B f_n. \qquad (3.53)$$

EXAMPLE 3.5.14 We illustrate the various possible situations before the proof; we take $A = 0$ and $B = 1$. If f_n are as in Example 3.5.1 then strict inequality occurs in (3.53). If $f_n = n^2 1_{(0,n^{-1})}$ then the left-hand side of (3.53) is zero and the right-hand side is ∞. If $f_n(x) = x^{-2}1_{(n^{-1},1)}$ then both sides are infinite. Finally if $f_n = 1_{(0,1)} - 2n1_{(0,n^{-1})}$ the left-hand side is 1 and the right-hand side is -1: inequality (3.53) fails because f_n are not non-negative.

† For the proof see Exercise 7.21.

Proof of Fatou's lemma Let $g_k = \inf\{f_i;\ n \le i \le k\}$, $h_n = \lim\limits_{k\to\infty} g_k = \inf\{f_i;\ i \ge n\}$. Since $0 \le \int_A^B g_k \le \int_A^B f_n$ all the functions h_n are integrable by the monotone convergence theorem. Since $\int_A^B h_n \le \int_A^B f_n$ we have

$$\lim_{n\to\infty} \int_A^B h_n \le \liminf_{n\to\infty} \int_A^B f_n. \tag{3.54}$$

If the limit on the left-hand side is ∞ then both sides in (3.53) become infinite†, otherwise by the monotone convergence theorem

$$\int_A^B \liminf_{n\to\infty} f_n = \int_A^B \lim_{n\to\infty} h_n = \lim_{n\to\infty} \int_A^B h_n.$$

This together with (3.54) gives (3.53).   •

REMARK 3.5.15 If the right-hand side of (3.53) is finite then Fatou's lemma asserts the integrability of $\liminf f_n$. This is often used in proving the integrability of a given function. The next example is typical in this regard.

EXAMPLE 3.5.16 If f is increasing, bounded and has a derivative‡ a.e. on $[a, b] \subset \mathbb{R}$ then f' is KH-integrable on $[a, b]$ and we have $\int_a^b f' \le f(b) - f(a)$. First we extend f by setting $f(x) = f(b)$ for $x > b$ and then define

$$g_n(x) = n\left[f(x + \frac{1}{n}) - f(x)\right].$$

It is easily checked that

$$\int_a^b g_n = n\int_b^{b+\frac{1}{n}} f - n\int_a^{a+\frac{1}{n}} f \le f(b) - f(a).$$

By Fatou's lemma

$$\int_a^b f' = \int_a^b \liminf_{n\to\infty} n\left[f(x + \frac{1}{n}) - f(x)\right] \le f(b) - f(a). \quad •$$

The assumption that f is increasing is essential; if $f(x) = \sin\frac{1}{x}$ then f' is not integrable on $[0, 1]$ because $\lim\limits_{x\to 0}\int_x^1 f'$ does not exist.

---

† The left-hand side because of the definition of an integral with value ∞.
‡ It is proved in Section 3.14 that an increasing function has a finite derivative a.e. and so this assumption can be omitted.

The next theorem is easy to apply and is probably the theorem most often used in integration.

THEOREM 3.5.17 (Lebesgue dominated convergence theorem)
If the functions f_n, G and g are KH-integrable on $[A, B] \subset \overline{\mathbb{R}}$,

$$g \leq f_n \leq G \tag{3.55}$$

for all $n \in \mathbb{N}$ and

$$\lim_{n \to \infty} f_n = f \quad a.e.$$

then f is KH-integrable and equation (3.35) holds.

Proof We apply Fatou's lemma to the sequences $n \mapsto f_n - g$ and $n \mapsto G - f_n$ and obtain successively

$$\int_A^B (f - g) \leq \liminf_{n \to \infty} \int_A^B (f_n - g) \leq \int_A^B (G - g),$$

$$\int_A^B f \leq \liminf_{n \to \infty} \int_A^B f_n, \tag{3.56}$$

$$\int_A^B (G - g) \leq \liminf_{n \to \infty} \int_A^B (G - f_n) \leq \int_A^B (G - f),$$

$$\int_A^B f \geq \limsup_{n \to \infty} \int_A^B f_n. \tag{3.57}$$

Combining (3.56) and (3.57) proves that the right-hand side of (3.35) exists and equals to the left-hand side.

COROLLARY 3.5.18 *If the inequality (3.55) is replaced by*

$$|f_n| \leq G \tag{3.58}$$

then (3.35) holds and f is absolutely KH-integrable.

COROLLARY 3.5.19 *If $[A, B] = [a, b] \subset \mathbb{R}$ and there is $C \in \mathbb{R}$ such that $|f_n(x)| \leq C$ for almost all $x \in [A, B]$ then f is absolutely KH-integrable and (3.34) holds.*

The inequality

$$g \leq f \leq G \tag{3.59}$$

alone is not sufficient to guarantee (3.35) as Example 3.5.1 shows. However, it will follow from Theorems 3.11.6 and 3.11.2 that if inequality

(3.55) in Theorem 3.5.17 is replaced by inequality (3.59) then the assertion that f is integrable remains valid.

EXAMPLE 3.5.20 The Γ function is defined for $t > 0$ by

$$\Gamma(t) = \int_0^\infty e^{-x} x^{t-1} dx.$$

We wish to prove the formula

$$\Gamma(t) = \lim_{n \to \infty} \frac{n! \, n^t}{t(t+1) \cdots (t+n)}.$$

It is easy to show by straightforward calculus methods that

$$\int_0^1 (1-y)^n y^{t-1} dy = \frac{n!}{t(t+1)\ldots(t+n)},$$

$$\int_0^n \left(1 - \frac{u}{n}\right)^n u^{t-1} du = n^t \int_0^1 (1-y)^n y^{t-1} dy.$$

Consequently it suffices to show

$$\Gamma(t) = \lim_{n \to \infty} \int_0^n \left(1 - \frac{u}{n}\right)^n u^{t-1} du.$$

Since

$$\lim_{n \to \infty} \left(1 - \frac{x}{n}\right)^n = e^{-x}$$

we need only an integrable non-negative function G such that $0 \le f_n \le G$, where

$$f_n(x) = \begin{cases} 0 & \text{if } x > n \\ (1 - \frac{x}{n})^n x^{t-1} & \text{if } 0 \le x \le n. \end{cases}$$

Recalling the inequality $\log(1 - x/n) \le -x/n$ gives $(1 - x/n)^n \le e^{-x}$ for $0 \le x \le n$. Consequently we can take $G(x) = e^{-x} x^{t-1}$.   •

The dominated convergence theorem is important not only for dealing with concrete problems but also as a tool for proving new theorems, this we shall see later in this book. Although f_n and f need not be absolutely integrable the dominated convergence theorem is really a convergence theorem about absolutely integrable functions since $G - f$ and $G - f_n$ are absolutely integrable.

EXAMPLE 3.5.21 Let f be non-absolutely integrable and $f_n = f/n$. Then (3.35) holds but there is no integrable G dominating all f_n. Indeed, if $f_n(x) \le G(x)$ almost everywhere in $[A, B]$ for all n then $G \ge f^+$ and G cannot be integrable. •

Section 3.7 and several theorems in Chapter 5 deal with the interchange of limit and integration for non-absolutely integrable functions with no monotonicity condition or majorizing functions present.

THEOREM 3.5.22 (Mean convergence, completeness) *Let the functions f_n, $n = 1, 2, \ldots$ be absolutely integrable and for every positive ε let there be N such that*

$$\int_a^b |f_n - f_m| < \varepsilon \qquad (3.60)$$

for $n, m > N$. Then

(i) *there exists an almost everywhere convergent subsequence,*

$$\lim_{n_k \to \infty} f_{n_k} = f$$

,

(ii) $\displaystyle \lim_{n \to \infty} \int_a^b |f_n - f| = 0,$

(iii) $\displaystyle \lim_{n \to \infty} \int_a^b f_n = \int_a^b f.$

Proof By assumption for every $k \in \mathbb{N}$ there is n_k such that

$$\int_a^b |f_{n_k} - f_n| < \frac{1}{2^k}, \qquad (3.61)$$

for $n \ge n_k$ and $n_k < n_{k+1}$. Since the series

$$\sum_{k=1}^\infty \int_a^b |f_{n_k} - f_{n_{k+1}}|$$

converges, by Beppo Levi the series

$$f_{n_1} + \sum_{k=1}^\infty (f_{n_k} - f_{n_{k+1}})$$

converges absolutely a.e. and this proves (i).

Next we apply Fatou's lemma to the sequence $j \mapsto |f_{n_k} - f_{n_{k+j}}|$ and obtain from (3.61)

$$\int_a^b |f_{n_k} - f| \le \frac{1}{2^k},$$

and this together with (3.61) gives

$$\int_a^b |f_n - f| \leq \frac{2}{2^k},$$

for $n \geq n_k$. This establishes (ii). Condition (iii) is a direct consequence
of (ii).   ●

If (ii) holds then we say that the sequence $\{f_n\}$ is *mean convergent* to
f or that f_n *converges* to f *in the mean*. Convergence in the mean allows
interchange of limit and integration but a mean convergent sequence of
absolutely integrable functions need not converge at any point. However,
the theorem implies that a mean convergent sequence contains an almost
everywhere convergent subsequence.

EXAMPLE 3.5.23 *Diminishing step travelling.* Write $n = 2^{i-1} +$
$k - 1$ where $k = 1, 2, \ldots, 2^{i-1}$ and $i = 1, 2, \ldots$.   Put $f_n(x) = 1$
when $x \in [(k-1)/2^{i-1},\ k/2^{i-1}]$ and 0 elsewhere. In other words,
the functions f_n, $n = 1, 2, \ldots$ take the value 1 on the intervals

$[0,\ 1]$, $[0,\ 1/2]$, $[1/2,\ 1]$, $[0,\ 1/4]$, $[1/4,\ 1/2]$, $[1/2, 3/4]$, $[3/4, 1]$, $\ldots$

and 0 elsewhere. Then $\{f_n\}$ is mean convergent to zero but the sequence
$\{f_n(x)\}$ diverges for every x in $[0, 1]$.

The set of all absolutely KH-integrable functions is often denoted by $\mathcal{L}$.
The property of $\mathcal{L}$ stated in Theorem 3.5.22 is referred to as completeness
of $\mathcal{L}$. This is because, in the language of functional analysis, the space
$\mathcal{L}$ with the distance $d(f, g)$ defined by $d(f, g) = \int_a^b |f - g|$ is complete
by Theorem 3.5.22.

3.6 Absolute continuity

There is another characterization of absolute integrability which uses a
subclass of functions of bounded variation, namely absolutely continuous
functions. A function F is said to be *absolutely continuous on* $[A, B]$ or
briefly AC if for every $\varepsilon > 0$ there is an η such that for every partial
division D with

$$\sum_D (v - u) < \eta \tag{3.62}$$

we have

$$\left| \sum_D F(u,v) \right| < \varepsilon. \tag{3.63}$$

Taking into account the obvious inequalities

$$\left| \sum_D F(u,v) \right| \le \sum_D |F(u,v)|$$

$$\le \sum_D [F(u,v)]^+ + \sum_D [F(u,v)]^-$$

it is easy to see that the meaning of the definition of absolute continuity is not changed if the absolute value in inequality (3.63) is put inside the summation sign. It is also clear that a function which is AC on $[A, B]$ is uniformly continuous there. The next example provides a function which is increasing and uniformly continuous on $[0,1]$ but not AC. It is even possible to define a continuous *strictly* increasing function which is not AC. (See [42] pp. 198–200.)

EXAMPLE 3.6.1 *Devil's stairs.* First we let $H(x) = 0$ for $x \le 0$ and $H(x) = 1$ for $x \ge 1$. Then we set $H(x) = \frac{1}{2}$ in the interval $(\frac{1}{3}, \frac{2}{3})$, and generally if $J \subset [0,1]$ is an interval of the complement of Cantor's discontinuum C and z its mid-point then we define $H(x) = z$ for $x \in J$. Then we extend H by setting $H(x) = \sup\{H(t); t \le x, t \notin C\}$. The function H is clearly increasing and assumes all values of the form $\frac{k}{2^n}$ with k, n non-negative integers and $k \le 2^n$. Consequently every sub-interval of $[0,1]$ contains values of H and from this the continuity of H follows. Indeed if H were discontinuous at c, with $0 \le c \le 1$, the interval $[\lim_{x \uparrow c} H(x), \lim_{x \downarrow c} H(x)]$ would be free of any value of H. Let $\pi \equiv \{[u, v]\}$ be the partial division of $[0,1]$ which consists of the 2^n closed intervals left in the n-th step of the construction of Cantor's set. The total length of these intervals is $(\frac{2}{3})^n$ and tends to zero with $n \to \infty$. Since H is constant on the contiguous intervals $\sum_\pi H(u,v) = H(1) - H(0) = 1$ and H cannot be AC.

THEOREM 3.6.2 *If F is AC on a closed and bounded $[a, b]$ then it is of bounded variation on $[a, b]$.*

Proof There exists a positive η such that

$$\sum_D |F(u,v)| < 1 \tag{3.64}$$

whenever the intervals (u, v) satisfy (3.62). Let

$$a = x_0 < x_1 < \cdots < x_N = b$$

be a division of $[a, b]$ into subintervals of length smaller than η. Then by equation (3.28)

$$\mathrm{Var}_a^b F \leq \sum_1^N \mathrm{Var}_{x_{i-1}}^{x_i} F < N. \quad \bullet$$

The definition of AC has some similarity with the definition of bounded variation. It is therefore no coincidence that the statement and proof of the next theorem are very similar to those of Theorem 3.3.2.

THEOREM 3.6.3 *If F and G are AC on a closed bounded interval $[a, b]$ then so are the functions $|F|$, $F \pm G$, FG, $\mathrm{Max}(F,G)$ and $\mathrm{Min}(F, G)$. If moreover $1/G > 0$ then F/G is also AC.*

Proof Since F and G are continuous we have $|F| \leq K$ and $|G| \leq K$ on $[a, b]$ for some K. We rewrite the inequalities from the proof of Theorem 3.3.2 thus:

$$\left|\, |F(v)| - |F(u)|\, \right| \leq \left|F(u, v)\right|,$$
$$|(F(v) \pm G(v)) - (F(u) \pm G(u))|$$
$$\leq |F(u, v)| + |G(u, v)|,$$

$$\left.\begin{array}{l}|F(v)G(v) - F(u)G(u)| \\ = |F(v)G(u, v) + G(u)F(u, v)| \\ \leq K\,|G(u, v)| + K\,|F(u, v)|, \end{array}\right\} \qquad (3.65)$$

$$|\mathrm{Max}\,(F(v),\, G(v)) - \mathrm{Max}\,(F(u),\, G(u))|$$
$$\leq |F(u, v)| + |G(u, v)|,$$
$$|\mathrm{Min}\,(F(v),\, G(v)) - \mathrm{Min}\,(F(u),\, G(u))|$$
$$\leq |F(u, v)| + |G(u, v)|.$$

If now

$$\sum_\pi |F(u,v)| < \frac{\varepsilon}{2K} \quad \text{and} \quad \sum_\pi |G(u,v)| < \frac{\varepsilon}{2K}$$

then it follows from (3.65) that

$$\sum_\pi |F(v)G(v) - F(u)G(u)| < \varepsilon.$$

This proves the absolute continuity of the product. The rest of the proof follows similarly. $\quad \bullet$

THEOREM 3.6.4 (*AC* **and absolute integrability**) *A KH-integrable f is absolutely integrable on* $[A, B] \subset \overline{\mathbb{R}}$ *if and only if the KH-primitive F,* $F(x) = \mathcal{KH} \int_c^x f$, *is absolutely continuous on* $[A, B]$ *and for every* $\varepsilon > 0$ *there is K such that*

$$\int_a^b |f| < \varepsilon \qquad (3.66)$$

for every bounded $[a, b]$ *satisfying* $[a, b] \subset [A, B]$ *and* $[a, b] \cap [-K, K] = \emptyset$.

COROLLARY 3.6.5 *A KH-integrable f is absolutely integrable on* $[a, b] \subset \mathbb{R}$ *if and only if the KH-primitive F,* $F(x) = \mathcal{KH} \int_c^x f$, *is absolutely continuous on* $[a, b]$.

Combining Theorem 3.4.1 and the corollary we have

COROLLARY 3.6.6 *If f is KH-integrable on* $[a, b] \subset \mathbb{R}$ *and its primitive F is of bounded variation, then F is absolutely continuous on* $[a, b]$.

Proof of Theorem 3.6.4 Let f be absolutely KH-integrable and $\varepsilon > 0$. By the dominated convergence theorem there is $N \in \mathbb{N}$ such that

$$\int_A^B |f - f^N| < \frac{\varepsilon}{2}.$$

For intervals (u, v) satisfying (3.62) we obtain

$$\int_u^v |f| \le \int_u^v |f - f^N| + \int_u^v |f^N|,$$

$$\sum_D \int_u^v |f| \le \frac{\varepsilon}{2} + N\eta.$$

This shows that F is AC. If f is absolutely integrable then condition (3.66) is satisfied because

$$\lim_{b \to B} \int_c^b |f| \quad \text{and} \quad \lim_{a \to A} \int_a^c |f| \qquad (3.67)$$

exist. Conversely, if F is AC then it is of bounded variation on every compact interval in $[A, B]$, and consequently f is absolutely integrable on every compact interval. Condition (3.66) guarantees the existence of the limits (3.67). Hence $|f|$ is integrable on $[A, B]$ by Theorem 2.9.3 and Remark 2.9.4. •

We return to AC and absolute integrability in Theorem 3.9.4.

3.7 Equiintegrability

We begin with a definition.

DEFINITION 3.7.1 *A family $\mathfrak{F}$ of KH-integrable functions is said to be equiintegrable on $[A, B]$ if for every positive ε there exists a gauge δ such that for every $f \in \mathfrak{F}$ and every δ-fine partition π*

$$\left| \sum_{\pi} f - \int_a^b f \right| < \varepsilon. \qquad (3.68)$$

For every KH-integrable f there is a δ such that (3.68) holds; for an equiintegrable family there is such a δ *common to all f.*

EXAMPLE 3.7.2 If $\mathfrak{F}$ is a family of functions with uniformly bounded variation on $[a, b] \subset \mathbb{R}$, i.e. there is a constant C such that for every $f \in \mathfrak{F}$ the variation $\mathrm{Var}_a^b f \leq C$, then $\mathfrak{F}$ is equiintegrable on $[a, b]$. Let $\varepsilon > 0$ be given, and choose $\delta \in \mathbb{R}_+$ such that $2C\delta < \varepsilon$. If $\pi \ll \delta$ then in every interval $[u, v]$ tagged by ξ we choose η in such a way that

$$\sup \{|f(\xi) - f(t)| \, ; \, t \in [u, v]\} < |f(\xi) - f(\eta)| + \frac{\varepsilon}{2(b - a)}$$

and we have

$$\left| \sum_{\pi} f(\xi)(v - u) - \int_a^b f \right| = \left| \sum_{\pi} \int_u^v [f(\xi) - f(t)] \, dt \right|$$

$$\leq \delta \sum_{\pi} |f(\xi) - f(\eta)| + \frac{\varepsilon}{2} \leq C\delta + \frac{\varepsilon}{2} < \varepsilon. \qquad \bullet$$

EXAMPLE 3.7.3 If $f_n(x) = 0$ for $x \neq 0$ and $f_n(0) = n$ then the family f_n is evidently not equiintegrable on $[-1, 1]$.

EXAMPLE 3.7.4 If f is KH-integrable on $[A, B]$ and

$$H_{c, \gamma}(x) = \begin{cases} 0 & \text{if } x < c \\ \gamma, \, 0 \leq \gamma \leq 1 & \text{if } x = c \\ 1 & \text{if } x > c \end{cases}$$

then the family of functions of the form $H_{c, \gamma} f$ with $c \in (A, B)$ and $0 \leq \gamma \leq 1$ is equiintegrable. Given $\varepsilon > 0$ there is $\delta_1 > 0$ such that† if

---

† Since f is integrable on $[A, B]$ the function $\int_c^x f$ is uniformly continuous on $[A, B]$ even if $[A, B]$ is infinite.

$|x - y| < \delta_1$ then

$$\left| \int_x^y f \right| < \frac{\varepsilon}{4}. \tag{3.69}$$

There exists also a gauge δ_2 such that

$$\sum_\sigma \left| f(\xi)(v - u) - \int_u^v f \right| < \frac{\varepsilon}{4} \tag{3.70}$$

for any δ_2-fine subpartition σ. Let

$$\delta(x) \subset \delta_2(x), \quad |\delta(x)| < \mathrm{Min}\left(\delta_1, \frac{\varepsilon}{8(1 + |f(x)|)} \right)$$

and

$$\pi \equiv \{ (\xi_i, [u_i, v_i]); i = 1, 2, \ldots, n \}$$

be a δ-fine partition of $[A, B]$; $c \in [u_k, v_k)$. Clearly

$$\left| \sum_\pi H_{c,\gamma}(\xi_i) f(\xi_i)(v_i - u_i) - \int_A^B H_{c,\gamma} f \right| \le |\gamma f(\xi_{k-1})(v_{k-1} - u_{k-1})|$$

$$+ |f(\xi_k)(v_k - u_k)| + \left| \int_c^{v_k} f \right| + \sum_{i=k+1}^n \left| f(\xi_i)(v_i - u_i) - \int_{u_i}^{v_i} f \right|.$$

The first or last term on the right-hand side is to be omitted if $k = 1$ or n, repectively. The terms on the right-hand side are less than $\varepsilon/4$, the first and second by the definition of δ, the third by (3.69) and the last by (3.70).   •

THEOREM 3.7.5 (Theorem on equiintegrability) *If* $\lim\limits_{n \to \infty} f_n(x) = f(x)$ *for every* $x \in [A, B] \subset \overline{\mathbb{R}}$ *and the sequence* $\{f_n\}$ *is equiintegrable on* $[A, B]$ *then* f *is integrable and (3.35) holds.*

Proof First we prove that $\lim\limits_{n \to \infty} \int_A^B f_n$ exists and is finite. For $\varepsilon > 0$ we find a gauge δ such that for all $n \in \mathbb{N}$ we have

$$\left| \sum_\pi f_n - \int_A^B f_n \right| < \frac{\varepsilon}{3} \tag{3.71}$$

whenever the partition π of $[A, B]$ is δ-fine. We fix a δ-fine partition

$$\pi \equiv \{ (\xi_i, [u_i, v_i]); i = 1, 2, \ldots, p \}$$

and find N such that

$$\left| \sum_{i=1}^{p} f_n(\xi_i)(v_i - u_i) - \sum_{i=1}^{p} f_m(\xi_i)(v_i - u_i) \right| < \frac{\varepsilon}{3}. \qquad (3.72)$$

This is possible since f_n converges to f and there are only p, i.e. finitely many, ξ_i. The sequence $\left\{ \int_A^B f_n \right\}$ is Cauchy since

$$\left| \int_A^B f_n - \int_A^B f_m \right| \le \left| \int_A^B f_n - \sum_\pi f_n \right|$$

$$+ \left| \sum_\pi f_n - \sum_\pi f_m \right| + \left| \sum_\pi f_m - \int_A^B f_m \right|.$$

The first and the last term are less than $\varepsilon/3$ by (3.71), the middle term by (3.72). Let $\lim_{n \to \infty} \int_A^B f_n = L$.

We now allow π to be any δ-fine partition. Sending $n \to \infty$ in (3.71) gives

$$\left| \sum_{i=1}^{p} f(\xi_i)(v_i - u_i) - L \right| < \varepsilon.$$

This proves the integrability of f and that $\int_A^B f = L$. $\qquad \bullet$

As a first application of this theorem we give a one-line proof of the theorem stating that there are no improper integrals in KH-integration, namely Theorem 2.8.3. We choose $c_n > a$, $c_n \to a$ and have

$$\lim_{n \to \infty} \int_{c_n}^b f = \lim_{n \to \infty} \int_a^b H_{c_n, 0} f = \int_a^b \lim_{n \to \infty} H_{c_n, 0} f = \int_a^b f.$$

The interchange of limit and integration is justified, since by Example 3.7.4, the family $\{ H_{c_n, 0} f; n = 1, 2, \dots \}$ is equiintegrable.

The following properties of equiintegrability are immediate consequences of the definition.

un Let $\mathfrak{F}_i$, $i = 1, 2$ be the families of functions constituted by restrictions of functions from $\mathfrak{F}$ to $[A, C]$ and $[C, B]$, respectively. If $\mathfrak{F}_i$ are equiintegrable for $i = 1, 2$ then $\mathfrak{F}$ is equiintegrable on $[A, B]$.

su If $\mathfrak{F}_1$ and $\mathfrak{F}_2$ are equiintegrable then so are $\mathfrak{F}_1 \cup \mathfrak{F}_2$ and the family of functions of the form $f_1 + f_2$ with $f_i \in \mathfrak{F}_i$;

m If $K \in \mathbb{R}_+$ and $\mathfrak{F}$ is equiintegrable then so is the family of functions of the form cf with $f \in \mathfrak{F}$ and $|c| \le K$;

cx Let $\mathfrak{H}$ be a family of functions h with the following property: there exist positive numbers α_i, $i = 1, 2, \ldots, n$ with $\sum_1^n \alpha_i = 1$ and functions $f_i \in \mathfrak{F}$ such that $h = \sum_1^n \alpha_i f_i$. If $\mathfrak{F}$ is equiintegrable then so is $\mathfrak{H}$.

We also need

cl Let $\bar{\mathfrak{F}}$ denote the set of all functions which can be uniformly approximated by functions from $\mathfrak{F}$. If $\mathfrak{F}$ is equiintegrable on $[A, B] \subset \mathbb{R}$ so is $\bar{\mathfrak{F}}$.

Proof Let $\varepsilon > 0$ and $\bar{f} \in \bar{\mathfrak{F}}$. There exists a gauge δ such that

$$\left| \sum_\pi f - \int_A^B f \right| < \frac{\varepsilon}{2} \tag{3.73}$$

whenever $f \in \mathfrak{F}$ and π is δ-fine. For $n \in \mathbb{N}$ we find $f_n \in \mathfrak{F}$ such that

$$|f_n - \bar{f}| < \frac{1}{n}. \tag{3.74}$$

By Theorem 3.7.5

$$\lim_{n \to \infty} \int_A^B f_n = \int_A^B \bar{f}. \tag{3.75}$$

For a δ-fine partition π and any $n \in \mathbb{N}$ we have by (3.73)

$$\left| \sum_\pi \bar{f} - \int_A^B \bar{f} \right| < \frac{\varepsilon}{2} + \left| \sum_\pi (\bar{f} - f_n) \right| + \left| \int_A^B f_n - \int_A^B \bar{f} \right|. \tag{3.76}$$

By (3.75) there exists n_0 such that for $n > n_0$ the second term in (3.76) is less than $\varepsilon/4$ and then, by (3.74), we find $n > n_0$ which will make the first term in (3.76) less than $\varepsilon/4$. ●

THEOREM 3.7.6 *If*

(i) $K \in \mathbb{R}_+$ *and* f *is KH-integrable on* $[A, B] \subset \bar{\mathbb{R}}$,

(ii) $\mathfrak{G}$ *is a family of increasing† functions with* $|g| \le K$ *for* $g \in \mathfrak{G}$

then the family of functions of the form gf, *with* $g \in \mathfrak{G}$, *is equiintegrable on* $[A, B]$.

Proof The family $\{fg; \ g \in \mathfrak{G}, \ g(a) = g(B)\}$ is obviously equiintegrable. By **su** we can assume that $g(A) < g(B)$. We can further assume, because of **su** and **m**, that

$$g(A) = 0 \qquad \text{and} \qquad g(B) = 1. \tag{3.77}$$

† Or decreasing.

We prove the theorem first with the additional assumption that the functions g are simple. Any such function can be represented as

$$g = \alpha_1 H_{c_1,\gamma_1} + \alpha_2 H_{c_2,\gamma_2} + \cdots + \alpha_n H_{c_n,\gamma_n}, \tag{3.78}$$

with $\alpha_i > 0$, $\sum_1^n \alpha_i = 1$, $0 \le \gamma_i \le K$ and

$$A \le c_1 < c_2 < \cdots < c_n \le B$$

The equiintegrability now follows from Example 3.7.4, **su** and **m**. In view of **cl** the proof will be completed when we show that every increasing g satisfying (3.77) can be uniformly approximated by simple increasing functions. Let $\varepsilon > 0$ and $\{[y, w]\}$ be an ε-fine division of $[0,1]$. The set $g^{-1}([y, w))$ could be empty, a singleton or an interval. For $x \in g^{-1}([y, w))$ let $\varphi(x) = y$ and $\varphi(x) = 1$ for $x \in g^{-1}(\{g(B)\})$. Then φ is increasing and has a finite range, and by construction $0 \le f(x) - \varphi(x) < \varepsilon$ on $[A, B]$.   •

COROLLARY 3.7.7 *If f is KH-integrable on $[A, B]$ and g monotonic and bounded then fg is KH-integrable on $[A, B]$.*

EXAMPLE 3.7.8 The family of functions of the form

$$f(x) = x^{-1} e^{-kx} \sin x$$

with $k > 0$ is equiintegrable on $[0, \infty]$ since $\int_0^\infty x^{-1} \sin x \, dx$ is finite. By Theorem 3.7.5

$$\lim_{k \to 0} \int_0^\infty e^{-kx} \frac{\sin x}{x} \, dx = \int_0^\infty \frac{\sin x}{x} \, dx.$$

3.7.1 The second mean value theorem

If the function f oscillates highly then the mean value theorem 2.5.27 provides weak results in comparison with our next theorem, which fruitfully uses the additional assumption that g is monotonic. Compare Example 3.7.10 below.

THEOREM 3.7.9 (The second mean value theorem) *If f is KH-integrable and g monotonic and bounded on $[a, b] \subset \mathbb{R}$ then gf is KH-integrable and there exists $\xi \in [a, b]$ such that*

$$\int_a^b gf = g(a) \int_a^\xi f + g(b) \int_\xi^b f. \tag{3.79}$$

Proof† The theorem is true if $g(a) = g(b)$. Hence we assume $g(a) \neq g(b)$. It is a matter of simple calculation to verify that if α, β are real constants and (3.79) holds with g replaced by $\hat{g}$ then it also holds for any functions of the form $g = \alpha + \beta\hat{g}$. It therefore suffices to prove the theorem for increasing functions g which satisfy $g(a) = 0$, $g(b) = 1$. First let g be a simple function with the representation (3.78), $F(x) = \int_\xi^b f$ and M, m the maximum or minimum of F, respectively. Then fg is integrable and

$$m \leq \int_a^b fg = \sum_1^n \alpha_i \int_{c_i}^b f \leq M. \tag{3.80}$$

Consequently, by continuity of F, there is a ξ such that

$$\int_a^b gf = F(\xi) = \int_\xi^b f = g(a) \int_a^\xi f + g(b) \int_\xi^b f. \tag{3.81}$$

For a general g there is a sequence of step functions g_n converging uniformly to g. Equation (3.81) holds with g and ξ replaced by g_n and ξ_n, i.e.

$$\int_a^b g_n f = \int_{\xi_n}^b f. \tag{3.82}$$

Passing to the limit through a subsequence for which $\xi_{n_k} \to \xi$ and using Theorems 3.7.6 and 3.7.5 to justify the interchange of limit and integration we obtain

$$\int_a^b gf = \lim_{n_k \to \infty} \int_a^b g_{n_k} f = \lim_{n_k \to \infty} \int_{\xi_{n_k}}^b f = \int_\xi^b f. \qquad \bullet$$

EXAMPLE 3.7.10 We show that $\int_2^\infty \sin x (\log x)^{-1} dx$ converges. For $a > 2$ we obtain by the second mean value theorem

$$\left| \int_a^b \frac{\sin x}{\log x} dx \right| = \left| \frac{1}{\log a} \int_a^\xi \sin x \, dx + \frac{1}{\log b} \int_\xi^b \sin x \, dx \right|$$

$$\leq \frac{2}{\log a} + \frac{2}{\log b} \leq \frac{4}{\log a}.$$

Consequently $\lim_{b \to \infty} \int_2^b \sin x (\log x)^{-1} dx$ exists and is finite. $\qquad \bullet$

† This due to Š. Schwabik (Czech lecture notes).

Using the first mean value theorem gives a poor result. We have

$$\int_a^b \frac{\sin x}{\log x}\, dx = \sin \xi \int_a^b \frac{1}{\log x}\, dx$$

and the integral on the right-hand side is 'big'.

3.8 Differentiation of integrals

In this section the Henstock lemma finds an important application in the proof of Theorem 3.8.2 below.

First we generalize the concept of length of an interval to sets which are finite unions of intervals. The length of $S = \bigcup_1^n I_j$, denoted by $\ell(S)$, is, by definition, the integral of the characteristic set of S. Clearly, if the intervals are non-overlapping then $\ell(S) = \sum_1^n |I_j|$ and always $\ell(S) \le \sum_1^n |I_j|$. If $S_1 \subset S_2$ then $\ell(S_1) \le \ell(S_2)$. For the proof of the next theorem we need the following lemma.

LEMMA 3.8.1 (Austin†) *Every finite family of intervals $\{I_k;\ k \in P\}$ contains a disjoint subfamily $\{I_k;\ k \in Q \subset P\}$ such that*

$$\frac{1}{3}\ell\left(\bigcup_{k \in P} I_k\right) \le \ell\left(\bigcup_{k \in Q} I_k\right). \tag{3.83}$$

Proof We may and shall assume without loss of generality that $P = \{1, 2, \ldots, n\}$ and

$$\ell(I_1) \ge \ell(I_2) \ge \ell(I_3) \ge \cdots \ge \ell(I_n).$$

We construct Q from P by removing some elements from it. Firstly we put $1 \in Q$. Then we discard all intervals which intersect I_1. In the next step we omit all intervals which intersect the largest interval left and so on. After a finite number of steps we obtain a disjoint subfamily, and since we keep at each step at least one third, the length of what is left must be at least one third of what we started with.   •

At the end of Section 2.6 we anticipated the following theorem.

THEOREM 3.8.2 *If f is KH-integrable on $[a, b]$ and F is defined by*

$$F(x) = \int_a^x f \tag{3.84}$$

† [3].

then

$$F'(x) = f(x) \qquad (3.85)$$

almost everywhere on $[a, b]$.

Proof Let E_μ be the set of all x such that every neighbourhood of x contains an interval $[u, v]$ with the property that

$$x \in (u, v) \quad \text{and} \quad |F(v) - F(u) - f(x)(v - u)| > \mu(v - u). \qquad (3.86)$$

Let $E = \bigcup E_{\frac{1}{n}}$. If $x \notin E$ then F has a derivative at x and equation (3.85) holds. Indeed, let $\tau > 0$ and choose $n \in \mathbb{N}$ with $1/n < \tau$. Since $x \notin E_{\frac{1}{n}}$ there is a neighbourhood U of x such that for any interval $[u, v] \subset U$ with $x \in (u, v)$ we have

$$|F(v) - F(u) - f(x)(u - v)| \leq \frac{1}{n}(v - u).$$

By continuity this inequality persists for $v = x$ or $u = x$. Hence

$$|F(v) - F(u) - f(x)(v - u)| < \tau(v - u)$$

for $(u, v) \subset U$. By Lemma 2.6.1 we have (3.85) for $x \notin E$.

It now suffices to show that for every positive μ the set E_μ is of measure zero. By the KH-integrability of f and by Henstock's lemma there exists, for every positive ε, a function $\delta : [a, b] \mapsto \mathbb{R}_+$ such that

$$\sum_\pi |F(v) - F(u) - f(x)(v - u)| < \frac{\varepsilon\mu}{6}, \qquad (3.87)$$

whenever π is a δ-fine subpartition of $[a, b]$. If $E_\mu = \emptyset$ there is nothing to prove. Otherwise for every $x \in E_\mu$ choose an interval $[u, v]$ such that inequality (3.86) holds and $[u, v] \subset (x - \delta(x), x + \delta(x))$. To emphasize the dependence of $[u, v]$ on x we write $[u_x, v_x]$. Now define $\delta_1 : E_\mu \to \mathbb{R}_+$ with the property that $[x - \delta_1(x), x + \delta_1(x)] \subset (u_x, v_x)$. By the covering lemma (Lemma 2.11.4) there is a countable system of closed non-overlapping intervals K_n with the following properties:

$$E_\mu \subset \bigcup_1^\infty K_n$$

and each K_n contains a point $x_n \in E_\mu$ such that

$$K_n \subset (x_n - \delta_1(x_n), x + \delta_1(x_n)) \subset (u_{x_n}, v_{x_n}).$$

Let

$$k = \sum_1^\infty |K_n| \quad \text{and} \quad k < 2 \sum_1^N |K_n|.$$

From the system $\{[u_{x_i}, v_{x_i}]; i = 1, 2, \ldots, N\}$ we find, by using Lemma 3.8.1, a finite disjoint system

$$\{[u_{x_i}, v_{x_i}]; \quad i \in Q\} \tag{3.88}$$

such that

$$\ell \left(\bigcup_{i \in Q} [u_{x_i}, v_{x_i}] \right) \geq \frac{1}{3} \ell \left(\bigcup_{i=1}^N [u_{x_i}, v_{x_i}] \right)$$

and obtain

$$\sum_{i \in Q} (v_{x_i} - u_{x_i}) = \ell \left(\bigcup_{i \in Q} [u_{x_i}, v_{x_i}] \right) \geq \frac{1}{3} \ell \left(\bigcup_{i=1}^N [u_{x_i}, v_{x_i}] \right)$$

$$\geq \frac{1}{3} \ell \left(\bigcup_{i=1}^N K_i \right) = \frac{1}{3} \sum_1^N |K_i| \geq \frac{k}{6}. \tag{3.89}$$

The intervals in (3.88) together with the points x_i form a δ-fine subpartition of $[a, b]$ and by inequalities (3.87), (3.86)

$$\mu \sum_{i \in Q} (v_{x_i} - u_{x_i}) \leq \sum_\pi |F(v_{x_i}) - F(u_{x_i}) - f(x_i)(v_{x_i} - u_{x_i})| < \frac{\varepsilon \mu}{6}. \tag{3.90}$$

Combining inequalities (3.89) and (3.90) gives $k < \varepsilon$.   •

3.9 Characterization of the KH-integral

In this section we prove the strongest version of the Fundamental Theorem, using the concept of negligible variation. We say that a function F defined on $[a, b]$ is *of negligible variation on a set* $S \subset [a, b]$ if for every positive ε there is a $\gamma : S \mapsto \mathbb{R}_+$ such that

$$\sum_\pi |F(v) - F(u)| < \varepsilon \tag{3.91}$$

whenever π is a γ-fine partial division on S. If S is countable and F continuous on S then F is of negligible variation on S. Also if F is AC on $[a, b] \subset \mathbb{R}$ and S is of measure zero then F is also of negligible variation on S. The following theorem is the strongest version of the Fundamental Theorem†.

THEOREM 3.9.1 (R. Bartle) *If*

(i) *F is continuous on $[a, b]$,*
(ii) *there exists $S \subset [a, b] \subset \mathbb{R}$ of measure zero and F is of negligible variation on S,*
(iii) *$F'(x)$ exists and is finite for $x \in [a, b]$ and $x \notin S$*

then F' is if KH-integrable on $[a, b]$ and

$$F(x) - F(a) = \mathcal{KH} \int_a^x F' \tag{3.92}$$

for every $x \in [a, b]$.

Proof Since F is continuous at a and b, it will remain of negligible variation on S if a and b are included in S. Hence we may and shall assume that $a, b \in S$. For every $x \in [a, b] \setminus S$ there is a $\delta(x)$ such that

$$|F(v) - F(u) - F'(x)(v - u)| < \varepsilon(v - u) \tag{3.93}$$

whenever $x - \delta(x) < u \le x \le v < x + \delta(x)$. Let γ be as above and define δ on S by setting it equal to γ. Let $f^*(x) = F'(x)$ on S and $f^*(x) = 0$ otherwise. If π is a tagged division of $[a, b]$ and π_1 and π_2 its parts tagged in $[a, b] \setminus S$ and S, respectively, then by (3.91) and (3.93)

$$\left| F(b) - F(a) - \sum_\pi f^*(x)(v - u) \right| = \left| \sum_\pi [F(v) - F(u) - f^*(x)(v - u)] \right|$$

$$\le \left| \sum_{\pi_1} [F(v) - F(u) - F'(x)(v - u)] \right| + \left| \sum_{\pi_2} [F(v) - F(u)] \right|$$

$$\le \varepsilon(b - a) + \varepsilon.$$

This proves (3.92) with F' replaced by f^*. ●

THEOREM 3.9.2 *If f is KH-integrable on $[a, b]$ and $F(x) = \int_a^x f$ then F is of negligible variation on every set S of measure zero.*

Proof Let

$$A_n = \{x; \, n - 1 \le |f(x)| < n, n \in \mathbb{N}\}.$$

---

† We obtained this theorem in a private communication from Professor R. Bartle.

It suffices to show that F is of negligible variation on $Z_n = S \cap A_n$ for every n. Firstly we have the inequality

$$\sum_\pi |F(u,v)| \leq \sum_\pi |F(u,v) - f(\xi)(v-u)| + \sum_\pi |f(\xi)(v-u)| \quad (3.94)$$

valid for any partial division π. By Henstock's lemma, for every positive ε there is a $\delta : [a,b] \mapsto \mathbb{R}_+$ such that

$$\sum_\pi |F(u,v) - f(\xi)(v-u)| < \frac{\varepsilon}{2} \quad (3.95)$$

whenever π is a δ-fine subpartition of $[a,b]$. There exists a countable system of open disjoint intervals I_j such that

$$Z_n \subset \bigcup_1^\infty I_j \quad \text{and} \quad \sum_1^\infty |I_j| < \frac{\varepsilon}{2n}.$$

For $x \in I_j$ we modify δ obtained from the Henstock lemma in such a way that $(x - \delta(x), x + \delta(x)) \subset I_j$. Let π be now a δ-fine tagged partial division on E. Then by (3.94) and by (3.95) we have

$$\sum_\pi |F(u,v)| \leq \frac{\varepsilon}{2} + n \sum_\pi (v-u) \leq \frac{\varepsilon}{2} + n \sum_1^\infty |I_j| < \varepsilon. \qquad \bullet$$

Combining Theorem 2.5.29, Theorem 3.8.2 and Theorem 3.9.2 we have

THEOREM 3.9.3 *Let f be KH-integrable on $[a,b]$ and*

$$F(x) = \mathcal{KH} \int_a^x f. \quad (3.96)$$

Then conditions (i), (ii) and (iii) from Theorem 3.9.1 are satisfied.

It is now easy to obtain an important characterization of absolute KH-integrability.

THEOREM 3.9.4 *A function f is absolutely KH-integrable on $[a,b]$ if and only if there is an AC function F with $F' = f$ almost everywhere on $[a,b]$.*

Proof Let f be absolutely KH-integrable and $F(x) = \int_a^x f$. The function F is AC by Theorem 3.6.4. Next $F' = f$ a.e. by Theorem 3.8.2. Conversely if F is AC then it is of negligible variation on every set of measure zero, and in particular on the set where $F' \neq f$. Hence $F(x) - F(a) = \int_a^x f$ by Theorem 3.9.1. Since F is AC it is of bounded variation and the function f is absolutely integrable by Theorem 3.4.1.$\bullet$

3.10 Lebesgue points, approximation by step functions

If f is integrable on $[a, b]$ then x is said to be a *Lebesgue point* of f if

$$\lim_{h \to 0} \frac{1}{h} \int_x^{x+h} |f(t) - f(x)| dt = 0. \qquad (3.97)$$

Equivalently x is a Lebesgue point of f if and only if for every positive ε there is a $\delta(x)$ such that

$$\int_u^v |f(t) - f(x)| dt < (v - u)\varepsilon \qquad (3.98)$$

whenever $x - \delta(x) < u \leq x \leq v < x + \delta(x)$. Obviously, if x is a Lebesgue point of f then the indefinite integral of f has a derivative $f(x)$ at x. For the function $f(t) = \sin t^{-1}$ for $t \neq 0$ and $f(0) = 0$ the point $x = 0$ is not a Lebesgue point† but for $F(t) = \int_{-1}^t \sin u^{-1} \, du$ we have $F'(0) = 0 = f(0)$. However, the set of Lebesgue points and the set where the indefinite integral has a derivative equal to the value of the integrand are of the same 'size', as the next theorem shows.

THEOREM 3.10.1   *If f is absolutely KH-integrable on $[a, b]$ then almost every point is a Lebesgue point.*

Proof   We can and shall assume that f is defined everywhere on $[a, b]$. Let r be a rational number. By Theorem 3.8.2

$$\lim_{h \to 0} \frac{1}{h} \int_x^{x+h} |f(t) - r| dt = |f(x) - r| \qquad (3.99)$$

for almost all $x \in [a, b]$. Let M_r be the set of those x for which (3.99) does not hold and $M = \bigcup_r M_r$; then M is of measure zero. Now we choose $x \in (a, b) \setminus M$ and a rational r such that $|f(x) - r| < \varepsilon/3$. Then

$$\left| \frac{1}{h} \int_x^{x+h} |f(t) - f(x)| dt \right| \leq \left| \frac{1}{h} \int_x^{x+h} |f(t) - r| dt \right| + \frac{\varepsilon}{3}.$$

Since $x \notin M_r$ we have (3.99) and consequently there is a $\delta > 0$ such that

$$\left| \frac{1}{h} \int_x^{x+h} |f(t) - r| dt \right| < |f(x) - r| + \frac{\varepsilon}{3},$$

for $0 < |h| < \delta$. Hence

$$\left| \frac{1}{h} \int_x^{x+h} |f(t) - f(x)| dt \right| < \varepsilon,$$

---

† Another simple example is in Exercise 3.12.

for $0 < |h| < \delta$ and $x \in (a,b) \setminus M$.   •

THEOREM 3.10.2 *If f is absolutely integrable on $[A, B]$ then for every positive ε there is a step function φ such that*

$$\int_A^B |f - \varphi| < \varepsilon. \tag{3.100}$$

Proof Since we can always find a bounded interval $[c, d]$ such that

$$\int_A^B |f| - \int_c^d |f| < \varepsilon,$$

it clearly suffices to prove the theorem if $[A, B]$ is a bounded interval. If x is a Lebesgue point there exists a $\delta(x)$ such that

$$\int_u^v |f(t) - f(x)| < (v - u)\frac{\varepsilon}{2(B - A)}, \tag{3.101}$$

whenever $x - \delta(x) < u \leq x \leq v < x + \delta(x)$. The set N of points which are not Lebesgue points is of measure zero and therefore we can assume without loss of generality that $f = 0$ on N. By Theorem 3.9.2 the indefinite integral of $|f|$ is of negligible variation on N and there is

$$\delta : N \mapsto \mathbb{R}_+$$

such that

$$\sum_\pi \int_u^v |f| < \frac{\varepsilon}{2}, \tag{3.102}$$

whenever π is δ-fine and tagged in N. Let $\pi \equiv \{(x, [u, v])\}$ be a δ-fine partition of $[A, B]$ and

$$\varphi(t) = f(x) \quad \text{for} \quad t \in (u, v).$$

Then

$$\int_A^B |f - \varphi| \leq \sum_\pi \int_u^v |f(t) - \varphi(t)| dt$$
$$\leq \sum_{x \notin N} \int_u^v |f(t) - f(x)| \, dt + \sum_{x \in N} \int_u^v |f(t)| \, dt$$
$$\leq \frac{\varepsilon}{2(B - A)} \sum_{x \notin N} (v - u) + \frac{\varepsilon}{2}.$$

Consequently we have (3.100).   •

REMARK 3.10.3 Since

$$\left| \int_a^b f - \int_a^b \varphi \right| \le \int_a^b |f - \varphi|,$$

Theorem 3.10.2 indicates that the integral of an absolutely KH-integrable function can be evaluated with arbitrary accuracy by the integral of a step function. The assumption that f is absolutely integrable in Theorem 3.10.2 is essential, since clearly integrability of $|f - \varphi|$ implies absolute integrability of f. Hence 'if' in Theorem 3.10.2 can be replaced by 'if and only if' and the theorem remains valid.

3.11 Measurable functions and sets

A function may fail to be integrable for two reasons. It can be 'too large' like $\mathbf{1}_{(0,\infty)}$ or $\sin x$ on $[0, \infty]$, or it can be very irregular. In this section we shall study that aspect of regularity which makes functions integrable. It is interesting to note that it is not easy to find a bounded function which fails to be integrable over a bounded interval. We deal with this question in Subsection 3.11.1.

DEFINITION 3.11.1

(i) *A* **function** $f : \mathbb{R} \mapsto \mathbb{R}$ *is said to be* **measurable** *if there exists a sequence of continuous functions* $\{\varphi_n\}$ *converging a.e. to* f;

(ii) *a* **set is measurable** *if and only if its characteristic function is measurable;*

(iii) *a* **function** f *is called* **measurable on a set** $S \subset \mathbb{R}$ *if* S *is measurable and if* f^*, *the function which equals* f *on* S *and zero outside* S, *is measurable.*

In future we shall slightly abuse the notation and denote the extension f^* by $f\mathbf{1}_S$ even if f is not defined everywhwere. The term comes from the way measurable sets were originally introduced by Lebesgue.† Obviously, measurable functions are related to continuous functions. The so-called Luzin theorem (Theorem 3.11.17) describes precisely how much continuity is preserved in any measurable function.

The following theorem is an obvious consequence of the dominated convergence theorem. Roughly speaking it says that a measurable function is integrable if it is not 'too large'.

---

† An alternative approach to measure and measurability, closer to Lebesgue's original ideas, is outlined in Section 3.13.

THEOREM 3.11.2 *If f is measurable on* (A, B) *and there is a KH-integrable g such that* $|f| \leq g$ *on* (A, B) *then f is (absolutely) KH-integrable on* $[A, B]$. *In particular, if f is bounded and measurable on a closed bounded interval* $[a, b]$, *then it is (absolutely) KH-integrable on* $[a, b]$.

An immediate consequence of the definition and Theorem 3.8.2 is

THEOREM 3.11.3 *If f is KH-integrable on* $[a, b]$ *then it is measurable on* $[a, b]$. *If f is KH-integrable on every bounded interval then it is measurable.*

The usual arithmetic and other frequently used operations preserve measurability.

THEOREM 3.11.4 *If* $\Phi : \mathbb{R}^n \mapsto \mathbb{R}$ *is continuous and* f_i, $i = 1, 2, \ldots, n$ *measurable on S then the composition* $\Phi \circ f_i$,

$$\Phi \circ f_i(x) = \Phi(f_1(x), f_2(x), \ldots, f_n(x))$$

is measurable on S. If f and g are measurable on S then so are the functions $|f|$, $f \pm g$, fg, $\mathrm{Max}(f, g)$, $\mathrm{Min}(f, g)$, f^+, f^-.

Proof All assertions are obvious from the definition.   •

A consequence of the previous three theorems is: If f is KH-integrable on a bounded interval $[a, b]$ and $f^{N,M}$ its truncation from below by M and from above by N then $f^{N,M}$ is absolutely KH-integrable. Also if f is absolutely KH-integrable on $[A, B] \subset \mathbb{R}$ and c measurable and bounded, $|c(x)| \leq K$ then cf is measurable and majorized by an integrable function, namely $K|f|$. Consequently it is absolutely KH-integrable. We state this as

THEOREM 3.11.5 *The product of an absolutely KH-integrable function with a bounded and measurable function is absolutely KH-integrable.*

Measurability is preserved under the limit passage.

THEOREM 3.11.6 *If the functions* f_n, $n = 1, 2, \ldots$ *are measurable on S and* $\lim_{n \to \infty} f_n = f$ *exists almost everywhere then f is measurable on S.*

Proof We can and shall assume that each f_n is zero outside of S. Let $\Phi : \mathbb{R} \mapsto (0, 1)$ be continuous and strictly increasing, Ψ its inverse. $\Phi \circ f_n$

are measurable and bounded and hence KH-integrable on every bounded interval. By the Lebesgue dominated convergence theorem $\Phi \circ f$ is KH-integrable on every bounded interval and therefore measurable. So is $f = \Psi \circ \Phi \circ f$. •

COROLLARY 3.11.7 *If the functions f_n, $n = 1, 2, \ldots$ are measurable then so are* $\sup \{f_n; \, n \in \mathbb{N}\}$ *and* $\inf \{f_n; \, n \in \mathbb{N}\}$.

COROLLARY 3.11.8 *A function f is measurable if and only if there exists a sequence of step functions or a sequence of integrable functions converging a.e. to f.*

Many properties of measurable sets follow simply from the already established properties of measurable functions.

THEOREM 3.11.9 (Properties of measurable sets)

(i) *If A, B are measurable then so are $A \cap B$, $A \cup B$ and $A \setminus B$.*

(ii) *Open sets and closed sets are measurable, in particular $\mathbb{R}$ and $\emptyset$ are measurable.*

(iii) *If A_n are measurable for $n = 1, 2, \ldots$ then so are $\bigcap_1^\infty A_n$ and $\bigcup_1^\infty A_n$.*

A consequence of (i) and (iii) is

(iv) *A set is measurable if and only if its intersections with bounded intervals are measurable.*

A *simple function* is defined as a function which has a finite range. In the theory of the integral simple functions are used extensively. Often something is proved for simple functions first and then a limit passage is used to extend the result to the general case. The next theorem describes measurability of functions in terms of simple functions and measurability of sets. It is also an approximation theorem.

THEOREM 3.11.10 *Let $f : \mathbb{R} \mapsto \mathbb{R}$. Then the following statements are equivalent.*

(I) *f is measurable;*

(II) *the set $\{x; f(x) > c\}$ is measurable for every $c \in \mathbb{R}$;*

(III) *all four sets*

$$S_> = \{x; f(x) > c\}, \tag{3.103}$$
$$S_\leq = \{x; f(x) \leq c\}, \tag{3.104}$$
$$S_< = \{x; f(x) < c\}, \tag{3.105}$$
$$S_\geq = \{x; f(x) \geq c\}, \tag{3.106}$$

are measurable for every c;

(IV) *there exists a sequence of simple measurable functions $\{f_n\}$ such that*

  (i) $\lim_{n\to\infty} f_n(x) = f(x)$ *for every x;*

  (ii) *if $|f| \leq C$ then $|f_n| \leq C$ for all n and the convergence is uniform;*

  (iii) *if f is non-negative then so are all f_n and $f_n \leq f_{n+1}$.*

Proof (I)$\Rightarrow$(II). Define

$$d_n(x) = \frac{\mathrm{Min}(f(x), c + \frac{1}{n}) - \mathrm{Min}(f(x), c)}{\frac{1}{n}}.$$

Then d_n are measurable for every n and as n goes to infinity d_n tends to the characteristic function of the set $S_>$.

(II)$\Rightarrow$(III). $S_\leq$ is the complement of $S_>$ and hence is measurable. Replacing f by $-f$ and c by $-c$ yields measurability of $S_<$ and $S_\geq$.

(III)$\Rightarrow$(IV). The sets

$$E_{k,n} = \left\{x; \frac{k}{2^n} \leq f(x) < \frac{k+1}{2^n}\right\}$$

are measurable for all $k \in \mathbb{Z}$ and $n \in \mathbb{N}$. Let $f_n(x) = k2^{-n}$ for $x \in E_{k,n}$. The functions f_n are measurable, simple and have all the required properties by construction.

(IV)$\Rightarrow$(I). This follows from Theorem 3.11.6.   •

The *Lebesgue measure $m(S)$* of a measurable set $S \subset \mathbb{R}$ is by definition

$$m(S) = \int_{-\infty}^{\infty} 1_S. \tag{3.107}$$

If the integral in (3.107) converges we call the *set S integrable*. In this book we shall call Lebesgue measure simply *measure*. Measure is a generalization of the concept of length of an interval to more general sets. We shall see later that in two or three dimensions measure is a generalization of area or volume, respectively. It is an interesting question whether or not one can generalize these concepts *to all sets in* $\mathbb{R}^n$ in such a way that some fundamental properties of length, area and volume are preserved. The answers are surprising and depend on the dimension and on whether or not we include property (3.109) in Theorem 3.11.11 below among 'fundamental properties'. We refer to [34] Chapter 3 §7 for discussion of these matters.

Properties of measure follow simply from the already established properties of the integral.

THEOREM 3.11.11 (Properties of measure)

(i) $0 \leq m(A) \leq \infty$ *for any measurable* A; $m(\emptyset) = 0$, $m(\mathbb{R}) = \infty$.

(ii) *If* A, B *are measurable then* $m(A \cup B) = m(A) + m(B) - m(A \cap B)$; *in particular, if* A *and* B *are disjoint then*

$$m(A \cup B) = m(A) + m(B).$$

If $A \subset B$ *then* $m(A) \leq m(B)$.

(iii) *If* A_n *are measurable for* $n = 1, 2, \ldots$ *then*

$$m(\bigcup_1^\infty A_n) \leq \sum_{n=1}^\infty m(A_n) \tag{3.108}$$

and if A_n *are pairwise disjoint then*

$$m(\bigcup_1^\infty A_n) = \sum_{n=1}^\infty m(A_n). \tag{3.109}$$

The properties in (iii) are referred to as *countable subadditivity* and *countable additivity*, respectively. Mathematicians of the nineteenth century had difficulties in accepting countable additivity of measure for various reasons. One of them was that countable additivity was not compatible with Riemann theory. It was the success of Lebesgue theory and its rapid penetration into other branches of mathematics which made countable additivity a cornerstone of modern integration theories.

The series in (3.108) and (3.109) are not ordinary series; some terms can be ∞. We make the following definition: The sum $\sum_1^\infty a_n$ of a series of non-negative terms a_n is the limit (possibly infinite) of partial sums $\sum_1^n a_i$ if all $a_i \in \mathbb{R}$ and is ∞ when some $a_i = \infty$. The following theorem follows from the countable additivity of measure or even more easily from the monotone convergence theorem.

THEOREM 3.11.12 *If* E_n *are measurable and*

$$E_1 \subset E_2 \subset \cdots \subset E_n \subset \cdots$$

then

$$\lim_{n\to\infty} m(E_n) = m(\bigcup_1^\infty E_n). \qquad (3.110)$$

If $m(E_1) < \infty$ and

$$E_1 \supset E_2 \supset \cdots \supset E_n \supset \cdots$$

then

$$\lim_{n\to\infty} m(E_n) = m\left(\bigcap_1^\infty E_n\right). \qquad (3.111)$$

REMARK 3.11.13 The assumption $m(E_1) < \infty$ is essential: if $E_n = (n,\infty)$ then the left side of (3.111) is ∞ and the right side is 0.

If A and B are sets then the *symmetric difference* of these sets is $(A \setminus B) \cup (B \setminus A)$ and is denoted by $A\triangle B$. The characteristic function of $A\triangle B$ is $|1_A - 1_B|$. If Z is a measurable set then there exists a sequence of step functions z_n converging to 1_Z a.e.. By redefining each $z_n(x)$ to be 1 or 0 depending on whether $z_n(x) \le 1/2$ or $> 1/2$ we make each z_n the characteristic function of a set Z_n which is a finite union of intervals and $|1_{Z_n} - 1_Z| \to 0$ a.e. If Z is a part of a bounded interval K then we can take all $Z_n \subset K$ and it then follows from the Lebesgue dominated convergence theorem that $m(Z\triangle Z_n) \to 0$. Thus we have proved

THEOREM 3.11.14 *If $Z \subset \mathbb{R}$ is bounded and measurable then, for every positive ε, there exists a finite union of intervals† S such that*

$$m(Z\triangle S) < \varepsilon. \qquad (3.112)$$

Now we take a finite union of closed intervals S for which (3.112) holds with ε replaced by $\varepsilon/2$. Using Theorem 2.11.3 we cover $Z\triangle S$ and all endpoints of intervals in S by a system of open intervals J_n, $G_0 = \bigcup_1^\infty J_n$ with $m(G_0 \setminus (S\triangle G_0)) < \varepsilon/2$. Then, for $F = S \setminus G_0$ and $G = S \cup G_0$, we have

$$F \text{ is closed, } F \subset Z, \qquad G \text{ is open, } Z \subset G \qquad (3.113)$$
$$m(G \setminus F) \;<\; \varepsilon. \qquad (3.114)$$

The last relations completely characterize measurable sets. More precisely we have

† Closed or open or of any kind.

THEOREM 3.11.15 *A set $Z \subset \mathbb{R}$ is measurable if and only if for every positive ε there are sets F and G such that (3.113) and (3.114) hold.*

Proof Let Z be measurable. We know that (3.113) and (3.114) hold if Z is bounded. Therefore there exist closed sets F_n and open sets G_n such that

$$F_n \subset [-n, n] \cap Z \subset G_n,$$
$$m(G_n \setminus F_n) < \frac{\varepsilon}{2^n},$$

for every $n \in \mathbb{N}$. The set $F = \bigcup_1^\infty F_n$ contains all its limit points, hence it is closed. $G = \bigcup_1^\infty G_n$ is obviously open and we have

$$F \subset Z \ \subset \ G$$
$$G \setminus F \ \subset \ \bigcup_1^\infty (G_n \setminus F_n)$$
$$m(G \setminus F) \ < \ \varepsilon \sum_1^\infty \frac{1}{2^n}.$$

Going in the opposite direction we choose a bounded interval K. By (iv) of Theorem 3.11.9 it suffices to show that $Z \cap K$ is integrable. The sets $F_K = F \cap K$ and $G_K = G \cap K$ are integrable, $G_K \setminus F_K \subset G \setminus F$ and therefore $m(G_K \setminus F_K) < \varepsilon$. The functions $\mathbf{1}_{G_K}$, $\mathbf{1}_{F_K}$ are integrable,

$$\mathbf{1}_{G_K} \geq \mathbf{1}_{Z \cap K} \geq \mathbf{1}_{F_K},$$

and consequently $Z \cap K$ is integrable by Theorem 2.5.16.   •

COROLLARY 3.11.16 *If $m(Z) < \infty$ then there exist a **compact** set F and an open set G such that (3.114) holds.*

THEOREM 3.11.17 (Luzin's theorem)   *Let S be a measurable set. The following statements are equivalent:*

(I) *f is measurable on S;*

(II) *for every positive ε there exists an open set O_ε with $m(O_\varepsilon) < \varepsilon$ such that the restriction of f to $S \setminus O_\varepsilon$ is continuous;*

(III) *for every positive ε there exists a closed set F with $m(S \setminus F) < \varepsilon$ and a function $f^* : \mathbb{R} \to \mathbb{R}$ continuous on all of $\mathbb{R}$ such that $f^*(x) = f(x)$ for $x \in F$.*

REMARK 3.11.18   The reader is advised to treat the continuity of the restriction with some caution. It means precisely this: for every $x \in S$ and every sequence $\{x_n\}$ with $x_n \in S$ and $x_n \to x$ we have $f(x_n) \to f(x)$. For example the restriction of the characteristic function of the rationals to the rationals (or irrationals) is continuous although the function itself is everywhere discontinuous.

REMARK 3.11.19   The Tietze theorem asserts that a function continuous on a closed set can be extended to a function continuous on the whole of $\mathbb{R}$. We need this theorem for proving (II)$\Rightarrow$(III). The Tietze Theorem in $\mathbb{R}^1$ is easy to prove: one extends the function by making it linear on the contiguous intervals of the closed set. For a Tietze theorem in a general topological setting we refer to [39] or [19].

Proof of Theorem 3.11.17   We carry out the proof of (I)$\Rightarrow$(II) in several steps.

(i) If f is a characteristic function of a set $Z \subset S$ then, by Theorem 3.11.15, there are F and G satisfying (3.113) and (3.114). If $G_\varepsilon = G \setminus F$ then f is continuous on G_ε because it is constant on F.

It follows easily that (I)$\Rightarrow$(II) holds for a function which is a linear combination of characteristic functions of measurable sets, i.e. for a simple measurable function.

(ii) If $|f| \leq 1$ then for each function f_n from Theorem 3.11.10 there is an open set O_n with $m(O_n) < \varepsilon/2^{-n}$ and the restriction of f_n to $\mathbb{R} \setminus O_n$ continuous. Let $G_\varepsilon = \bigcup_1^\infty O_n$. Since the restriction of f_n to the complement of G_ε is continuous and $\{f_n\}$ converges uniformly to f, we have proved the theorem in this case.

(iii) We take a strictly increasing function Φ which maps $\mathbb{R}$ onto $(-1, 1)$ and denote its inverse by Ψ. Applying (ii) to $\Phi \circ f$ gives us G_ε, $m(G_\varepsilon) < \varepsilon$ with the restriction of this function to the complement of G_ε continuous. $f = \Psi \circ \Phi \circ f$ has also this property.

(II)$\Rightarrow$(III). There exists an open set O, with $m(O) < \varepsilon/2$, such that the restriction of f to $S \setminus O$ is continuous. By Theorem 3.11.15 there are a closed set F and an open set G such that $F \subset S \setminus O \subset G$ and $m(G \setminus F) < \varepsilon/2$. Clearly the restriction of f to F is continuous and $m(S \setminus F) < \varepsilon$ since $S \setminus F \subset (S \setminus O) \cup (G \setminus F)$. Using the Tietze theorem the function f can be extended from the set F to a function continuous on all of $\mathbb{R}$.

(III)$\Rightarrow$(I). Assuming (III) we obtain for every $n \in \mathbb{R}$ a continuous function f_n and a set $F_n \subset S$ such that $f_n = f$ on F_n and $m(S \setminus F_n) < 2^{-n}$. We denote $Z_n = \bigcap_n^\infty F_i$, $M = \bigcup_1^\infty Z_n$. If $x \in Z_n$ for some n then $f_n(x) \to f(x)$, and consequently this convergence takes place for every $x \in M$. It remains to show that $S \setminus M$ is of measure zero. The following set relations are fairly obvious:

$$S \setminus M \subset S \setminus \bigcup_1^\infty Z_i$$

$$= \bigcap_1^\infty (S \setminus Z_i) \subset S \setminus Z_n$$

$$= S \setminus \bigcap_n^\infty F_i = \bigcup_n^\infty (S \setminus F_i).$$

Consequently

$$m(S \setminus M) \le \sum_n^\infty \frac{1}{2^i} = \frac{1}{2^{n-1}}.$$

This shows $S \setminus M$ to be of measure zero.   ●

THEOREM 3.11.20 (Egoroff's theorem) *Let $\{f_n\}$ be a sequence of functions measurable on S, $m(S) < \infty$. If $f_n(x) \to f(x)$ almost everywhere in S, then for every $\eta > 0$ there is an open set G with $m(G) < \eta$ such that f_n converges to f uniformly on $S \setminus G$.*

Proof Since a set of measure zero can be covered by an arbitrarily small open set, we may assume that $f_n(x) \to f(x)$ everywhere on S. Given x and $\varepsilon > 0$, there is an integer N such that

$$|f_n(x) - f(x)| < \varepsilon \quad \text{whenever} \quad n \ge N.$$

Let X_N denote the set of x such that the above inequality holds. Obviously, X_N depends on ε, $X_i \subset X_{i+1}$ and $\bigcup_1^\infty X_i = S$. Consequently $S \setminus X_i \supset S \setminus X_{i+1}$ and $\bigcap_1^\infty S \setminus X_i = \emptyset$. Therefore we can, by Theorem 3.11.12, choose an N such that $m(S \setminus X_N)$ is arbitrarily small.

Take $\{\varepsilon_i\}$ decreasing to 0 and $\eta > 0$. Then by what we have proved above there exist N_i and Y_i such that $m(S \setminus Y_i) < \eta 2^{-i}$ and

$$|f_n(x) - f(x)| < \varepsilon_i \quad \text{whenever} \quad n \ge N_i \quad \text{and} \quad x \in Y_i.$$

Next, take $Y = \bigcap_{i=1}^\infty Y_i$. Then, obviously, $S \setminus Y = \bigcup_1^\infty (S \setminus Y_i)$ and $m(S \setminus Y) < \eta$ by Theorem 3.11.12. For every $\varepsilon > 0$ we find $\varepsilon_i < \varepsilon$ and

$N_\varepsilon = N_i$ such that

$$|f_n(x) - f(x)| < \varepsilon_i < \varepsilon \quad \text{for all} \quad x \in Y \quad \text{whenever} \quad n \geq N_\varepsilon.$$

That is, f_n converges to f uniformly on Y. We can choose an open set $G \supset S \setminus Y$ such that $m(G) < \eta$ and the proof is complete.   •

REMARK 3.11.21   The assumption that $m(S) < \infty$ is essential as the example of $S = \mathbb{R}$, $f_n = \mathbf{1}_{(n,\infty)}$ shows.

REMARK 3.11.22   An example in [48] shows, perhaps surprisingly, that Egoroff's theorem ceases to be true if the limit as $n \to \infty$ in Egoroff's theorem is replaced by the limit of a continuously changing variable.

3.11.1 A non-measurable set

If S is a set and

$$S_a = \{x + a; \ x \in S\}$$

then S_a is measurable if and only if S is and then $m(S) = m(S_a)$. We denote by $E\{a\} = \{x; x - a \in \mathbb{Q}, -1 \leq x \leq 1\}$. It is easy to see that two sets $E\{a\}$, $E\{b\}$ are either disjoint or identical. Now we make up a set S which is constituted by exactly one element from each set $E\{a\}$. Let $r_1, r_2, \ldots$ be a sequence which contains only rationals from $[-2, 2]$ and each rational exactly once. We claim that

$$[-1, 1] \subset \bigcup_{n=1}^{\infty} S_{r_n} \subset [-3, 3], \tag{3.115}$$

$$\text{for} \quad i \neq j \quad S_{r_{n_i}} \cap S_{r_{n_j}} = \emptyset. \tag{3.116}$$

Since S contains an element from $E\{x\}$ for every $x \in [-1, 1]$ there are $s \in S$ and $q \in \mathbb{Q}$ such that $s - x = q$, i.e. $x \in S_{-q}$. Since obviously $|q| \leq 2$, we have that $-q = r_n$ for some n. The second inclusion is obvious. It follows from (3.115) and (3.116) that

$$2 \leq \sum_{1}^{\infty} m(S_{r_n}) \leq 6.$$

This together with $m(S_{r_n}) = m(S)$ leads to a contradiction. The assumption that S is measurable is not tenable — S is non-measurable.

If S is not measurable and bounded then $\mathbf{1}_S$ cannot be KH-integrable.

The existence of a non-measurabale set provides an example of a bounded function which fails to be KH-integrable over a closed bounded interval.

The sentence: '*we make up a set which is constituted by exactly one element in each set $E\{a\}$.*' is not accepted by all mathematicians as a legitimate definition of a set. Its ultimate validity hinges on the following statement — the so-called axiom of choice.

Axiom of choice *If $\mathfrak{F}$ is a family of sets then there exists a function*

$$g : \mathfrak{F} \mapsto \bigcup_{S \in \mathfrak{F}} S$$

such that $g(S) \in S$ for every $S \in \mathfrak{F}$.

Most mathematicians accept set theory as a foundation stone for all mathematics and the axiom of choice is one of the axioms of set theory. The reason why the axiom of choice is controversial is that it leads to some truly amazing, almost paradoxical, results; one such theorem, the so-called Hausdorff decomposition of a sphere, is discussed in [34]. In this book we accept the axiom of choice without hesitation, but we remark that an attempt has been made to build measure theory (without the axiom of choice) such that every set in $\mathbb{R}$ is measurable.

3.12 The McShane integral

The definition of the KH-integral is a small but very significant modification of the definition of the Riemann integral. There is yet another variation of the Riemann definition similar in spirit to the Kurzweil–Henstock one which leads to an absolutely convergent integral. This development is due to McShane [31] and [32]. In his theory the tags are allowed to float outside the intervals they tag. The theory is slightly simpler for a compact interval but we shall expand it to include infinite intervals as well. The reader should be familiar with the conventions regarding the infinite numbers ∞ and $-\infty$ we made in Section 2.10.

Let us remind ourselves that we agreed to call a map δ from a closed interval $[A, B]$ into the set of open intervals in $\overline{\mathbb{R}}$ a gauge if $\xi \in \delta(\xi)$ for every $\xi \in [A, B]$ and $\delta(\xi)$ was a *bounded* open interval for every $\xi \in \mathbb{R} \cap [A, B]$. Similarly as in the definition of a partition in Section 1.2, an M-partition of a closed interval in $\overline{\mathbb{R}}$ is a set of pairs (ξ_k, I_k), $k = 1, \ldots, n$, such that the closed intervals I_k are non-overlapping with $\bigcup_1^n I_k = [A, B]$, but for an M-partition the points ξ_k are merely in $[A, B]$.† Here ξ_k is still called the tag of I_k. In contrast to partitions,

---

† The reader should note that ξ_k might be outside I_k.

where a point can be a tag of at most two intervals, for an M-partition a point can tag many intervals. The effect of this on a Riemann sum can be seen in Figure 3.1. Exactly as in Section 2.10 we shall say that

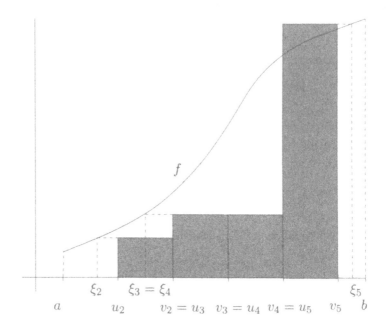

Fig. 3.1. Riemann sum for an M-partition

an M-partition $\pi \equiv \{(\xi_k, J_k)\}$ of $[A, B]$ is δ-fine if δ is a gauge defined on $[A, B]$ and $J_k \subset \delta(\xi_k)$. The geometrical meaning of δ-fine for an M-partition is indicated in Figure 3.2 where $\delta(\xi) = (\xi - \delta(\xi), \xi + \delta(\xi))$. We

Fig. 3.2. δ-fine M-partition

denote by $\pi \ll \delta$ that π is δ-fine. As before the unbounded intervals of a δ-fine partition are tagged by either ∞ or $-\infty$. We make the same convention as in Section 2.10, namely that every function to be integrated is automatically defined (redefined) at ∞ and $-\infty$ as zero. The Riemann sum $\sum_\pi f = \sum_\pi f(x_k)|J_k|$ for a partition π of a possibly unbounded interval $[A, B]$ is always meaningful because $f(x_k)|J_k| = 0.\infty = 0$ for unbounded J_k.

DEFINITION 3.12.1 *A number I is the McShane integral of f from A to B (or over [A, B]), in symbols $\mathcal{M}\int_A^B f$, if for every positive ε there is a gauge δ such that for every δ-fine M-partition π*

$$\left|\sum_\pi f - I\right| < \varepsilon. \tag{3.117}$$

We may use other notation, like $\mathcal{M}\int_a^b f$ or $\mathcal{M}\int_a^b f(x)dx$ or $\mathcal{M}\int_I f$ for the McShane integral. If the McShane integral of f exists then f is called McShane integrable or briefly M-integrable. Since every partition is also an M-partition every McShane integrable function f is also KH-integrable and

$$\mathcal{KH}\int_A^B f = \mathcal{M}\int_A^B f. \tag{3.118}$$

The Cauchy convergence principle for the M-integral reads:

THEOREM 3.12.2 (Bolzano–Cauchy condition for the McShane integral.) *A function f is M-integrable if and only if for every positive ε there is a gauge δ such that*

$$\left|\sum_{\pi_1} f - \sum_\pi f\right| < \varepsilon \tag{3.119}$$

whenever the M-partitions π_1 and π are δ-fine.

Proof If f is M-integrable then (3.119) clearly holds. Going in the other direction we denote by

$$I_\delta = \sup\left\{\sum_\pi f; \pi \ll \delta\right\} \text{ and } I = \inf\{I_\delta; \delta \text{ a gauge}\}.$$

If $δ$ is chosen so that (3.119) holds with $ε/2$ then

$$\sum_{\pi_1} f - \frac{\varepsilon}{2} < \sum_\pi f < \sum_{\pi_1} f + \frac{\varepsilon}{2}.$$

Keeping π_1 fixed and taking the supremum over all $π$ with $π \ll δ$ we obtain

$$\sum_{\pi_1} f - \frac{\varepsilon}{2} \le I_\delta \le \sum_{\pi_1} f + \frac{\varepsilon}{2}.$$

Consequently

$$\sum_{\pi_1} f - \frac{\varepsilon}{2} \le I \le \sum_{\pi_1} f + \frac{\varepsilon}{2}.$$

Since π_1 is an arbitrary δ-fine M-partition f is M-integrable and I is the M-integral of f. •

Only minor changes are needed for many proofs of theorems for the KH-integral to become valid for the McShane integral. All theorems in Section 2.5 remain valid if the KH-integral is replaced by the McShane integral and the interval $[a, b]$ by $[A, B] \subset \overline{\mathbb{R}}$. The concept of a sub-partition can be modified in an obvious way to M-subpartition and the Henstock lemma proved almost verbatim for the M-subpartitions and the McShane integral. Very importantly the monotone convergence theorem, Fatou's lemma and the Lebesgue dominated convergence theorem are valid for the M-integral.

There are, however, important differences. For the M-integral the fundamental theorem does not hold as stated either in Theorem 2.6.2 or in Theorem 3.9.1 but becomes valid with the additional assumption that F' is M-integrable. The second version of the fundamental theorem as stated in Theorem 2.6.8 is true and has the same proof for the M-integral. Theorems 2.8.3 and 2.9.3 are false for the M-integral. The integral $\mathcal{M} \int_a^b f$ can be evaluated as a limit

$$\lim_{c \downarrow a} \mathcal{M} \int_c^b f = \mathcal{M} \int_a^b f \quad \text{or} \quad \lim_{c \uparrow b} \mathcal{M} \int_a^c f = \mathcal{M} \int_a^b f$$

if f is integrable on $[a, b]$. However, f may fail to be M-integrable on $[a, b]$ even if it is integrable e.g. on $[c, b]$ for every c with $a < c < b$ and the first limit above exists. All these comments will be obvious after we prove Theorem 3.12.5 which asserts the equivalence of absolute KH-integrability and M-integrability. One very important difference between KH and M-integration is the next lemma.

LEMMA 3.12.3 (Absolute integrability lemma) *If f is M-integrable then for every positive ε there is a gauge δ such that for two δ-fine M-partitions*

$$\{(x_k, I_k); \ k = 1, 2 \dots, s\}, \tag{3.120}$$

$$\{(z_j, J_j); \ j = 1, 2 \dots, t\} \tag{3.121}$$

it is true that

$$\sum_{k=1}^{s} \sum_{j=1}^{t} |f(x_k) - f(z_j)| \, |I_k \cap J_j| < \varepsilon. \tag{3.122}$$

Proof For a given positive ε let δ be as in Theorem 3.12.2. Define

$\xi_{kj} = x_k$ and $\zeta_{kj} = z_j$ if $f(x_k) \geq f(z_j)$, otherwise let $\xi_{kj} = z_k$ and $\zeta_{kj} = x_j$. This choice causes

$$|f(x_k) - f(z_j)| = f(\xi_{kj}) - f(\zeta_{kj}) \qquad (3.123)$$

for $k = 1, \dots, s$ and $j = 1, \dots, t$. Both sets of pairs†

$$\{(\xi_{kj}, I_k \cap J_j);\ k = 1, \dots, s, j = 1, \dots, t\},$$
$$\{(\zeta_{kj}, I_k \cap J_j);\ k = 1, \dots, s, j = 1, \dots, t\}$$

are δ-fine M-partitions, and consequently

$$\left| \sum_{k=1}^{s} \sum_{j=1}^{t} f(\xi_{kj}) \, |I_k \cap J_j| - \sum_{k=1}^{s} \sum_{j=1}^{t} f(\zeta_{kj}) \, |I_k \cap J_j| \right| < \varepsilon.$$

By (3.123) this implies (3.122). ●

COROLLARY 3.12.4 *If f is M-integrable and L a function satisfying the inequality $|L(y) - L(z)| \leq |y - z|$ on the the range of f then $L \circ f$ is M-integrable.*

Proof For the two partitions (3.120) and (3.121) we have

$$\left| \sum_{k=1}^{s} L(f(x_k)) I_k - \sum_{j=1}^{s} L(f(y_j)) I_k \right| \leq \sum_{k=1}^{s} \sum_{j=1}^{t} |f(x_k) - f(z_j)| \, |I_k \cap J_j|.$$

By the Cauchy convergence principle $L \circ f$ is M-integrable. ●

A catch phrase for Corollary 3.12.4 is: A Lipschitz function of an M-integrable function is M-integrable.

A suitable choice of L in the above corollary yields important results for M-integrability. For $L(u)$ equal to $|u|$ or u^+ or u^- or $L(u) = \mathrm{Max}(u, N)$ for $u \geq 0$ and $L(-u) = L(u)$ we obtain, for an M-integrable f, the integrability of $|f|$ or f^+ or f^- or f^N.

For the next theorem we need the concept of semicontinuity. A function $U : S \mapsto [-\infty, \infty)$ is said to be *upper semicontinuous at a point x on a set S* if for every $c > U(x)$ there is a δ such that $U(y) < c$ whenever $|x - y| < \delta$ and $y \in S$. A function L is *lower semicontinuous at a point x on a set S* if $-L$ is upper semicontinuous at x on S. A function is said to be *upper (lower) semicontinuous on a set S* if it is upper (lower) semicontinuous at every point of the set. A function is upper (lower)

---

† Some of these sets may be empty and some intervals degenerate but that does not influence the argument.

semicontinuous if it is upper (lower) semicontinuous on $\mathbb{R}$. The definition of semicontinuity looks very much like the definition of continuity but only half of the crucial inequality is retained. Many theorems about continuity also hold for semicontinuity. We need the following facts: the sum of two upper (lower) semicontinuous function is upper (lower) semicontinuous, and the multiple of an upper (lower) semicontinuous function by a non-negative constant is upper (lower) semicontinuous†. Semicontinuous functions have some important properties which continuous functions do not have: the characteristic function of an open set is lower semicontinuous, the characteristic function of a closed set is upper semicontinuous, the supremum of a set of lower semicontinuous functions is lower semicontinuous and the sum of a convergent series of non-negative lower semicontinuous functions is also lower semicontinuous. With this preparation we are ready for the theorem of this section.

THEOREM 3.12.5 (Vitali–Carathéodory)   *The following three statements are equivalent*

(A) *f is M-integrable on $[A, B]$;*

(B) *f is absolutely KH-integrable on $[A, B]$;*

(C) *for every positive ε there are absolutely KH-integrable functions U and L such that $U \leq f \leq L$, U is upper semicontinuous and bounded above on $[A, B] \cap \mathbb{R}$, L is lower semicontinuous and bounded below on $[A, B] \cap \mathbb{R}$, and*

$$\mathcal{KH} \int_A^B (L - U) < \varepsilon. \tag{3.124}$$

REMARK 3.12.6   The name Vitali–Carathéodory theorem is usually used for a similar theorem in Lebesgue theory.

COROLLARY 3.12.7   *After the theorem has been proved the KH-integral in (3.124) can be replaced by the M-integral and the theorem remains valid.*

Proof of Theorem 3.12.5   If f is M-integrable then so is $|f|$ and hence (A)$\Rightarrow$(B).

We prove the implication (B)$\Rightarrow$(C) first for $f \geq 0$. By Theorem 3.11.10 part (IV) there is a sequence of non-negative simple measurable functions $\{f_n\}$ converging everywhere increasingly to f. The difference

---

† Obviously multiplication by a negative constant changes a lower semicontinuous function to a upper semicontinuous one.

$f_n - f_{n-1}$, being a simple non-negative measurable function, can be represented as $k_1 1_{E_1} + k_2 1_{E_2} + k_3 1_{E_3} + \cdots + k_p 1_{E_p}$ with $k_i > 0$ and E_i measurable. Consequently there are measurable sets S_n and positive constants c_n such that $f(x) = \sum_1^\infty c_n 1_{S_n}(x)$ for every $x \in [A, B] \cap \mathbb{R}$. f is integrable and therefore

$$\sum_1^\infty c_n m(S_n) = \int_A^B f < \infty. \tag{3.125}$$

Since S_n are integrable there are compact sets K_n and open sets G_n such that

$$K_n \subset S_n \subset G_n$$
$$c_n m(G_n \setminus K_n) < \frac{\varepsilon}{2^{n+2}}, \quad n = 1, 2, \ldots \tag{3.126}$$

The series in (3.125) converges; hence we can find N such that

$$\sum_{N+1}^\infty c_n m(S_n) < \frac{\varepsilon}{4}. \tag{3.127}$$

Define

$$U = \sum_1^N c_n 1_{K_n}, \qquad L = \sum_1^\infty c_n 1_{G_n}.$$

It is easy to see that $U \leq f \leq L$, U is upper semicontinuous and L is lower semicontinuous. We also have

$$
\begin{aligned}
L - U &= \sum_1^N c_n 1_{G_n \setminus K_n} + \sum_{N+1}^\infty c_n 1_{G_n} \\
&\leq \sum_1^\infty c_n 1_{G_n \setminus K_n} + \sum_{N+1}^\infty c_n 1_{S_n}.
\end{aligned}
$$

This together with (3.126) and (3.127) implies (3.124). Since U is measurable and $0 \leq U \leq f$ the function U is absolutely KH-integrable and in view of (3.124) so is L. In the general case find U and L for f^+ and f^- and call them U_+, L_+ and U_-, L_-, respectively. Then $U = U_+ - L_-$ and $L = L_+ - U_-$ are well defined and finite a.e. It is easy to check that they have all the required properties.

(C)$\Rightarrow$(A). We can and shall assume that $f(A) = f(B) = 0$. For given

$\varepsilon > 0$ we choose a strictly positive function† $\mathcal{E}$ such that $\mathcal{KH}\int_A^B \mathcal{E} < \dfrac{\varepsilon}{3}$. There exists a gauge δ with the following properties:

(i) $\delta(x) \subset (A, B)$ for every $x \in (A, B)$ and consequently the intervals of any δ-fine partition which contain A and B are tagged by A and B, respectively;

(ii)

$$\int_{\delta(A)} (|U| + |L|) < \frac{\varepsilon}{3}, \qquad \int_{\delta(B)} (|U| + |L|) < \frac{\varepsilon}{3}.$$

(iii) for every $x \in [A, B] \cap \mathbb{R}$ and every $t \in \delta(x)$

$$U(t) < f(x) + \mathcal{E}(x) \quad \text{and} \quad L(t) > f(x) - \mathcal{E}(x). \tag{3.128}$$

Let $\{(x_i, I_i); i = 1, 2, \dots, n + m + p\}$ be a δ-fine partition‡ of $[A, B]$ with $x_i = A$ for $i = n + 1, \dots, n + m$ and $x_i = B$ for $i = n + m + 1, \dots, n + m + p$. Integrating inequalities (3.128) over I_k and summing gives

$$\sum_1^n \mathcal{KH}\int_{I_k} U - \frac{\varepsilon}{3} \le \sum_1^{n+m+p} f(x_k)|I_k| \le \sum_1^n \mathcal{KH}\int_{I_k} L + \frac{\varepsilon}{3}.$$

It follows from (ii) that

$$\sum_{n+1}^{n+m+p} \mathcal{KH}\int_{I_k} U - \frac{2\varepsilon}{3} < 0 \quad \text{and} \quad \sum_{n+1}^{n+m+p} \mathcal{KH}\int_{I_k} L + \frac{2\varepsilon}{3} > 0.$$

Consequently

$$\mathcal{KH}\int_A^B U - \varepsilon < \sum_1^{n+m+p} f(x_k)|I_k| < \mathcal{KH}\int_A^B L + \varepsilon.$$

Since

$$\mathcal{KH}\int_A^B U \le \mathcal{KH}\int_A^B f \le \mathcal{KH}\int_A^B L,$$

it follows that

$$\left| \sum_1^{n+m+p} f(x_k)|I_k| - \mathcal{KH}\int_A^B f \right| < \varepsilon.$$

This proves M-integrability of f and equation (3.118).   •

----

† If the interval $[A, B]$ is finite it suffices to take $\mathcal{E}$ constant equal to $\varepsilon 3^{-1}(B - A)^{-1}$. A function like $\mathcal{E}$ is often used instead of a constant ε to modify proofs in order to make them valid for an infinite interval.

‡ Keep in mind that a point might tag many intervals.

3.12.1 A short proof

There is a very short proof of the equivalence of absolute KH-integrability and McShane integrability, which uses the results of Section 3.11 on measurability, the monotone convergence theorem and the dominated convergence theorem for the McShane integral. A proof with minimal prerequisites is given in [47]. We proved Theorem 3.12.5 because part (C) is important in itself. Assume f is non-negative: otherwise we can consider f^+ and f^- instead of f. Let $f_n(x) = 0$ if either $x \notin [-n, n]$ or $f_n(x) > n$. Otherwise let $f_n(x) = f(x)$. If f is absolutely KH-integrable then f_n is bounded and measurable. By the dominated convergence theorem for the McShane integral it is M-integrable on $[A, B]$ since the integration is effectively over a bounded interval. Hence $\mathcal{M}\int_A^B f_n = \mathcal{KH}\int_A^B f_n$. Employing the monotone convergence theorem first for the KH-integral and then for the M-integral gives

$$\mathcal{KH}\int_A^B f = \lim_{n\to\infty} \mathcal{KH}\int_A^B f_n = \lim_{n\to\infty} \mathcal{M}\int_A^B f_n = \mathcal{M}\int_A^B f.$$

3.13 The Lebesgue integral

The aim of this section is to give a brief exposition of Lebesgue integration, not independently but by using our knowledge of KH-theory. The reader familiar with the Lebesgue integral and interested only in the relation between the KH and Lebesgue integrals can proceed directly to the closing Subsection 3.13.2.

There are several ways introducing the Lebesgue integral. The traditional one starts with measure theory, and this will be our approach here. The so-called Riesz definition is explained in Subsection 3.13.1.

The outer Lebesgue measure of a set $S \subset \mathbb{R}$, denoted by $\mu_e(S)$, is defined as

$$\mu_e(S) = \inf \{\sum_{k=1}^{\infty} |I_k| \, ; \, I_k \text{ intervals, } S \subset \bigcup_{k=1}^{\infty} I_k\}.$$

It follows easily from the definition that

$$\mu_e(A) \le \mu_e(B) \quad \text{if} \quad A \subset B. \tag{3.129}$$

It is also clear from the definition that

$$\mu_e(S_1 \cup S_2) \le \mu_e(S_1) + \mu_e(S_2) \tag{3.130}$$

and more generally

$$\mu_e(\bigcup_1^\infty S_k) \le \sum_1^\infty \mu_e(S_k). \qquad (3.131)$$

The next step is to select a class of sets such that equality holds in (3.130) and (3.131) if the unions are disjoint. This class, which we shall temporarily call L-measurable, is defined as a class of sets S with the following property: for every positive ε there is a closed set F and an open set G such that

$$F \subset S \subset G \quad \text{and} \quad \mu_e(G \setminus F) < \varepsilon. \qquad (3.132)$$

The set $G \setminus F$ being open is a countable union of open intervals, and consequently

$$\mu_e(G \setminus F) = m(G \setminus F).$$

Hence by Theorem 3.11.15 a set is L-measurable if and only if it is measurable in the sense of Definition 3.11.1. Therefore we can use the properties of measurable sets stated in Theorem 3.11.9 for L-measurable sets. In fact there is no need any more to make a distinction between L-measurable and measurable sets. A *Lebesgue measure* $\mu(S)$ of a measurable set S is by definition $\mu(S) = \mu_e(S)$. At this stage, in a proper development of Lebesgue theory, one establishes those properties of the Lebesgue measure which for the measure generated by the KH-integral were stated in Theorem 3.11.11. We show instead that the Lebesgue measure and the measure generated by the KH-integral coincide. Let S be a measurable set and I_k, $k = 1, 2, \ldots$ a disjoint system of open intervals with $S \subset G = \bigcup_1^\infty |I_k|$. Then

$$m(S) = \mathcal{KH}\int_{\mathbf{R}} \mathbf{1}_S \le \mathcal{KH}\int_{\mathbf{R}} \mathbf{1}_G = \sum_1^\infty |I_k|. \qquad (3.133)$$

Consequently $m(S) \le \mu_e(S) = \mu(S)$. For proving the reverse inequality we can assume $m(S) < \infty$ and choose G with the additional property that $m(G) < m(S) + \varepsilon$. This is possible because of Theorem 3.11.15. It follows that $\mu(S) = \mu_e(S) \le m(S) + \varepsilon$ and since ε is arbitrary $\mu(S) \le m(S)$. Now that we have proved that Lebesgue measure and the measure generated by the KH-integral are identical we shall drop the notational distinction and denote measure simply by m.

DEFINITION 3.13.1 *We denote by f^* the function which equals f on S and zero outside S. The KH-integral of a function f over a set S*

is defined as $\mathcal{KH} \int_{-\infty}^{\infty} f^*$ and is denoted by $\mathcal{KH} \int_S f$. If the integral is finite then f is KH-integrable over S.

If S is an interval this definition is consistent with the definitions stated earlier. Also if f is defined on $[A, B]$ and $S \subset [A, B]$ then $\int_S f = \int_A^B \mathbf{1}_S f$. If f is non-negative and measurable on a measurable set S then $\mathcal{KH} \int_S f$ always exists†. This follows from Theorem 3.11.10 and the monotone convergence theorem.

The *Lebesgue integral* $\mathcal{L} \int_S f$ of a *measurable* function f over a *measurable* set S is defined in three steps.

DEFINITION 3.13.2

(i) *If f is a non-negative simple function, $f = \sum_1^n c_i \mathbf{1}_{E_i}$ with disjoint‡ E_i and positive c_i, then*

$$\mathcal{L} \int_S f \stackrel{\text{def}}{=} \sum_1^n c_i m(E_i \cap S). \qquad (3.134)$$

(ii) *If f is measurable and non-negative then*

$$\mathcal{L} \int_S f \stackrel{\text{def}}{=} \sup \left\{ \mathcal{L} \int_S \varphi; 0 \le \varphi \le f, \varphi \text{ simple} \right\}. \qquad (3.135)$$

(iii) *If f is measurable then*

$$\mathcal{L} \int_S f \stackrel{\text{def}}{=} \mathcal{L} \int_S f^+ - \mathcal{L} \int_S f^-. \qquad (3.136)$$

provided the right-hand side is defined.

If the integral in (3.136) exists and is finite then we say that f is Lebesgue integrable *on S.*

The main result of this section is

THEOREM 3.13.3 *The equality*

$$\mathcal{L} \int_S f = \mathcal{KH} \int_S f$$

is valid if and only if one of the following conditions is satisfied:

(a) *the integral on the left exists;*

---

† Although it could be ∞.
‡ The definition stays correct if the E_i are not required to be disjoint but then one has to prove that the integral is independent of the way in which f is represented as a linear combination of characteristic functions.

(b) *the integral on the right exists and one of the integrals $\mathcal{KH} \int_S f^+$, $\mathcal{KH} \int_S f^-$ is finite.*

Proof We go step by step through the definition of the Lebesgue integral.

(i) If f is simple and non-negative then the right-hand side of (3.134) is the KH-integral of f.

(ii) Continuing with the second step we have first

$$\mathcal{L} \int_S \varphi = \mathcal{KH} \int_S \varphi \leq \mathcal{KH} \int_S f,$$

and by (3.135)

$$\mathcal{L} \int_S f \leq \mathcal{KH} \int_S f.$$

By Theorem 3.11.10 there exists an increasing sequence of simple non-negative measurable functions φ_n converging a.e. to f. By (3.135)

$$\mathcal{L} \int_S \varphi_n = \mathcal{KH} \int_S \varphi_n \leq \mathcal{L} \int_S f.$$

However, by the monotone convergence theorem for the KH-integral

$$\mathcal{KH} \int_S \varphi_n = \mathcal{KH} \int_{-\infty}^{\infty} 1_S \varphi_n \to \mathcal{KH} \int_{-\infty}^{\infty} f^* = \mathcal{KH} \int_S f.$$

Consequently

$$\mathcal{KH} \int_S f \leq \mathcal{L} \int_S f.$$

(iii) If $\mathcal{L} \int_S f$ exists then by what we have just proved the KH-integrals of f^+ and f^- exist and

$$\mathcal{L} \int_S f = \mathcal{L} \int_S f^+ - \mathcal{L} \int_S f^- = \mathcal{KH} \int_S f^+ - \mathcal{KH} \int_S f^- = \mathcal{KH} \int_S f. \tag{3.137}$$

If one of the KH-integrals of f^+ or f^- is finite and the KH-integral of f exists then the validity of (3.137) is obtained by reading it from right to left. $\bullet$

3.13.1 F. Riesz' definition

Now we explain another way of introducing the Lebesgue integral; this approach is due to F. Riesz. We again give the definition in several steps.

DEFINITION 3.13.4

(i) *The integral is first defined for step functions, in the obvious way.*

(ii) *Sets of measure zero (and the concept of almost everywhwere) are defined as in Section 2.11.*

(iii) *A function f is said to belong to class $\mathfrak{L}^+$ if there exists an increasing sequence of step functions $\{\varphi_n\}$ converging a.e. to f and such that the integrals of φ_n are bounded. The FR-integral† of $f \in \mathfrak{L}^+$ is defined by*

$$\mathcal{FR}\int_a^b f \overset{\text{def}}{=} \lim_{n\to\infty} \int_a^b \varphi_n. \tag{3.138}$$

The class $\mathfrak{L}$ is the family of functions f which can be represented as $f = f_1 - f_2$ with f_1 and f_2 in $\mathfrak{L}^+$. The integral of $f \in \mathfrak{L}$ is

$$\mathcal{FR}\int_a^b f \overset{\text{def}}{=} \mathcal{FR}\int_a^b f_1 - \mathcal{FR}\int_a^b f_2.$$

It is immediate that $f \in \mathfrak{L}$ is absolutely KH-integrable and

$$\mathcal{FR}\int_a^b f = \mathcal{KH}\int_a^b f. \tag{3.139}$$

This means that the FR-integral is well defined, a fact which in Riesz' approach must be established independently. In order to show that (3.139) holds if f is absolutely KH-integrable we choose, for every $n \in \mathbb{R}$, a step function ψ_n such that

$$\mathcal{KH}\int_a^b |f - \psi_n| < \frac{1}{2^n}.$$

This is possible by Theorem 3.10.2. Since $\mathcal{KH}\int_a^b |\psi_n - \psi_{n-1}| < 2^{-n+1}$ the series $\sum |\psi_n - \psi_{n-1}|$ is convergent a.e. By Fatou's lemma $\psi_n \to f$

---

† This notation is used only in this subsection and is not found anywhere in the literature. FR comes from Frederic Riesz.

a.e. Let us define

$$\varphi_n^1 \;=\; \psi_1^+ + \sum_1^n (\psi_i - \psi_{i-1})^+,$$

$$\varphi_n^2 \;=\; \psi_1^- + \sum_1^n (\psi_i - \psi_{i-1})^-.$$

Clearly $f^j = \lim_{n\to\infty} \varphi_n^j$ belongs to $\mathcal{L}^+$ for $j = 1, 2$ and $f = f_1 - f_2$. We have proved

THEOREM 3.13.5   *The integral $\mathcal{FR}\int_a^b f$ exists if and only if f is absolutely KH-integrable and then equation (3.139) holds.*

3.13.2 Quick proofs

Lebesgue integrability is characterized by the same conditions as absolute KH-integrability was either in Theorem 3.9.4 or in Theorem 3.12.5 part (C). Consequently these concepts of integrability are equivalent. This fact can also be proved by the argument given in Subsection 3.12.1 with the M-integral replaced by the Lebesgue integral.

3.14 Differentiation almost everywhere

At the end of the eighteenth and the beginning of the nineteenth century it was widely believed that a continuous function must be differentiable except possibly on a small set. This belief was shown to be completely false when the German mathematician K. Weierstrass produced an example of a function which is continuous everywhere but differentiable nowhere. The example is simple enough, namely the function is given by the formula

$$f(x) = \sum_{n=1}^{\infty} b^n \cos(a^n \pi x),$$

with $b < 1$ and $ab > 1 + 3\pi/2$. However, the proof is not easy and many mathematicians have produced their own examples of non-differentiable continuous functions which have additional features (e.g. having a finite or infinite derivative neither from the left nor from the right anywhere) or for which the proofs are elegant. The simplest example is in [29]. However, the intuition that most functions we encounter in 'real life' are differentiable at many points has a sound basis: it is difficult to picture a continuous function which is not monotonic on some interval of its domain of definition and on that interval the function is, according to a theorem proved by Lebesgue, differentiable except on a set of measure zero. Most of this section is devoted to the proof of Theorem 3.14.4, which asserts that a function of bounded variation is differentiable almost everywhere. Quite a bit of preparation is needed for this theorem and we start with several lemmas.

Let us recall the definition of the Dini derivates of a function F at x:

$$D^+F(x) = \limsup_{h\downarrow 0} \frac{F(x+h)-F(x)}{h},$$

$$D^-F(x) = \limsup_{h\uparrow 0} \frac{F(x+h)-F(x)}{h},$$

$$D_+F(x) = \liminf_{h\downarrow 0} \frac{F(x+h)-F(x)}{h},$$

$$D_-F(x) = \liminf_{h\uparrow 0} \frac{F(x+h)-F(x)}{h}.$$

The Dini derivates always exist but one has to keep in mind that they might be infinite. A finite derivative of F exists at x if and only if

$$-\infty < D^+F(x) \le D_-F(x) \le D^-F(x) \le D_+F(x) \le D^+F(x) < \infty. \tag{3.140}$$

The third and fifth inequalities in this chain are always satisfied. The next two lemmas deal with the other inequalities.

LEMMA 3.14.1 *If F is of bounded variation on $[a, b]$ then the set*

$$E_\infty = \big\{x; D^+F(x) = \infty \quad \text{or} \quad D^+F(x) = -\infty\big\}$$

is of measure zero.

Proof For every positive ε and every $x \in E_\infty$ there is a positive h_x such that

$$|F(x+h_x) - F(x)| > M(\varepsilon)h_x, \tag{3.141}$$

where $M(\varepsilon) = \dfrac{12\mathrm{Var}_a^b F}{\varepsilon}$. By Lemma 2.11.4 there are countable systems of points x_i and non-overlapping intervals K_i such that

$$x_i \in K_i \tag{3.142}$$

$$K_i \subset (x_i - h_{x_i}, x_i + h_{x_i}) \tag{3.143}$$

$$E_\infty \subset \bigcup_{i=1}^{\infty} K_i \tag{3.144}$$

for $i = 1, 2, \ldots$ We choose N such that

$$\sum_{i=1}^{\infty} |K_i| < 2 \sum_{i=1}^{N} |K_i|. \tag{3.145}$$

By using Lemma 3.8.1 on the system $(x_i - h_{x_i}, x_i + h_{x_i})$ with $i = 1, 2 \ldots, N$ and renumbering if necessary we find a finite disjoint subsystem of intervals $(x_i - h_{x_i}, x_i + h_{x_i})$ with $i = 1, \ldots, p$ such that

$$\sum_{i=1}^{N} h_{x_i} \leq 3 \sum_{i=1}^{p} h_{x_i}. \tag{3.146}$$

It follows that

$$\sum_{i=1}^{\infty} |K_i| < 2 \sum_{i=1}^{N} |K_i| \leq 12 \sum_{i=1}^{p} h_{x_i} \leq \frac{12}{M(\varepsilon)} \sum_{i=1}^{p} |F(x_i + h_{x_i}) - F(x_i)| \leq \varepsilon.$$

This shows that E_∞, being covered by $\bigcup_1^\infty K_i$, is of measure zero.   •

LEMMA 3.14.2 *If for **every** function F of bounded variation the inequality*

$$D^+ F(x) \leq D_- F(x) \tag{3.147}$$

holds almost everywhere then every function of bounded variation has a finite derivative almost everywhere.

Proof Since $D^- F(x) = -D_-(-F)(x)$ and $D_+ F(x) = -D^+(-F)(x)$ we can apply inequality (3.147) to $-F$ and have

$$D^- F(x) \leq D_+ F(x) \text{ a.e.} \tag{3.148}$$

and inequalities (3.140) follow.   •

LEMMA 3.14.3 *Let L be a continuous piecewise linear function on the interval $[y, Y]$ and $\{J_k; k = 1, \ldots, s\}$ be a finite system of closed* *disjoint intervals such that*

$$L' < -\alpha \quad \text{on} \quad \bigcup_{k=1}^{s} J_k \quad \text{if} \quad L(y) \leq L(Y) \qquad (3.149)$$

and

$$L' > \alpha \quad \text{on} \quad \bigcup_{k=1}^{s} J_k \quad \text{if} \quad L(y) \geq L(Y). \qquad (3.150)$$

Then

$$\text{Var}_a^b L \geq |L(Y) - L(y)| + 2\alpha \sum_{k=1}^{s} |J_k|. \qquad (3.151)$$

Proof It is sufficient to consider the case (3.149) because the assertion is not changed if L is replaced by $-L$. Clearly

$$\text{Var}_y^Y L = \int_y^Y |L'| \geq \int_y^Y L' + \sum_{i=1}^{s} \int_{J_k} (|L'| - L')$$

$$\geq L(Y) - L(y) + 2\alpha \sum_{i=1}^{p} |J_k|,$$

since $L(Y) \geq L(y)$ and $|L'| - L' \geq 2\alpha$. •

We are now ready for the main theorem of this section.

THEOREM 3.14.4 (The Lebesgue differentiation theorem) *If F is of bounded variation on $[a, b]$ then F has a finite derivative a.e. on $[a, b]$.*

Proof By Lemmas 3.14.1 and 3.14.2 it suffices to show that the inequality (3.147) holds a.e. Hence we want to show that the set

$$E_-^+ = \{x; D_-F(x) < D^+F(x)\}$$

is of measure zero. This set E_-^+ is a countable union of sets of the form

$$\left\{x; D_-F(x) < r_n - \frac{1}{k} < r_n + \frac{1}{k} < D^+F(x)\right\},$$

where $r_n \in \mathbb{Q}$ and $k \in \mathbb{N}$. The proof that each of these sets is of measure zero can be reduced to the case $r_n = 0$. Indeed by adding a suitable linear function to F the modified function will be still of bounded variation and will have $r_n = 0$.

Let ε and α be positive and

$$E_\alpha = \left\{ x; D_- F(x) < -\alpha < \alpha < D^+ F(x) \right\}.$$

There is a division D given by $D \equiv a = y_0 < y_1 < \cdots < y_m = b$ such that

$$\sum_{i=1}^m |F(y_i) - F(y_{i-1})| + \frac{\varepsilon \alpha}{6} > \mathrm{Var}_a^b F. \tag{3.152}$$

We denote by E_D the set $E_\alpha \setminus \{y_0, y_1, \ldots, y_m\}$. For every $x \in E_D$ there is an interval (u_x, v_x) such that $x \in (u_x, v_x) \subset [y_k, y_{k+1}]$ for some k and

$$F(x) - F(u_x) < -\alpha(x - u_x), \tag{3.153}$$

$$F(v_x) - F(x) > \alpha(v_x - x). \tag{3.154}$$

If $F(y_k) \leq F(y_{k+1})$ we set $J_x = [u_x, x]$, and $J_x = [v_x, x]$ if $F(y_k) > F(y_{k+1})$. We also denote $|J_x| = h_x$. For the set E_D we now find similarly as in the proof of Lemma 3.14.1 the intervals K_i, points x_i and numbers N and p such that (3.142), (3.143), (3.145) and (3.146) are satisfied and (3.144) holds with E_∞ replaced by E_D. Now we define a continuous piecewise linear function L which agrees with F at the points y_k and at the endpoints of intervals J_{x_n} with $n = 1, \ldots, p$. Applying Lemma 3.14.3 to each interval $[y_{k-1}, y_k]$ and summing for $k = 1, \ldots, m$ we have

$$\mathrm{Var}_a^b F \geq \mathrm{Var}_a^b L \geq \sum_{i=1}^m |F(y_{i-1}) - F(y_i)| + 2\alpha \sum_{i=1}^p h_{x_i}. \tag{3.155}$$

Combining this inequality with inequality (3.152) gives

$$\frac{\varepsilon}{6} \geq \sum_{i=1}^p h_{x_i} \geq \frac{1}{3} \sum_{i=1}^N h_{x_i} \geq \frac{1}{6} \sum_{i=1}^\infty |K_i|. \tag{3.156}$$

This means that E_D and consequently E_α is of measure zero.   •

The Lebesgue theorem on differentiability of functions of bounded variation is an important theorem, but it plays a far less central rôle in our approach to integration because we did not need it for the proof of Theorem 3.8.2. For the proof of the next theorem it is indispensable.

THEOREM 3.14.5 *A function F is AC on $[a, b]$ if and only if there exists an absolutely KH-integrable function f such that*

$$F(x) = F(a) + \mathcal{KH} \int_a^x f.$$

Proof The if part is covered by Theorem 3.6.4. If F is AC then it is of bounded variation and has a finite derivative F' a.e. By Theorem 3.9.1

$$F(x) - F(a) = \mathcal{KH} \int_a^x F'. \qquad \bullet$$

The next theorem shows that term by term differentiation of series is possible for series with monotonic terms.

THEOREM 3.14.6 (Fubini's differentiation theorem) *If the series $\sum_1^\infty h_n(x) = s(x)$ converges for every x in $[a,b] \subset \mathbb{R}$ and for each $n \in \mathbb{N}$ the function h_n is increasing (or for each n decreasing) then*

(i) *$t(x) = \sum_1^\infty h'_n(x)$ converges a.e. on $[a,b]$*
(ii) *$s'(x) = t(x)$ a.e. on $[a,b]$.*

Proof The assertion involves only derivatives. We can therefore assume without loss of generality that for all $n \in \mathbb{N}$ we have $h_n(a) = 0$. Let $s_n(x) = \sum_{k=1}^n h_k(x)$ and $t_n(x) = \sum_{k=1}^n h'_n(x)$. By the Lebesgue differentiation theorem there is a set N of measure zero such that for $x \notin N$ all the functions h_n, s_n, s, $s - s_n$ have finite non-negative derivatives and it follows that

$$t(x) = \lim_{n \to \infty} s'_n(x) \le s'(x). \qquad (3.157)$$

This proves (i). Let us choose an increasing sequence of natural numbers n_k such that $s(b) - s_{n_k}(b) < 2^{-k}$ and hence $0 \le s(x) - s_{n_k}(x) < 2^{-k}$ for all $x \in [a,b]$. The series $\sum_{k=1}^\infty (s(x) - s_{n_k}(x))$ converges everywhere on $[a,b]$ and we can use the already established part (i) of the theorem to conclude that the series $\sum_{k=1}^\infty (s'(x) - s'_{n_k}(x))$ converges a.e. on $[a,b]$. Consequently $\lim_{k \to \infty} s'_{n_k}(x) = s(x)$ a.e. on $[a,b]$. This combined with (3.157) proves (ii). $\qquad \bullet$

3.15 Exercises

EXERCISE 3.1 ①① *Suppose for every positive ε there exist functions M_P and m_P which satisfy the following conditions:*
(a) *M_P is continuous on $[a,b]$ and $f(x) \le \underline{D}M_P(x)$ for nearly all $x \in [a,b]$, i.e. for all $x \in [a,b]$ except a countable number of points,*
(b) *m_P is continuous on $[a,b]$ and $f(x) \ge \overline{D}m_P(x)$ for nearly all $x \in [a,b]$,*

(c) $M_P(b) - M_P(a) - [m_P(b) - m_P(a)] < \epsilon$.

Show that f is Perron integrable on $[a, b]$.

EXERCISE 3.2 Modify the construction of the Perron integral of f by considering only P-major and P-minor functions which are of bounded variation on $[a, b]$. Show that if (3.6) holds then the function f is absolutely integrable on $[a, b]$.

EXERCISE 3.3 Let p and q be positive integers, for example, $p = 1$, $q = 1$ or $p = 1$, $q = 2$. Show that

$$\int_0^1 \frac{x^{p-1} dx}{1 + x^q} = \frac{1}{q} - \frac{1}{p+q} + \frac{1}{p+2q} - \frac{1}{p+3q} + \cdots$$

[Hint: Let $f_n(x) = x^{p-1}(1 - x^q + x^{2q} - x^{3q} + \cdots + x^{(2n-2)q} - x^{(2n-1)q})$, and apply the monotone convergence theorem.]

EXERCISE 3.4 Let $F(x) = x^2 \sin x^{-2}$ when $x \neq 0$ and $F(0) = 0$. Show that F is not of bounded variation on $[0, 1]$. Hence deduce F is not absolutely continuous on $[0, 1]$.

EXERCISE 3.5 ⓘShow that if f is continuous and g is of bounded variation on $[a, b]$ then the Stieltjes integral $\int_a^b f \, dg$ exists and the integration-by-parts formula in Exercise 2.19 holds.

EXERCISE 3.6 ⓘⓘ Use Exercise 2.20 and Exercise 3.5 to show that if f is of bounded variation and g is KH-integrable on $[a, b]$ then the product fg is KH-integrable on $[a, b]$.

EXERCISE 3.7 ⓘWe know from Example 3.5.7 that if f is absolutely integrable on $[a, b]$ then so is the truncated function f^N and $\int_a^b f^N \to \int_a^b f$, where $f^N(x) = f(x)$ when $|f(x)| \leq N$, $f^N(x) = N$ when $f(x) > N$ and $f^N(x) = -N$ when $f(x) < -N$. Show further that

$$N.m\{x \in [a, b]; |f(x)| > N\} \to 0 \text{ as } N \to \infty.$$

EXERCISE 3.8 Fatou's lemma is stated in Theorem 3.5.13. Do we have a similar result involving

$$\int_a^b \limsup_{n \to \infty} f_n \geq \limsup_{n \to \infty} \int_a^b f_n \ ?$$

Explain.

EXERCISE 3.9 *Show that Arzelà's theorem 1.5.6 is a consequence of the Lebesgue dominated convergence theorem.*

EXERCISE 3.10 *Let f_n, $n = 1, 2, \ldots$ be KH-integrable on $[a, b]$. Show that if $\lim\limits_{m,n \to \infty} \int_a^b |f_m - f_n| = 0$ then there is a subsequence $\{f_{n_k}\}$ which is dominated in the sense that $g \le f_{n_k} \le G$ for all n and for some KH-integrable functions g and G. Give an example to show that we cannot prove that the sequence $\{f_n\}$ itself is dominated.*

EXERCISE 3.11 *Prove that a function F is the KH-primitive of a function of bounded variation on $[a, b]$ if and only if F satisfies the following condition:*

$$\sum_{i=0}^{n-2} \left| \frac{F(x_{i+2}) - F(x_{i+1})}{x_{i+2} - x_{i+1}} - \frac{F(x_{i+1}) - F(x_i)}{x_{i+1} - x_i} \right| + \left| \frac{F(x_n) - F(x_{n-1})}{x_n - x_{n-1}} \right|$$

is bounded for all divisions $a = x_0 < x_1 < \cdots < x_n = b$. Such a function is said to be of bounded slope variation.

EXERCISE 3.12 *Let $f(0) = 0$ and the graph of f on the interval $[(n+1)^{-1}, n^{-1}]$ be formed by an isosceles triangle of height 2, the triangle pointing upwards if n is odd and downwards if n is even. Show that, for $F(x) = \int_0^x f$, the derivative $F'(0) = 0$ and that 0 is not a Lebesgue point for f. [Hint: For $\hat{h} = p^{-1}$ with $p \in \mathbb{N}$ we have $|F(\hat{h})| \le \hat{h}^2(1+\hat{h})^{-1}$ and $\int_0^{\hat{h}} |f| = \hat{h}$. The required results follow now easily.]*

EXERCISE 3.13 ⓘⓘ *Show that the following conditions are equivalent: (a) The function F is the indefinite integral of a function f which is McShane integrable on $[a, b]$; (b) for every positive ε there exists a positive K such that for any partial division D of $[a, b]$ we have $|\sum_D F(u, v)| \le \varepsilon + K \sum_D |v - u|$ where $F(u, v) = F(v) - F(u)$; (c) the function F is absolutely continuous on $[a, b]$. [Hint: To prove that (c) implies (a), we prove that every absolutely continuous function can be written as the difference of two increasing functions and use the fact that every increasing function is differentiable almost everywhere.]*

EXERCISE 3.14 *Let $f(x) = (-1)^{n+1}$ when $x \in (\frac{1}{n+1}, \frac{1}{n}]$ for $n = 1, 2, \ldots$, and $f(0) = 0$. Determine whether f is integrable on $[0, 1]$ and whether it is absolutely integrable on $[0, 1]$. [Hint: compare Exercsise 2.10.]*

EXERCISE 3.15 Let f be KH-integrable on $[a, b]$. Let $\mathfrak{F}$ be the family of functions of the form $f\mathbf{1}_{[\alpha,\beta]}$ with $[\alpha\,\beta] \subset [a, b]$. Prove that $\mathfrak{F}$ is equiintegrable.

EXERCISE 3.16 ⓘDirichlet's test. Let f be KH-integrable on $[a, x]$ for every $x > a$ and the function $F(x) = \int_a^x f$ be bounded on $[a, \infty)$. If g is decreasing and $\lim_{x \to \infty} g(x) = 0$ show that fg is KH-integrable on $[a, \infty]$. [Hint: use the second mean value theorem similarly as in Example 3.7.10.]

EXERCISE 3.17 ⓘGive an example of a bounded function f which is not KH-integrable on $[0, 1]$ although $|f|$ is. [Hint: Use the existence of a non-measurable set.]

EXERCISE 3.18 Prove by definition that a continuous function on $[a, b]$ is McShane integrable there.

EXERCISE 3.19 ⓘLet f_n, $n = 1, 2, \ldots$ be KH-integrable on $[a, b]$ with the primitives F_n, $n = 1, 2, \ldots$, the sequence $f_n \to f$ almost everywhere in $[a, b]$, and $F_n(x) \to F(x)$ for every $x \in [a, b]$. Then in order that f is KH-integrable on $[a, b]$ with the primitive F, it is necessary and sufficient that for every positive ε there exists $M : [a, b] \mapsto \mathbb{N}$ such that for infinitely many $m(\xi) \geq M(\xi)$ there is $\delta : [a, b] \to R_+$ such that for any δ-fine partition π of $[a, b]$ we have

$$\left| \sum_\pi F_{m(\xi)}(u, v) - F(a, b) \right| < \varepsilon.$$

This is known as the basic convergence theorem. [Hint: Define $\delta_n : [a, b] \mapsto \mathbb{R}_+$ such that $\sum_\pi |f_n(\xi)(v - u) - F_n(u, v)| < \varepsilon 2^{-n}$ for all δ_n-fine partitions π. Assume $f_n \to f$ everwhere. Choose $M(\xi)$ such that $|f_{m(\xi)}(\xi) - f(\xi)| < \varepsilon$ for all $m(\xi) \geq M(\xi)$. Then put $\delta(\xi) = \delta_{m(\xi)}(\xi)$. If f is KH-integrable then modify δ so that $\sum |f(\xi)(v - u) - F(u, v)| < \varepsilon$ for any δ-fine partition π and the required inequality follows. Conversely, rearrange the above inequalities.]

EXERCISE 3.20 ⓘProve the monotone convergence theorem 3.5.2 using the proof of the basic convergence theorem (see Exercise 3.19). [Hint: Follow the hint of Exercise 3.19. Let q be the minimum of $m(\xi)$ while $(\xi, [u, v]) \in \pi$. Since the sequence $\{f_n\}$ is monotone we have

$F_q(a, b) \leq \sum_\pi F_{m(\xi)}(u, v) \leq L$, where $L = \lim\limits_{n \to \infty} F_n(a, b)$. We can choose $m(\xi)$ sufficiently large so that $L - F_q(a, b) < \varepsilon$. Hence the required inequality in Exercise 3.19 holds and the result follows. Note that this proof works for the KH-integral only whereas the proof of Theorem 3.5.2 works for the McShane integral as well.]

EXERCISE 3.21 ⓘⓘ *Show that a non-negative function f is integrable to I on $[a, b]$ if and only if for every positive ε and positive η there exist an open set G and a positive constant δ such that the Lebesgue measure $m(G) < \eta$ and such that for any partition $\pi = \{(\xi, [u, v])\}$ with $0 < v - u < \delta$ we have*

$$\left| \sum_{\pi^*} |f(\xi)(v - u) - I| \right| < \varepsilon,$$

where π^* is a tagged partial division obtained from π by omitting those pairs $(\xi, [u, v])$ for which $\xi \in G$.

EXERCISE 3.22 *Prove that if a function f is integrable on every measurable subset X of $[a, b]$ then f is absolutely integrable on $[a, b]$.*

EXERCISE 3.23 ⓘ*Prove that a function f is absolutely integrable on $[a, b]$ if and only if there exist real numbers $c_1, c_2, \ldots$ and subintervals $I_1, I_2, \ldots$ of $[a, b]$ such that*

$$f = \sum_{i=1}^{\infty} c_i \mathbf{1}_{I_i} \text{ almost everywhere, } \sum_{i=1}^{\infty} |c_i| m(I_i) \text{ converges.}$$

The above holds true with intervals replaced by measurable sets X_i.

EXERCISE 3.24 *Let $X \subset [a, b]$. The characteristic function of X is integrable to I on $[a, b]$ if and only if for every postive ε there exists $\delta : [a, b] \to \mathbb{R}_+$ such that for any δ-fine M-partition π we have $|\sum_\pi \mathbf{1}_X - I| < \varepsilon$. Use the above fact to give an alternative proof of Theorem 3.11.15.*

EXERCISE 3.25 ⓘⓘ *Let f_n, $n = 1, 2, \ldots$ be integrable on $[a, b]$. Prove that if $f_n(x) \uparrow f(x) \in \mathbb{R}$ everywhere as $n \to \infty$ and the sequence $\{\int_a^b f_n\}$ is bounded then the sequence $\{f_n\}$ is equiintegrable on $[a, b]$. [Hint: Define δ as in the hint of Exercise 3.19 with $\delta_{n+1} \leq \delta_n$ for all n. For any δ-fine partition π of $[a, b]$, write π_1 to be the part of π such that $n \leq m(\xi)$ and $\pi_2 = \pi - \pi_1$. Then $|\sum_{\pi_1} f_n(\xi)(v - u) - F_n(u, v)| < \varepsilon 2^{-n}$*

and

$$\left|\sum_{\pi_2}[f_n(\xi)(v-u)-F_n(u,v)]\right| \leq \left|\sum_{\pi_2}[f_n(\xi)-f_{m(\xi)}(\xi)](v-u)\right|$$

$$+\left|\sum_{\pi_2}[f_{m(\xi)}(\xi)(v-u)-F_{m(\xi)}(u,v)]\right| + \left|\sum_{\pi_2}[F_{m(\xi)}(u,v)-F_n(u,v)]\right|$$

in which the last term is small by means of the hint of Exercise 3.20.]

EXERCISE 3.26 Let f_n, $n = 1,2,\ldots$ and g be integrable on $[a,b]$. Prove that if $|f_n| \leq g$ and $f_n \to f$ everywhere as $n \to \infty$, then the sequence $\{f_n\}$ is equiintegrable on $[a,b]$. [Hint: Follow the hint of Exercise 3.25 except $|\sum_{\pi_2}[F_{m(\xi)}(u,v)-F_n(u,v)]| \leq |\sum_{\pi_2}[F_{m(\xi)}(u,v)-F(u,v)]| + |\sum_{\pi_2}[F(u,v)-F_n(u,v)]|$ in which both terms are small by means of the basic convergence theorem (see Exercise 3.19) and the mean convergence theorem.]

EXERCISE 3.27 Give an example of a sequence $\{f_n\}$ which is equiintegrable and which converges nowhere. [Hint: Take $f_n(x) = (-1)^n$ for all $x \in [a,b]$.]

EXERCISE 3.28 Let $f_n(x) = n$ when $x \in (\frac{1}{n+1}, \frac{1}{n})$ for $n = 1,2,\ldots$ and 0 otherwise. Show that the sequence $\{f_n\}$ is mean convergent but not dominated by any integrable functions.

EXERCISE 3.29 A function f is uniformly continuous on $[a,b]$ if and only if for every x_n, $y_n \in [a,b]$, $x_n - y_n \to 0$ implies $f(x_n) - f(y_n) \to 0$. We define a function f to be uniformly if there exists a sequence $\{\nu_n\}$ of tag functions (see Exercise 2.21) such that for any ν_n-fine interval I_n we have $\omega(f, \nu_n(I_n)) \to 0$ where ω denotes the oscillation of f on $\nu_n(I_n)$. Prove that a uniformly continuous function on $[a,b]$ is uniformly ν-continuous there, but not conversely.

4

The SL-integral

4.1 The strong Luzin condition

In this chapter we offer an alternative development of the KH-integral. The definition is perhaps less natural but its advantage is that sets of measure zero are automatically neglected and some proofs become easier and shorter.

The Russian mathematician N.N. Luzin recognized the importance the condition which now bears his name and is also referred to as *condition N*. A function F satisfies *condition N* on a set S if $F(E)$ is of measure zero† for every $E \subset S$ of measure zero. Most functions encountered in applications satisfy condition N. The function H from Example 3.6.1 is continuous and increasing but it maps Cantor's discontinuum which is of measure zero onto a set which is obtained from $[0, 1]$ by removing a countable set. It is even possible to define a continuous *strictly* increasing function which does not satisfy the Luzin condition. (See [42] pp. 198–200.) We need a condition stronger than N.

DEFINITION 4.1.1 (Definition of *SL*) *A function F is said to satisfy the strong Luzin condition, or briefly SL, on a set $S \subset \mathbb{R}$ if for every set $E \subset S$ of measure zero and every positive ε there exists $\gamma : S \to \mathbb{R}_+$ such that if $\pi \equiv \{(\xi, [u, v])\}$ is any γ-fine partial division tagged in E‡*

$$\sum_\pi |F(u, v)| < \varepsilon. \tag{4.1}$$

A function F is said to satisfy SL on an interval $[A, B] \subset \overline{\mathbb{R}}$ if it satisfies

---

† As usual $F(E) = \{y : y = F(x), \ x \in E\}$.
‡ We use the notation $F(u, v) = F(v) - F(u)$. This should not be confused with $F([u, v])$.

SL on (*A*, *B*) and there are finite limits

$$\lim_{x \downarrow A} F(x) = F(A), \qquad \lim_{x \uparrow B} F(x) = F(B). \qquad (4.2)$$

A function F satisfies the strong Luzin condition on a set $S \subset \mathbb{R}$ if and only if it has a negligible variation on every set E of of measure zero, $E \subset S$. If F is AC on $[a, b] \subset \mathbb{R}$ then it is SL on $[a, b]$.

REMARK 4.1.2 The definition of SL on $[A, B] \subset \overline{\mathbb{R}}$ is motivated by the fact that if F is continuous on $[a, b] \subset \mathbb{R}$ and satisfies SL on (a, b) then it satisfies SL on $[a, b]$.

REMARK 4.1.3 The meaning of the definition is not changed if inequality (4.1) is replaced by

$$\left| \sum_{\pi} F(u, v) \right| < \varepsilon.$$

This follows from

$$\left| \sum_{\pi} F(u, v) \right| \le \sum_{\pi} |F(u, v)| = \sum_{\pi} F^+(u, v) + \sum_{\pi} F^-(u, v). \qquad (4.3)$$

It is advantageous to make the following **convention.** If F is SL on (A, B), the values $F(A)$, $F(B)$ are not defined but the limits

$$\lim_{x \downarrow A} F(x), \quad \lim_{x \uparrow B} F(x) \qquad (4.4)$$

exist and are finite then we shall always agree to extend the definition of F at A or B by the respective limits (4.4), hence making F satisfy the SL condition on $[A, B]$.

The following properties follow easily from the definition:

(S) If F_1 and F_2 satisfy SL so does $F_1 + F_2$.

(M) If F satisfies SL and c is a constant then cF satisfies SL.

(C) If F satisfies SL on $[A, B]$ then F is continuous on $[A, B]$.

(U) If $S_k, k = 1, 2, \ldots$, are disjoint and F is SL on each S_k then f is SL on $\cup_{k=1}^{\infty} S_k$. Since every countable union can be written as countable union of disjoint sets the assumption that S_k are disjoint becomes superfluous.

REMARK 4.1.4 It is easy to see that F is SL on each set

$$\{x; \, n - 1 \le |F'(x)| < n, n \in \mathbb{N}\}.$$

It follows from **(U)** that F is SL on every set where it has a *finite* derivative.

THEOREM 4.1.5 *If F satisfies SL on $[a, b]$ then it also satisfies N.*

Proof Given ε positive and a set E of measure zero there exists a $\gamma : [a, b] \to \mathbb{R}_+$ such that

$$\sum_\pi |F(u, v)| < \frac{\varepsilon}{5}. \qquad (4.5)$$

whenever $\pi \equiv \{(\xi, [u, v])\}$ is a γ-fine partial division† tagged in E. By Lemma 2.11.4 there is a countable system of closed non-overlapping intervals $[a_i, b_i]$ covering E and each of these intervals contains a point $c_i \in E$ such that $(c_i - \gamma(c_i), c_i + \gamma(c_i)) \supset [a_i, b_i]$. By Weierstrass' theorem there are intervals $[\alpha_i, \beta_i]$ contained in $[a_i, b_i]$ such that

$$\sup\{F; [a_i, b_i]\} - \inf\{F; [a_i, b_i]\} = |F(\beta_i) - F(\alpha_i)| \qquad (4.6)$$

We denote by A, B, C the sets of those positive integers $i \leq N$ for which $\alpha_i \leq c_i \leq \beta_i$, $a_i \leq c_i < \alpha_i$ and $\beta_i < c_i \leq b_i$. Since

$$|F(\beta_i) - F(\alpha_i)| \leq |F(\beta_i) - F(a_i)| + |F(a_i) - F(\alpha_i)|$$
$$|F(\beta_i) - F(\alpha_i)| \leq |F(\beta_i) - F(b_i)| + |F(b_i) - F(\alpha_i)|$$

we have by (4.5)

$$\sum_{i=1}^{N} |F(\alpha_i, \beta_i)| \leq \sum_{i \in A} |F(\alpha_i, \beta_i)|$$
$$+ \sum_{i \in B} |F(\beta_i, a_i)| + \sum_{i \in B} |F(a_i, \alpha_i)|$$
$$+ \sum_{i \in C} |F(\beta_i, b_i)| + \sum_{i \in C} |F(b_i, \alpha_i)| < \varepsilon.$$

By letting $N \to \infty$ we obtain

$$\sum_{1}^{\infty} |F(\alpha_i, \beta_i)| \leq \varepsilon.$$

This proves $F(E)$ is of measure zero. ●

This theorem makes it clear that the function from Example 3.6.1 does not satisfy the strong Luzin condition on $[0, 1]$.

† For the definition of partial tagged divisions see Section 3.2.

REMARK 4.1.6 If a function F continuous and increasing on $[a, b]$ satisfies condition N then it also satisfies condition SL. Let ε be positive and $N \subset [a, b]$ be of measure zero. Then, by assumption, $F(N)$ is also of measure zero. There exists a system of open disjoint intervals J_k with $k \in \mathbb{N}$ such that $F(N) \subset \bigcup_1^\infty J_k$ and $\sum_1^\infty |J_k| < \varepsilon$. If $x \in N$ then there is a unique k_x such that $F(x) \in J_{k_x}$. Consequently, by continuity of F there is $\delta(x) > 0$ with the property that $F(t) \in J_{k_x}$ whenever $t \in (x - \delta(x), x + \delta(x))$. If $\pi \equiv \{(x_i, [u_i, v_i])\}$ is now a δ-fine tagged partial division on N then

$$\sum_\pi |F([u_i, v_i])| = \sum_\pi |F(u, v)| \leq \sum_{k=1}^\infty |J_k| < \varepsilon. \qquad \bullet$$

There is a much stronger theorem called the Banach–Zarecki theorem (see e.g. [34] p. 250) which states that if F is only of bounded variation and has the Luzin property then F is AC.

For a function f we shall denote by N_f the set of zeros of f, i.e. $N_f = \{x; f(x) = 0\}$. A function δ whose range is $[0, \infty)$ will be called a *tight gauge* if N_δ is of measure zero. We know that for a tight gauge δ there need not be a partition of $[a, b]$ which is δ-fine. However, for a tight gauge, δ-fine subpartitions always exist even if $N_\delta \neq \emptyset$.

DEFINITION 4.1.7 (Definition of the SL-integral) *A function f is said to be SL-integrable on $[A, B]$ if there exists a function F which is SL and has the property that for every positive ε there is a tight gauge δ such that*

$$\sum_\pi |f(\xi)(v - u) - F(u, v)| < \varepsilon \qquad (4.7)$$

for every δ-fine tagged partial division of $[A, B]$. The number $F(B) - F(A)$ is then the SL-integral of f and is denoted by $SL \int_A^B f$. The function F itself is called the SL-primitive of f, or simply the primitive.

We show that the definition is meaningful by proving that the function F is uniquely determined up to an additive constant.

LEMMA 4.1.8 *Let f be SL-integrable on $[A, B]$. If F_1 and F_2 are functions associated with f according to Definition 4.1.7 then $F_1(\beta) - F_1(\alpha) = F_2(\beta) - F_2(\alpha)$ for every $[\alpha, \beta] \subset (A, B)$.*

Proof Since F_1 and F_2 are SL so is $F = F_1 - F_2$. Let δ be the tight gauge associated with f, F_1, F_2 as in the definition of the SL-integral. Let γ be associated with F according to the SL condition. Define $\Delta(x) = \delta(x)$ for $x \notin N_\delta$ and $\Delta(x) = \gamma(x)$ for $x \in N_\delta$. Let π be a Δ-fine partition of $[\alpha, \beta]$ and π_1 and π_2 subpartitions of π for which $\xi \in N_\delta$ and $\xi \notin N_\delta$, respectively. Then

$$|F(\beta) - F(\alpha)| = \left| \sum_\pi (F(v) - F(u)) \right|$$

$$\leq \left| \sum_{\pi_1} (F(v) - F(u)) \right| + \left| \sum_{\pi_2} (F_1(v) - F_1(u) - f(\xi)(v - u)) \right|$$

$$+ \left| \sum_{\pi_2} (f(\xi)(v - u) - F_2(v) - F_2(u)) \right| < 3\varepsilon.$$

Since ε is arbitrary $F(\beta) - F(\alpha) = 0$.   ●

EXAMPLE 4.1.9 If $f = 0$ almost everywhere then

$$SL \int_a^b f = 0.$$

To prove this it is sufficient to choose $F = 0$ and $\delta(x) = 0$ for $x \notin N_f$.

It is now clear that a change of a function on a set of measure zero affects neither the value nor the existence of the SL-integral.

4.2 SL-integration

The basic properties of the SL-integral follow easily from the definition.

(M) If f is SL-integrable with primitive F and c is a constant then cf is SL-integrable with primitive cF.

(A) If f_1 and f_2 are SL-integrable with primitives F_1 and F_2 then $f_1 + f_2$ is SL-integrable with primitive $F_1 + F_2$.

(I) If f and g are SL-integrable and $f(x) \leq g(x)$ almost everywhere on $[a, b]$ then

$$SL \int_a^b f \leq SL \int_a^b g.$$

In particular if

$$m \leq f(x) \leq M$$

almost evewhere on $[a, b]$ then

$$m(b - a) \leq \mathcal{SL} \int_a^b f \leq M(b - a).$$

(S) A function f which is SL-integrable on $[A, B]$ is also SL-integrable on any subinterval of $[A, B]$.

(C) If f is SL-integrable and $F(x) = \mathcal{SL} \int_a^x f$ then F is continuous.

Proofs for the SL-integral are sometimes almost verbatim repetition of the corresponding proofs for the KH-integral. Examples of this are Theorems 2.6.8 and 3.8.2. The use of a tight gauge enables us to neglect sets of measure zero with ease and therefore many proofs for the SL-integral are simpler than the ones for the KH-integral. Examples of this are the next three theorems.

THEOREM 4.2.1 *If $-\infty < a < c < B \leq \infty$, f is SL-integrable on $[a, c]$ and*

$$\lim_{c \uparrow B} \mathcal{SL} \int_a^c f \quad \text{exists and equals} \quad J$$

then f is SL-integrable on $[a, B]$ and $\mathcal{SL} \int_a^B f = J$.

Proof Set $F(x) = \mathcal{SL} \int_a^x f$ and $F(B) = J$. Clearly F is SL on $[a, B]$. Let $n \mapsto c_n$ be a strictly increasing sequence with $x_1 = a$ and $x_n \to B$. For every positive ε there exists a tight gauge δ_n such that

$$\left| \sum_{\pi_n} (f(\xi)(v - u) - F(u, v)) \right| < \frac{\varepsilon}{2^n}$$

whenever π_n is a δ_n-fine tagged partial division of $[x_n, x_{n+1}]$. By re-defining δ_n, if necessary, we can achieve that

$$\delta_n(x_n) = \delta_n(x_{n+1}) \quad = \quad 0 \quad \text{for } n \in \mathbb{N},$$
$$(x - \delta_n(x), x + \delta_n(x)) \quad \subset \quad (x_n, x_{n+1}).$$

Let $\delta(x) = \delta_n(x)$ if $x \in [x_n, x_{n+1})$ and $\delta(B) = 0$. If $\pi \equiv \{(\xi, [u, v])\}$ is a δ-fine partial division tagged in $[a, B]$ and π_n that part of π which is tagged in (x_n, x_{n+1}) then

$$\left|\sum_\pi (f(\xi)(v-u) - F(u,v))\right| \le \left|\sum_{n=1}^\infty \sum_{\pi_n} (f(\xi)(v-u) - F(u,v))\right|$$

$$< \sum_{n=1}^\infty \frac{\varepsilon}{2^n} = \varepsilon. \quad \bullet$$

The Fundamental Theorem for the SL-integral is powerful and easy to prove.

THEOREM 4.2.2 *If F is SL and $F' = f$ almost everywhere on $[A, B]$ then*

$$SL\int_A^B f = F(B) - F(A). \tag{4.8}$$

Proof Given $\varepsilon > 0$ we choose a positive function $\mathcal{E} : [A,B] \to \mathbb{R}_+$ such that† $\int_A^B \mathcal{E} < \varepsilon/2$ and then a $\delta_0 : [A, B] \to \mathbb{R}_+$ such that for every δ_0-fine subpartition π we have

$$\sum_\pi \mathcal{E}(\xi)(v-u) < \varepsilon. \tag{4.9}$$

If $F'(x) \ne f(x)$ or $F'(x)$ does not exist set $\delta_1(x) = 0$. If $F'(\xi) = f(\xi)$ find $\delta_1(\xi)$ such that

$$|f(\xi)(v-u) - F(u,v)| < \mathcal{E}(\xi)(v-u), \tag{4.10}$$

for $\xi - \delta_1(\xi) < u \le \xi \le v < \xi + \delta_1(\xi)$. Set $\delta = \text{Min}(\delta_0, \delta_1)$. If π is a δ-fine tagged partial division of $[A, B]$ then by (4.9) and (4.10)

$$\left|\sum_\pi [f(\xi)(v-u) - F(u,v)]\right| < \varepsilon. \quad \bullet$$

Since F is SL on every countable set on which it is continuous, we have by Remark 4.1.4:

COROLLARY 4.2.3 *If $F' = f$ except on a countable set and F is continuous on $[A, B]$ then (4.8) holds.*

REMARK 4.2.4 The condition that F is SL is essential. For F from Example 3.6.1 we have $\int_0^1 F' = 0 \ne F(1) - F(0) = 1$.

† e.g. $\mathcal{E}(x) = \varepsilon/2\pi(1+x^2)$.

THEOREM 4.2.5 (Integration by parts) *If F and G are SL,*
$F'(x) = f(x)$, $G'(x) = g(x)$ *a.e. on* $[A, B]$ *then*

$$\mathcal{SL} \int_A^B (Fg + Gf) = F(B)G(B) - F(A)G(A). \qquad (4.11)$$

Proof The function FG is SL, as is easily seen from the identity

$$FG(u,v) = F(v)G(u,v) + G(u)F(u,v)$$

and the fact that both functions F, G are bounded. Now the theorem
follows immediately from Theorem 4.2.2 because

$$[F(x)G(x)]' = F(x)g(x) + G(x)f(x). \qquad \bullet$$

Since Theorem 3.8.2 is valid for the SL-integral also we have

COROLLARY 4.2.6 *If f, g are SL-integrable and F and G their*
corresponding SL-primitives then formula (4.11) holds.

Remark 2.7.2 concerning the lack of integrability of Fg also applies here
in full. Generally speaking one is **not** allowed to write $\int_A^B (Fg + Gf) = \int_A^B Fg + \int_A^B Gf$ in (4.11).

EXAMPLE 4.2.7 Taking $F(x) = \sin x$ and $g(x) = x^{-1}$, and applying
Theorem 4.2.5 together with the integrability of $x^{-2}\cos x$ on $[1, \infty]$ we
obtain the existence of

$$\int_0^\infty \frac{\sin x}{x} dx.$$

We now turn our attention to the substitution formula

$$\mathcal{SL} \int_{\varphi(A)}^{\varphi(B)} f = \mathcal{SL} \int_A^B (f \circ \varphi)\varphi'. \qquad (4.12)$$

If φ is one-to-one it is possible to show, similarly as in Theorem 2.7.8,
that the existence of one integral in equation (4.12) implies the existence
of the other and the equality of the two integrals. However there is a
slight complication if f is not defined everywhere. For instance, if

$$
\begin{aligned}
f(x) &= \frac{x}{|x|}, \\
\varphi(t) &= t + |t|, \\
(A, B) &= (-1, 1),
\end{aligned}
$$

then the left-hand side of (4.12) is well defined and equals 2, but the
right-hand side has no meaning since $f \circ \varphi$ is not defined on $[-1, 0]$.

However, if we redefine $(f \circ \varphi)\varphi'(t)$ to be zero on $(-1, 0)$ then the right-hand side of (4.12) becomes

$$\int_{-1}^{1} (f \circ \varphi)\varphi' = \int_{-1}^{0} 0 \, dt + \int_{0}^{1} 2 \, dt = 2$$

and the validity of (4.12) is restored. For the rest of this section we set

$$(f \circ \varphi)\varphi'(t) = 0$$

whenever $\varphi'(t) = 0$ (even if $f \circ \varphi(t)$ is not defined).

Similarly as in the proof of Theorem 2.7.8 there is a correspondence between subpartitions of $[A, B]$ and $[\varphi(A), \varphi(B)]$. If e.g. φ is continuous and strictly increasing then we have π and $\varphi \circ \pi$ given by $\pi \equiv \{(t, [u, v])\}$ and $\varphi \circ \pi \equiv \{(\varphi(t), [\varphi(u), \varphi(v)])\} \equiv \{(x, [w, z])\}$, respectively. Also every subpartition ω of $[\varphi(A), \varphi(B)]$ is $\varphi \circ \pi$ for some subpartition π of $[A, B]$, namely $\pi = \varphi_{-1} \circ \omega$.

We begin our discussion of the monotonic case with a special case of formula (4.12) in which $f = 1_N$, the characteristic function of a null set N. Formula (4.12) in this case reads

$$\mathcal{SL} \int_{A}^{B} 1_N(\varphi(t))\varphi'(t) \, dt = 0. \tag{4.13}$$

This holds if and only if $\varphi' = 0$ a.e. on $\varphi_{-1}(N)$. We now state this special case as

LEMMA 4.2.8 *If φ is differentiable almost everywhere and strictly increasing on $[A, B]$, and if $N \subset$ range of φ is of measure zero then φ' is zero a.e. on $\varphi_{-1}(N)$.*

Proof Let $\varepsilon > 0$ and $\mathcal{E}$ as in (4.9). For the proof of (4.13) we set $\delta(t) = 0$ if φ is not differentiable at t. Otherwise find δ such that

$$|\varphi(u, v) - \varphi'(t)(v - u)| < \mathcal{E}(t)(v - u), \tag{4.14}$$

for $t - \delta(t) < u \leq t \leq v < t + \delta(t)$. Since N is of measure zero there is a system of disjoint open intervals $\{I_k\}$ such that

$$N \subset \bigcup_{1}^{\infty} I_k, \tag{4.15}$$

$$\sum_{1}^{\infty} |I_k| < \varepsilon. \tag{4.16}$$

If $t \in \varphi_{-1}(N)$ then $\varphi(t) \in I_k$ for some k and we diminish $\delta(t)$, if necessary, to have $\varphi(w) \in I_k$ for $t - \delta(t) < w < t + \delta(t)$. If π is a δ-fine tagged partial division of $[A, B]$ then

$$0 \leq \sum_\pi \mathbf{1}_N(\varphi(t))\varphi'(t)(v - u) = \sum_{\varphi(t)\in N} \varphi'(t)(v - u)$$

$$< \sum_{\varphi(t)\in N} \varphi(v, u) + \sum_\pi \mathcal{E}(t)(v - u) < \sum_{k=1}^\infty \sum_{\varphi(t)\in I_k} \varphi(u, v) + \varepsilon$$

$$< \sum_{k=1}^\infty |I_k| + \varepsilon < 2\varepsilon. \qquad \bullet$$

THEOREM 4.2.9 (Integration by substitution) *Let $[A, B] \subset \overline{\mathbb{R}}$, and let the function φ be differentiable a.e., continuous, SL and strictly monotonic on $[A, B]$. Then if one side of equation (4.12) exists then so does the other and the equality holds.*

Proof Let $\varepsilon > 0$ and $\mathcal{E}$ be as in (4.9). Assume that f is integrable, let F be its SL-primitive and N be the set where F' does not exist or $F' = \pm\infty$ or does not equal f. Set $M = \varphi_{-1}(N)$. Let $\eta : N \to \mathbb{R}_+$ be such that if ω is a tagged partial division on N with $\omega \ll \eta$ then $\sum_\omega |F(w, z)| < \varepsilon$. By continuity of φ there is, for every $t \in M$, a positive $\delta(t)$ such that $|\varphi(\tau) - \varphi(t)| < \eta(\varphi(t))$ for $|t - \tau| < \delta(t)$. Consequently, if π_M is a tagged partial division on M and $\pi_M \ll \delta$ then

$$\sum_{\pi_M} |(F \circ \varphi)(u, v)| < \varepsilon. \qquad (4.17)$$

If $t \in M$ and $\varphi'(t) > 0$ or $\varphi'(t)$ does not exist let $\delta(t) = 0$. By Lemma 4.2.8 the set of zeros of δ is of measure zero. For $t \in (A, B) \setminus M$ there is a positive $\delta(t)$ such that

$$|(F \circ \varphi)(u, v) - f(\varphi(t))\varphi'(t)(v - u)| < \mathcal{E}(t)(v - u) \qquad (4.18)$$

whenever $t - \delta(t) < u \leq t \leq v < t + \delta(t)$. Let π be δ-fine, π_M the largest partial division tagged in M with $\pi_M \subset \pi$. Denote $\pi_c = \pi \setminus \pi_M$. Then by inequalities (4.17) and (4.18)

$$\sum_{\pi} |(F \circ \varphi)(u, v) - f(\varphi(t))\varphi'(t)(v - u)|$$

$$\leq \sum_{\pi_c} |(F \circ \varphi)(u, v) - f(\varphi(t))\varphi'(t)(v - u)| + \sum_{\pi_M} |(F \circ \varphi)(u, v)|$$

$$\leq \sum_{\pi_c} \mathcal{E}(t)(v - u) + \varepsilon < 2\varepsilon.$$

It is easy to see that $F \circ \varphi$ is SL because both F and φ are SL and φ is strictly increasing. That proves the change of variable formula 'from left to right'.

Going in the other direction let H be the SL primitive of $(f \circ \varphi)\varphi'$. Denote $\mathbf{F} = H \circ \varphi_{-1}$. If $\omega = \varphi \circ \pi$ is any tagged partial division of $[\varphi(A), \varphi(B)]$ then

$$\sum_{\omega} |\mathbf{F}(w, z) - f(x)(w - z)| \leq \sum_{\pi} |H(u, v) - f(\varphi(t))\varphi'(t)(v - u)|$$

$$+ \sum_{\pi} |f(\varphi(t))\varphi(u, v) - f(\varphi(t))\varphi'(t)(v - u)| . \quad (4.19)$$

There are tight gauges δ_1 and δ_2 such that

$$\sum_{\pi_1} |H(u, v) - f(\varphi(t))\varphi'(t)(v - u)| < \frac{\varepsilon}{2} \qquad (4.20)$$

and

$$\sum_{\pi_2} |f(\varphi(t))\varphi(u, v) - f(\varphi(t))\varphi'(t)(v - u)| < \frac{\varepsilon}{2} \qquad (4.21)$$

whenever $\pi_1 \ll \delta_1$ or $\pi_2 \ll \delta_2$, respectively. The existence of δ_1 is obvious from the fact that H is the SL-primitive. Next δ_2 is defined at every point of differentiability of φ by requiring

$$|\varphi(u, v) - \varphi'(t)(v - u)| < \frac{\mathcal{E}(t)(v - u)}{1 + |f(\varphi(t))|} \qquad (4.22)$$

and setting $\delta_2 = 0$ on the rest of $[A, B]$. Define $\Delta = \text{Min}(\delta_1, \delta_2)$ and

$$\eta_\Delta(x) = \eta_\Delta(\varphi(t)) = \text{Min}[\varphi(t) - \varphi(t - \Delta(t)), \varphi(t + \Delta(t)) - \varphi(t)]. \qquad (4.23)$$

Since φ is strictly increasing and as an SL function satisfies condition N the function η_Δ is a tight gauge on $[\varphi(A), \varphi(B)]$. If $\omega \ll \eta_\Delta$ then by

(4.19), (4.20) and (4.21)

$$\sum_\omega |\mathbf{F}(w,z) - f(x)(w-z)| < \varepsilon. \qquad (4.24)$$

To complete the proof it suffices to show that $\mathbf{F}$ is SL on $[\varphi(A), \varphi(B)]$. Let $Z \subset [\varphi(A), \varphi(B)]$ be of measure zero, $T = \varphi_{-1}(Z)$ and $S \subset T$ where $\varphi' = 0$ and $\delta_1 > 0$. The set $T \setminus S$ is of measure zero, hence there is $\gamma : T \setminus S \to \mathbb{R}_+$ such that

$$\sum_{\pi_0} |H(u,v)| < \varepsilon \qquad (4.25)$$

whenever π_0 is a γ-fine partial division tagged in $T \setminus S$. For $t \in S$ we can let $\gamma(t) = \delta_1(t)$. Define η_γ by equation (4.23) with Δ replaced by γ. Clearly $\eta_\gamma > 0$. If $\omega \ll \eta_\gamma$ and $\omega = \varphi \circ \pi$ then by (4.20) and (4.25)

$$\sum_\omega |\mathbf{F}(w,z)| = \sum_\pi |H(u,v)|$$

$$= \sum_{t\in T\setminus S} |H(u,v)| + \sum_{t\in S} |H(u,v) - f(\varphi(t))\varphi'(t)(v-u)| < \varepsilon + \varepsilon. \quad \bullet$$

REMARK 4.2.10 The condition that φ satisfies SL is essential: the theorem fails without it even if $f = 1$. In this case Theorem 4.2.9 reduces to the Fundamental Theorem and Remark 4.2.4 applies.

4.3 Limit and SL-integration

The theory for interchanging limit and SL-integration is very similar to that for KH-integration. However, in our presentation here we make the concept of SL-equiintegrability (defined below) our starting point and we deduce the other theorems, like the monotone convergence theorem, from it. The same path could have been taken with the KH-integral with equiintegrability playing the rôle of the foundation stone for the interchange of limit and integration. We hope that readers would be able if they so wished to adapt the presentation for the SL-integral below to the KH-integral.

DEFINITION 4.3.1 *A family of functions $\mathfrak{F}$ is said to satisfy the uniform strong Luzin condition, or briefly USL, on an interval $[A, B] \subset \overline{\mathbb{R}}$ if the following conditions are satisfied:*

(a) *for every positive ε and for every set E of measure zero, $E \subset (A, B)$, there exists $\gamma : (A, B) \mapsto \mathbb{R}_+$ such that inequality (4.1)* *holds for all $F \in \mathfrak{F}$ whenever π is a γ-fine partial division tagged in E;*

(b) *the limits (4.2) exist uniformly for $F \in \mathfrak{F}$.*

It is easy to see that if $F_n(x) \to F(x)$ for every $x \in [A, B]$ and $\{F_n\}$ is USL then F satisfies SL. Clearly there is a γ such that

$$\sum_\pi |F_n(u, v)| < \frac{\varepsilon}{2}$$

holds for all $n \in \mathbb{N}$ if π is γ-fine and is tagged in E. Sending $n \to \infty$ shows that condition (4.1) holds. By (b) there is a such that for all $n \in \mathbb{N}$

$$|F_n(x) - F_n(y)| < \varepsilon$$

whenever x and y are in (A, a). Letting $n \to \infty$ shows that the Cauchy condition for the existence of a limit of F at A is satisfied. The existence of $\lim\limits_{x \to B} F(x)$ follows similarly.

DEFINITION 4.3.2 *A family of SL-integrable functions $\mathfrak{f}$ is said to be SL-equiintegrable on $[A, B]$ if*

(i) *the family $\mathfrak{F}$, consisting of functions which are SL-primitives to functions in $\mathfrak{f}$, satisfies USL on $[A, B]$;*

(ii) *for every positive ε there exists a tight gauge δ such that inequality (4.7) holds for all $f \in \mathfrak{f}$ and their SL-primitives F whenever π is a δ-fine tagged partial division of $[A, B]$.*

The main theorem of this section is

THEOREM 4.3.3 (SL-equiintegrability) *If the sequence $\{f_n\}$ is SL-equiintegrable, F_n are the SL-primitives of f_n and $f_n \to f$ almost everywhere on $[A, B]$ then $F_n(c, x)$, with $c \in (A, B)$, converge for every $x \in (A, B)$ to $F(x)$. The function F is the SL-primitive of f on $[A, B]$.*

Proof For a given positive ε we choose, similarly as in the proof of Theorem 4.2.2, functions $\mathcal{E} > 0$† and δ_e such that

$$\sum_\pi \mathcal{E} < \frac{\varepsilon}{4}, \tag{4.26}$$

---

† This notation is kept throughout this section.

whenever π is a δ_e-fine tagged partial division of $[A, B]$. Let $\mathcal{N}$ be the set where f_n do not converge to f. If $x \notin \mathcal{N}$ then there is $m(x)$ such that

$$|f_n(x) - f_m(x)| < \mathcal{E}(x) \tag{4.27}$$

for $n, m > m(x)$. If $\pi_1 \equiv \{(x_i, [u_i, v_i])\}$ is a δ_e-fine tagged partial division and $m_0 > m(x_i)$ for all tags x_i then by (4.27) and (4.26)

$$\sum_{\pi_1} |f_n(x_i) - f_m(x_i)| \, (v_i - u_i) < \varepsilon \tag{4.28}$$

for $n, m > m_0$. Let δ be chosen according to condition (ii) in the Definition 4.3.2 with $\mathfrak{F} = \{f_n; n \in \mathbb{N}\}$. Let us denote, as usual, $N_\delta = \{x; \delta(x) = 0\}$. Choose γ according to condition (a) in Definition 4.3.1 with $E = \mathcal{N} \cup N_\delta$. Define $\Delta(x) = \mathrm{Min}\,(\delta(x), \delta_e(x))$ if $x \notin \mathcal{N} \cup N_\delta$ and $\Delta(x) = \gamma(x)$ otherwise. Let π be a Δ-fine partition of $[c, x]$ (or $[x, c]$), π_0 that part of π which is tagged in $\mathcal{N} \cup N_\delta$ and $\pi_1 = \pi \setminus \pi_0$. For $m, n > m_0$ we have

$$|F_m(c, x) - F_n(c, x)| = \left| \sum_\pi [F_m(u, v) - F_n(u, v)] \right|$$

$$\leq \sum_{\pi_0} |F_m(u, v)| + \sum_{\pi_0} |F_n(u, v)| \tag{4.29}$$

$$+ \left| \sum_{\pi_1} [F_m(u, v) - f_m(x)(v - u)] \right| \tag{4.30}$$

$$+ \left| \sum_{\pi_1} [F_n(u, v) - f_n(x)(v - u)] \right| \tag{4.31}$$

$$+ \sum_{\pi_1} |f_m(x) - f_n(x)|(v - u). \tag{4.32}$$

The terms in (4.29) are less than ε by USL, the sums (4.30) and (4.31) by SL-equiintegrability and the last sum (4.32) by (4.28). Consequently the sequence $\{F_n(c, x)\}$ is Cauchy and $F_n(c, x) \to F(x)$. Further F is SL because F_n are USL. We now redefine δ by setting it zero on $\mathcal{N}$. Then we have

$$\left| \sum_\pi [F_m(u, v) - f_m(x)(v - u)] \right| < \varepsilon$$

for all m whenever π is a δ-fine tagged partial division of $[A, B]$. Sending $m \to \infty$ gives

$$\left| \sum_\pi [F(u,v) - f(x)(v-u)] \right| \leq \varepsilon. \qquad \bullet$$

THEOREM 4.3.4 (Monotone convergence theorem) *If*

(i) *the sequence $\{f_n(x)\}$ is monotonic for almost all $x \in [A, B] \subset \overline{\mathbb{R}}$,*

(ii) *the functions f_n are SL-integrable and the sequence $\left\{ \mathcal{SL} \int_A^B f_n \right\}$ is bounded, $\left| \mathcal{SL} \int_A^B f_n \right| < K$,*

(iii) $\lim\limits_{n\to\infty} f_n = f$ *is finite a.e.*

then f is SL-integrable and

$$\int_A^x f = \lim_{n\to\infty} \int_A^x f_n, \qquad (4.33)$$

for all $x \in [A, B]$.

Proof We show the SL-equiintegrability of the sequence $\{f_n\}$. The assertion then follows from Theorem 4.3.3. By considering $-f_n$ or $f_n - f_1$ instead of f_n, if need be, we can achieve that the sequence $\{f_n\}$ is increasing and $f_n \geq 0$. We denote

$$F_n(x) = \mathcal{SL} \int_A^x f_n.$$

Clearly $F_n(x) \leq F_n(B)$, the sequence $\{F_n(B)\}$ is increasing and bounded, and hence the limit of $F_n(x)$ exists for every $x \in [A, B]$; let us denote it by $F(x)$. Given ε we can find n_0 such that

$$F(B) - F_n(B) < \frac{\varepsilon}{4}, \qquad (4.34)$$

for $n \geq n_0$. Next we denote by $\mathcal{N}$ the set where f_n do not converge to f and for $x \notin \mathcal{N}$ find $n(x) \geq n_0$ such that

$$f(x) - f_n(x) < \mathcal{E}(x), \qquad (4.35)$$

for $n > n(x)$. By the definition of the SL-integral there is a tight gauge δ_n such that

$$\sum_\pi |F_n(u,v) - f_n(x)(v-u)| < \frac{\varepsilon}{4.2^n} \qquad (4.36)$$

whenever the tagged partial division $\pi \ll \delta_n$. The functions δ_n can be chosen to form a decreasing sequence and we assume that is done. We define

$$\delta(x) = \mathrm{Min}(\delta_{n(x)}(x), \delta_e(x)) \quad \text{if} \quad x \notin \mathcal{N},$$

and zero otherwise. We now complete the proof in three steps.

Step 1. The family $\mathfrak{F} = \{F_n; n \geq n_0\}$ satisfies condition (a) in Definition 4.3.1.
Firstly we have

$$|F_n(u,v)| \leq |F_{n_0}(u,v)| + |F_n(u,v) - F_{n_0}(u,v)|. \tag{4.37}$$

Consequently for any tagged partial division π

$$\sum_\pi |F_n(u,v)| \leq \sum_\pi |F_{n_0}(u,v)| + \sum_\pi [F_n(u,v) - F_{n_0}(u,v)],$$

$$\sum_\pi |F_n(u,v)| \leq \sum_\pi |F_{n_0}(u,v)| + F(B) - F_{n_0}(B)$$

$$\leq \sum_\pi |F_{n_0}(u,v)| + \frac{\varepsilon}{4}. \tag{4.38}$$

There is a γ_{n_0} from the SL-condition of F_{n_0}. By (4.38) this γ_{n_0} can serve as the required γ in condition (a) for the whole family $\mathfrak{F}$.

Step 2. Condition (b) from Definition 4.3.1 is satisfied. It follows from inequality (4.37) that

$$|F_n(u,v)| \leq |F_{n_0}(u,v)| + F(B) - F_{n_0}(B) \leq |F_{n_0}(u,v)| + \frac{\varepsilon}{4}.$$

This however indicates that the Cauchy condition for the existence of a uniform limit at A or B is implied by the Cauchy condition for the existence of these limits for the function F_{n_0}.

Step 3. Condition (ii) from Definition 4.3.2 is satisfied. For the rest of the proof we fix m, $m \in \mathbb{N}$, $m \geq n_0$. Let $\pi_w \equiv \{(x_i, [u_i, v_i])\}$ be a δ-fine partial division of $[A, B]$, and π that part of π_w which is tagged in the set where $m > n(x_i)$ and $\pi_0 = \pi_w \setminus \pi$. Denote

$$t_i = f_m(x_i)(v_i - u_i) - F_m(u_i, v_i).$$

Since the sequence $\{\delta_n\}$ is decreasing $\pi_0 \ll \delta_m$ and we have

$$\left|\sum_{\pi_w} t_i\right| \leq \left|\sum_\pi t_i\right| + \left|\sum_{\pi_0} t_i\right| \leq \left|\sum_\pi t_i\right| + \frac{\varepsilon}{4.2^m}. \tag{4.39}$$

To estimate the right-hand side of (4.39) we interpose the following two sums:

$$\sum_\pi f_{n(x_i)}(x_i)(v_i - u_i) \quad \text{and} \quad \sum_\pi F_{n(x_i)}(u_i, v_i).$$

The idea is to show that the first one is close to the Riemann sum of f_m and the second to the other term in the sum of (4.39). We then conclude the proof by proving that their difference is small. We have

$$\left| \sum_\pi [f_m(x_i)(v_i - u_i) - F_m(u_i, v_i)] \right|$$

$$\leq \quad \sum_\pi |f_m(x_i) - f_{n(x_i)}(x_i)| \, (v_i - u_i) \qquad (4.40)$$

$$+ \quad \left| \sum_\pi [f_{n(x_i)}(x_i)(v_i - u_i) - F_{n(x_i)}(u_i, v_i)] \right| \qquad (4.41)$$

$$+ \quad \left| \sum_\pi [F_{n(x_i)}(u_i, v_i) - F_m(u_i, v_i)] \right|. \qquad (4.42)$$

For the sum in (4.40) we obtain by (4.35)

$$0 \leq \sum_\pi |f_m(x_i) - f_{n(x_i)}(x_i)| \, (v_i - u_i) \leq \sum_\pi \mathcal{E} < \frac{\varepsilon}{4}. \qquad (4.43)$$

The sum in (4.42) is also easily estimated

$$0 \leq \sum_\pi [F_m(u_i, v_i) - F_{n(x_i)}(u_i, v_i)] \quad \leq \quad \sum_\pi [F(u_i, v_i) - F_{n_0}(u_i, v_i)]$$

$$\leq \quad F(B) - F_{n_0}(B) < \frac{\varepsilon}{4}. \quad (4.44)$$

It remains to estimate (4.41). The $n(x_i)$ are not necessarily distinct; let $i_1, i_2, \ldots, i_\kappa$ be the distinct i such that $n(x_i) = \kappa$, i.e.

$$\{i; n(x_i) = \kappa\} = \{i_1, i_2, \ldots, i_\kappa\}.$$

We group together the terms with the same $n(x_i) = \kappa$ and estimate the sum using condition (4.7) for the definition of the SL-integral of f_κ.

$$\left| \sum_{j=1}^\kappa [f_k(x_{i_j})(v_{i_j} - u_{i_j}) - F_k(u_{i_j}, v_{i_j})] \right| < \frac{\varepsilon}{4.2^k}.$$

Consequently

$$\left| \sum_\pi \left[f_{n(x_i)}(x_i)(v_i - u_i) - \int_{u_i}^{v_i} f_{n(x_i)} \right] \right| < \sum_1^\infty \frac{\varepsilon}{4.2^k} = \frac{\varepsilon}{4}. \qquad (4.45)$$

Collecting (4.43), (4.44) and (4.45) and applying them to (4.40),(4.41) and (4.42) gives

$$\left| \sum_\pi [f_m(x_i)(v_i - u_i) - F_m(u_i, v_i)] \right| < \varepsilon. \qquad \bullet$$

THEOREM 4.3.5 *If*

(i) *the sequence $\{f_n(x)\}$ is monotonic for almost all $x \in [A, B] \subset \overline{\mathbb{R}}$,*

(ii) *the functions f_n are SL-integrable and the sequence $\left\{ \mathcal{SL}\int_A^B f_n \right\}$ is bounded*

then $\lim_{n\to\infty} f_n = f$ is finite a.e.

The proof of Theorem 3.5.8 given in Section 3.5 for the KH-integral applies verbatim here.

REMARK 4.3.6 (Absolute integrability) Absolute integrability for the SL-integral can be treated similarly as for the KH-integral. Here we only mention that if f, g are absolutely SL-integrable then so are $f + g$, $f - g$, Max(f, g) and Min(f, g).

The monotone convergence theorem can be rephrased for series as was done for the KH-integral. We shall not pursue this here. In most theories of the integral the dominated convergence theorem is an immediate consequence of the monotone convergence theorem. This is in particular true for the SL-integral.

THEOREM 4.3.7 (Dominated convergence theorem) *If the functions f_n, g, G are SL-integrable and*

$$g \le f_n \le G \qquad (4.46)$$

on $[A, B]$ for all $n \in \mathbb{N}$ then

$$\int_A^B \liminf_{n\to\infty} f_n \le \liminf_{n\to\infty} \int_A^B f_n \qquad (4.47)$$

$$\le \limsup_{n\to\infty} \int_A^B f_n \le \int_A^B \limsup_{n\to\infty} f_n. \qquad (4.48)$$

If moreover $\lim_{n\to\infty} f_n(x)$ exists a.e. then

$$\int_A^B \lim_{n\to\infty} f_n = \lim_{n\to\infty} \int_A^B f_n. \qquad (4.49)$$

Proof We prove only (4.47); inequality (4.48) can be proved similarly (or by considering $-f_n$) and (4.49) then follows.

$$\text{Let } g_k = \inf\{f_i - g;\, n \leq i \leq k\} + g \;=\; \inf\{f_i;\, n \leq i \leq k\},$$

$$h_n = \lim_{k \to \infty} g_k \;=\; \inf\{f_i;\, i \geq n\}.$$

All the functions g_k are integrable, and since $\int_A^B g \leq \int_A^B g_k \leq \int_A^B G$ all the functions h_n are integrable by the monotone convergence theorem. Obviously $\int_A^B h_n \leq \int_A^B f_n$ and therefore

$$-\infty < \int_A^B g \leq \lim_{n \to \infty} \int_A^B h_n \leq \liminf_{n \to \infty} \int_A^B f_n \leq \int_A^B G. \qquad (4.50)$$

Applying the monotone convergence theorem once more gives

$$\int_A^B \liminf_{n \to \infty} f_n = \int_A^B \lim_{n \to \infty} h_n = \lim_{n \to \infty} \int_A^B h_n.$$

This together with (4.50) gives (4.47).   ●

4.4 Equivalence with the KH-integral

Our goal in this final section is to establish the equivalence of SL- and KH-integration. We state this result as the next two theorems.

THEOREM 4.4.1 *If f is KH-integrable on $[A, B]$ then it is SL-integrable and*

$$\mathcal{KH} \int_A^B f = \mathcal{SL} \int_A^B f. \qquad (4.51)$$

Proof Denote $F(x) = \mathcal{KH} \int_A^x f$. By Henstock's lemma the function F satisfies the inequality (4.7) from the definition of the SL-integral. The limit relations

$$\lim_{c \downarrow A} F(c) = F(A) \qquad \lim_{c \uparrow B} F(c) = F(B)$$

hold by Theorem 2.8.3 and Remark 2.8.4 or by Theorem 2.9.3 and Remark 2.9.4. By **(U)** of Section 4.1 it suffices to show that F is SL on every bounded interval $K \subset [A, B]$. On that part of K where F' does not exist or is infinite F is of negligible variation by Theorem 3.9.2. On the rest of K it is SL by Remark 4.1.4.   ●

THEOREM 4.4.2 *If f is SL-integrable on $[A, B]$ then it is KH-integrable and equation (4.51) holds.*

Proof It is sufficient to prove the theorem if $[A, B] = [a, b] \subset \mathbb{R}$; the general case then follows from Theorems 2.9.3 and Remark 2.9.4. Let $\varepsilon > 0$, and denote by F_{SL} the SL-primitive of f. We can find a tight gauge δ_1 such that

$$\sum_{\pi_1} |F_{SL}(u, v) - f(\xi)(u - v)| < \frac{\varepsilon}{3}, \tag{4.52}$$

for every tagged partial division π_1 on $[a, b]$, $\pi_1 \ll \delta_1$. Let $Z = N_{\delta_1}$ where N_{δ_1} is the set of zeros of δ_1. Define $f_0 = f \mathbf{1}_Z$. This function is zero almost everywhere, and consequently KH-integrable with $\mathcal{KH} \int_a^b f_0 = 0$. Hence there exists $\delta_2 : Z \to \mathbb{R}_+$ such that for any partial division π_2 tagged in Z with $\pi_2 \ll \delta_2$

$$\left| \sum_{\pi_2} f_0(\xi)(v - u) \right| = \left| \sum_{\pi_2} f(\xi)(v - u) \right| < \frac{\varepsilon}{3}. \tag{4.53}$$

For the primitive F_{SL} there is a positive γ such that

$$\sum_{\pi_2} |F_{SL}(u, v)| < \frac{\varepsilon}{3}, \tag{4.54}$$

whenever $\pi_2 \ll \gamma$ is a tagged partial division on Z. We define

$$\delta(t) = \begin{cases} \delta_1(t) & \text{if } t \notin Z, \\ \text{Min}(\delta_2(t), \gamma(t)) & \text{if } t \in Z. \end{cases}$$

If $\pi = \pi_1 \cup \pi_2$, where π_2 is the largest part of π tagged in Z and $\pi \ll \delta$, then by (4.52), (4.53) and (4.54)

$$\left| F_{SL}(a, b) - \sum_{\pi} f(\xi)(u - v) \right| = \left| \sum_{\pi} (F_{SL}(u, v) - f(\xi)(u - v)) \right|$$

$$\leq \sum_{\pi_1} |(F_{SL}(u, v) - f(\xi)(u - v))| + \sum_{\pi_2} |F_{SL}(u, v)| + \left| \sum_{\pi_2} f(\xi)(u - v) \right|$$

$$< \frac{\varepsilon}{3} + \frac{\varepsilon}{3} + \frac{\varepsilon}{3}.$$

This shows that f is KH-integrable and $\mathcal{KH} \int_a^b f = F_{SL}(a, b)$. ●

COROLLARY 4.4.3 *If f is SL-integrable on $[a, b]$ and $F(x) = \int_a^x f$ then $F'(x) = f(x)$ a.e. on $[a, b]$.*

Combining this corollary with Theorem 4.2.2 we have

THEOREM 4.4.4 *A function F is the SL-primitive on* $[a, b]$ *of another function f if and only if F is SL on* $[a, b]$ *and* $F' = f$ *a.e. on* $[a, b]$.

4.5 Exercises

EXERCISE 4.1 ⓘ*Prove directly without use of Theorem 3.9.2 or Theorem 3.8.2 that the KH-primitive of a KH-integrable function f is SL. [Hint: Use* (**U**) *of section 4.1 to reduce the proof to the case of bounded f and then use Henstock's lemma.]*

EXERCISE 4.2 *Show that the SL condition in Theorem 4.4.4 cannot be replaced by the Luzin condition. More precisely, if F satisfies the Luzin condition and* $F' = f$ *a.e. on* $[a, b]$ *then* $F(b) - F(a)$ *is not be uniquely determined.*

EXERCISE 4.3 *Show that if f is SL-integrable on* $[a, b]$ *then there is a sequence* $\{f_n\}$ *of continuous functions such that* $f_n \to f$ *almost everywhere and* $\int_a^b f_n \to \int_a^b f$ *as* $n \to \infty$. *[Hint: Put* $f_n(x) = n[F(x + \frac{1}{n}) - F(x)]$ *where F is the primitive of f.]*

EXERCISE 4.4 *Show that if* $\{f_n\}$ *is equiintegrable on* $[a, b]$ *and if* $\{f_n\}$ *is uniformly bounded then their primitives* $F_1, F_2, \ldots$ *satisfy the uniform strong Luzin condition on* $[a, b]$.

EXERCISE 4.5 *Show that if* $\{f_n\}$ *is equiintegrable on* $[a, b]$ *and* $f_n \to f$ *everywhere then their primitives* $F_1, F_2, \ldots$ *satisfy the uniform strong Luzin condition on* $[a, b]$. *[Hint: If E is of measure zero, so is its subset* E_i *where* $|f_n(x)| \le i$ *for all* $x \in E_i$ *and all n. Then apply the proof of Exercise 4.4.]*

EXERCISE 4.6 *Show that if* $\{f_n\}$ *is equiintegrable on* $[a, b]$ *and* $f_n \to f$ *everywhere then f is integrable on* $[a, b]$ *and* $\int_a^b f_n \to \int_a^b f$. *[Hint: Use Exercise 4.5 and the SL-integrability.]*

EXERCISE 4.7 *If the pointwise convergence everywhere of* f_n *in Exercise 3.25 is replaced by almost everywhere, then the sequence* $\{f_n\}$ *is SL-equiintegrable on* $[a, b]$. *Similarly for Exercise 3.26 we can prove* that the sequence $\{f_n\}$ is SL-equiintegrable on $[a, b]$ if the pointwise convergence everywhere of f_n is replaced by almost everywhere.

EXERCISE 4.8 Let $[a_k, b_k]$ be a sequence of non-overlapping subintervals of $[a, b]$. Give an example of a function f such that f is SL-integrable on each $[a_k, b_k]$ but not on the union $\bigcup_{k=1}^{\infty}[a_k, b_k]$.

EXERCISE 4.9 Let $[a_k, b_k]$ be a sequence of non-overlapping subintervals of $[a, b]$, and f SL-integrable on each $[a_k, b_k]$. Write $H(u, v) = \sum_{k=1}^{\infty} \int_{[a_k, b_k] \cap [u, v]} f$ if it exists, and X the complement of $\bigcup_{k=1}^{\infty}(a_k, b_k)$. Show that f is SL-integrable on the union $\bigcup_{k=1}^{\infty}[a_k, b_k]$ with the primitive H if and only if the function H satisfies the following condition: for every $\varepsilon > 0$ there is $\delta : [a, b] \to \mathbb{R}_+$ such that for any δ-fine partition π of $[a, b]$ we have $\sum_{\pi^*} |H(u, v)| < \varepsilon$, where π^* is the part of π having tags in X.

EXERCISE 4.10 Let f be defined as in Exercise 4.9, and $f_n(x) = f(x)$ for $x \in [a_k, b_k]$, $k = 1, 2, \ldots, n$ and 0 otherwise. Show that if f is SL-integrable on the union $\bigcup_{k=1}^{\infty}[a_k, b_k]$ with the primitive H and if $\sum_{k=1}^{\infty} \omega(H; [a_k, b_k])$ is finite then $\{f_n\}$ is equiintegrable on $[a, b]$.

The following exercises come from [24].

EXERCISE 4.11 Let Π be a collection of partitions π of $[a, b]$ and $\sqsupseteq$ an order defined in Π. We write $\pi_2 \sqsupseteq \pi_1$ and say π_2 is finer than π_1. The pair $(\Pi, \sqsupseteq)$ is called a directed set if (1) $\pi \sqsupseteq \pi$ for all $\pi \in \Pi$; (2) if $\pi_1, \pi_2, \pi_3 \in \Pi$ with $\pi_1 \sqsupseteq \pi_2$ and $\pi_2 \sqsupseteq \pi_3$ then $\pi_1 \sqsupseteq \pi_3$; (3) if $\pi_1, \pi_2 \in \Pi$ with $\pi_1 \sqsupseteq \pi_2$ and $\pi_2 \sqsupseteq \pi_1$ then $\pi_1 = \pi_2$; and (4) for every $\pi_1, \pi_2 \in \Pi$ there exists $\pi_3 \in \Pi$ such that $\pi_3 \sqsupseteq \pi_1$ and $\pi_3 \sqsupseteq \pi_2$. We define $\pi_2 \geq \pi_1$ and say π_2 is finer than π_1 in the Riemann sense if for each $(\eta, [s, t]) \in \pi_2$ we have $[s, t] \subset [u, v]$ for some $(\xi, [u, v]) \in \pi_1$ and when $[s, t] = [u, v]$ we have $\eta = \xi$. Show that $(\Pi, \geq)$ is a directed set.

EXERCISE 4.12 Let $(\Pi, \geq)$ be the directed set in Exercise 4.11 where $\pi_2 \geq \pi_1$ means π_2 is finer than π_1 in the Riemann sense. The family $\{\sum_{\pi} f\}_{\pi \in \Pi}$ of Riemann sums is called the generalized sequence. Show that the Riemann integral of f on $[a, b]$ is the Moore–Smith limit of the generalized sequence $\{\sum_{\pi}\}_{\pi \in \Pi}$, that is, for every positive ε there is

$\pi_0 \in \Pi$ such that for any $\pi \geq \pi_0$ we have

$$|\sum_\pi f - I| < \varepsilon.$$

EXERCISE 4.13 *Let Π be the family of all δ-fine partitions π of $[a, b]$ for some positive function δ. For $\pi_1, \pi_2 \in \Pi$, define $\pi_2 \geq \pi_1$ and say π_2 is finer than π_1 in the Henstock sense using δ if for every $(\eta, [s, t]) \in \pi_2$ we have $[s, t] \subset [u, v]$ for some $(\xi, [u, v]) \in \pi_1$, and $\{\xi : (\xi, [u, v]) \in \pi_1\} \subset \{\eta : (\eta, [s, t]) \in \pi_2\}$. Show that $(\Pi, \geq)$ is a directed set. The H_1-integral of f is defined to be the Moore–Smith limit of the generalized sequence $\{\sum_\pi f\}_{\pi \in \Pi}$ of Riemann sums using the above directed set Π. More precisely, f is H_1-integrable on $[a, b]$ if there is a directed set $(\Pi, \geq)$ as defined above by using δ and for every positive ε there is $\pi_0 \in \Pi$ such that for any $\pi \in \Pi$ and $\pi \geq \pi_0$ we have $|\sum_\pi f - I| < \varepsilon$. State and prove the corresponding H_1-equiintegrability theorem for the H_1-integral. (See Theorem 4.3.3.)*

EXERCISE 4.14 *A function f is said to be H_1-integrable on X if $f1_X$ is H_1-integrable on $[a, b]$. Let $X_1 \subset X_2$ be closed subsets of $[a, b]$ and f be H_1-integrable on X_1 using δ_1 and on X_2 with $f(x) = 0$ outside X_2. Show that if the primitive F of f on $[a, b]$ is absolutely continuous there then f is H_1-integrable on X_2 using δ where $\delta(x) = \delta_1(x)$ when $x \in X_1$. [Hint: Assume f is H_1-integrable on X_2 using δ_2 with $\delta_2 \leq \delta_1$. Put $\delta(x) = \delta_1(x)$ when $x \in X_1$ and δ_3 otherwise where $\delta_3 \leq \delta_2$ and $(x - \delta_3(x), x + \delta_3(x)) \cap X_1 = \emptyset$ when $x \notin X_1$.]*

EXERCISE 4.15 *Show that if f is continuous on $[a, b]$ then f is H_1-integrable on every closed subset X of $[a, b]$. [Hint: Use the Cauchy principle for the H_1-integral involving Riemann sums.]*

EXERCISE 4.16 *Let f be H_1-integrable on each closed set X_n with primitive F_n. Show that if f is non-negative on $[a, b]$ and $F_n(a, b) \to I$ then f is H_1-integrable on the union $X = \bigcup_{n=1}^\infty X_n$. [Hint: Assume $X_n \subset X_{n+1}$ and f is H_1-integrable on X_n using δ_n such that for any δ_n-fine partition π we have $\sum_{\pi^*} |f(\xi)(v - u) - F_n(u, v)| < 2^{-n}$ where π^* is the part of π with tags in X_n. Define $\delta(x) = \delta_n(x)$ when $x \in X_n - X_{n-1}$ for all n with $X_0 = \emptyset$ and arbitrary otherwise. Apply Exercise 4.14 if necessary. For $\varepsilon > 0$ choose N so that $I - F_N(a, b) < \varepsilon$ and $\sum_{n=N+1}^\infty 2^{-n} < \varepsilon$. Then choose π_N so that for any δ-fine π finer* than π_N in the Henstock sense we have

$$\sum_{\pi^*} |f(\xi)(v-u) - F_N(u,v)| < \varepsilon$$

where π^* is the part of π with tags in X_N. Using this π_N as π_0 in Exercise 4.13 we prove that f is H_1-integrable on X.]

EXERCISE 4.17 Let f be non-negative on $[a,b]$. Show that f is SL-integrable on $[a,b]$ if and only if there is an H_1-integrable function g such that $f = g$ almost everywhere. [Hint: We can construct, by Luzin's theorem (Theorem 3.10.17), a sequence $\{X_n\}$ of closed sets such that f is continuous on each closed set X_n with $m([a,b] - \bigcup_{n=1}^{\infty} X_n) = 0$. Further the conditions in Exercise 4.16 hold and f is H_1-integrable on $\bigcup_{n=1}^{\infty} X_n$. The converse is trivial. (In fact, the result is also true without the condition that f is non-negative. We have to wait until after Chapter 5.)]

5

Generalized AC Functions

5.1 Prologue

Classical integration theory often refers to two different ways of defining an integral. One is called descriptive and the other constructive. For example, the Newton definition is descriptive whereas the Riemann definition is constructive. The Kurzweil–Henstock integral is also constructive. A constructive definition usually begins with a function f, then by some process involving sums and limits arrives at $\int_a^b f$, the definite integral of f. If we write

$$F(x) = \int_a^x f(t)dt \quad \text{for} \quad a \leq x \leq b,$$

then F is the indefinite integral of f. A descriptive definition starts with a primitive function F satisfying certain condition or conditions and f is in some sense a derivative of F. For example, in the definition of the Newton integral $F'(x) = f(x)$ for all x and there is no additional condition. However, generally speaking, the additional conditions are the most important part of a descriptive definition. For instance the Lebesgue indefinite integral of f can be defined as a function F such that

$$F'(x) = f(x) \qquad \text{for almost all } x, \tag{5.1}$$

and F is absolutely continuous. By Theorem 3.9.1 the indefinite KH-integral of f can be defined as a continuous F for which (5.1) holds and which is of negligible variation on the set of non-differentiability. Example 3.6.1 shows that these two latter descriptive definitions become meaningless without the additional conditions. In this chapter, we shall give yet another descriptive definition of the Kurzweil–Henstock integral. First of all, we shall characterize the indefinite integral of a KH-integrable function by showing that it is generalized absolutely continuous in some sense. Later we shall use this characterization to formulate a convergence theorem, known as the controlled convergence theorem.

In a sense, the KH-integral is a countable extension of the Riemann integral. We recall that a number I is the KH-integral of f on $[a, b]$ if for every $\varepsilon > 0$ there exists $\delta : [a, b] \to R_+$ such that for every δ-fine partition π

$$\left| \sum_\pi f - I \right| < \varepsilon.$$

Let $\{\delta_n\}$ be a decreasing null sequence, and define $\delta^* : [a, b] \to R_+$ as follows:

$$
\begin{aligned}
\delta^*(x) &= \delta_1 \quad \text{when} \quad \delta(x) \geq \delta_1, \\
&= \delta_n \quad \text{when} \quad \delta_{n-1} > \delta(x) \geq \delta_n,
\end{aligned}
$$

$n = 2, 3, \cdots$. Then f is also KH-integrable on $[a, b]$ using δ^* in place of δ. In other words, we are using a countable number $\delta_1, \delta_2, \ldots$ above in place of a single constant δ as in the definition of the Riemann integral. Therefore we may regard the KH-integral as a
countable extension of the Riemann integral.

Again, we recall that if f is absolutely KH-integrable on $[a, b]$ then its primitive F is absolutely continuous. Now, to characterize the primitive F of a KH-integrable function f, we seek for a countable extension of the concept of absolute continuity. We give two examples now to motivate and a formal definition will be given in Section 5.3.

EXAMPLE 5.1.1   Let $f(x) = F'(x)$ for $x \in [0, 1]$ where $F(0) = 0$ and $F(x) = x^2 \sin x^{-2}$ when $x \neq 0$. Obviously, f is KH-integrable on $[0, 1]$ but F is not of bounded variation on $[0, 1]$ and therefore not absolutely continuous there. However, F has the following property: $[0, 1] = (\bigcup_i [\frac{1}{i+1}, \frac{1}{i}]) \cup \{0\}$ and F is absolutely continuous on $[\frac{1}{i+1}, \frac{1}{i}]$ for each i. That is, F is 'generalized absolutely continuous'. We shall make precise the definition of this concept later.

EXAMPLE 5.1.2   Let f be given as in Example 5.1.1 and $f_n(x) = f(x)$ when $x \in [\frac{1}{n}, 1]$ and 0 otherwise. It is easy to verify that $f_n(x) \to$

$f(x)$ as $n \to \infty$ for every $x \in [0, 1]$ and

$$\int_0^1 f_n(x)dx \to \int_0^1 f(x)dx \quad \text{as} \quad n \to \infty.$$

Yet none of the convergence theorems of Section 3.5 applies. The sequence $\{f_n\}$ is not dominated by any KH-integrable function on $[0, 1]$. However it is so on $[\frac{1}{n+1}, \frac{1}{n}]$ for each n. In what follows, we shall give a countable extension of the dominated convergence theorem which will take care of situations such as the above.

5.2 Uniformly AC functions

The concept of absolute continuity is also useful for interchange of limit and integration. A family $\mathfrak{F}$ of functions is said to be uniformly absolutely continuous on $[a, b]$ if for every positive ε there is a positive η with the following property: if D is a partial division of $[a, b]$ with

$$\sum_D (v - u) < \eta$$

then

$$\sum_D |F(u, v)| < \varepsilon \tag{5.2}$$

holds for all $F \in \mathfrak{F}$. The next theorem uses uniform absolute continuity and Egoroff's theorem (Theorem 3.11.20) which guarantees that a sequence of integrable functions converges uniformly except on an open set G of arbitrarily small measure.

THEOREM 5.2.1 *If the following conditions are satisfied:*

(i) $f_n(x) \to f(x)$ *almost everywhere in* $[a, b]$ *where each* f_n *is KH-integrable on* $[a, b]$,

(ii) *the primitives* F_n *of* f_n *are uniformly absolutely continuous on* $[a, b]$,

then f *is KH integrable on* $[a, b]$ *and*

$$\int_a^b f_n(x)dx \to \int_a^b f(x)dx \quad \text{as} \quad n \to \infty.$$

Proof Since each f_n is KH-integrable on $[a, b]$, Henstock's lemma implies that for every $\varepsilon > 0$ there exists $\delta_n : [a, b] \to R_+$ such that for any δ_n-fine partition π

$$\sum_\pi |f_n(x)(v - u) - F_n(u, v)| < \varepsilon.$$

By assumption F_n are uniformly absolutely continuous on $[a, b]$, that is, for every $\varepsilon > 0$ there is $\eta > 0$, independent of n, such that for any partial division π of $[a, b]$

$$\sum_\pi |v - u| < \eta \quad \text{implies} \quad \sum_\pi |F_n(u, v)| < \varepsilon.$$

Apply Egoroff's theorem (Theorem 3.11.20) and there are an open set G, with $m(G) < \eta$, and an integer N such that

$$|f_n(x) - f_m(x)| < \varepsilon \quad \text{whenever} \quad n, m \geq N \quad \text{and} \quad x \in [a, b] - G.$$

We may assume $(\xi - \delta_n(\xi), \xi + \delta_n(\xi)) \subset G$ when $\xi \in G$. Then for any division π of $[a, b]$ take a δ_n-fine and δ_m-fine partition π_1 finer than π and write $\pi_1 = \pi_2 \cup \pi_3$ in which the intervals of π_2 are tagged outside G and the intervals of π_3 in G. Then for $n, m \geq N$ we have

$$\sum_{\pi_1} |F_n(u, v) - F_m(u, v)| \leq \sum_{\pi_2} |F_n(u, v) - f_n(x)(v - u)|$$

$$+ \sum_{\pi_2} |f_n(x) - f_m(x)|(v - u) \quad + \quad \sum_{\pi_2} |f_m(x)(u, v) - F_m(u, v)|$$

$$+ \sum_{\pi_3} |F_n(u, v)| \quad + \quad \sum_{\pi_3} |F_m(u, v)|$$

$$< \quad 4\varepsilon + \varepsilon(b - a).$$

This implies

$$\text{Var}_a^b (F_n - F_m) \leq \varepsilon(4 + b - a).$$

In view of Theorem 3.4.1 we obtain

$$\int_a^b |f_n(x) - f_m(x)| dx \leq \varepsilon(4 + b - a) \quad \text{for} \quad n, m \geq N.$$

By the mean convergence theorem, the result follows.   •

COROLLARY 5.2.2   *If the conditions of Theorem 5.2.1 hold, then f is absolutely KH-integrable on $[a, b]$ and*

$$\int_a^b |f_n(x) - f(x)| dx \to 0 \quad \text{as} \quad n \to \infty.$$

COROLLARY 5.2.3   *If the conditions of Theorem 5.2.1 hold, then there are a function g, KH-integrable on $[a, b]$, and a subsequence $\{f_{n(i)}\}$ of $\{f_n\}$ such that $|f_{n(i)}(x)| \leq g(x)$ for almost all $x \in [a, b]$ and for all $n(i)$.*

REMARK 5.2.4   In fact, according to Theorem 3.6.4 all the functions $f_1, f_2, \ldots$ in Theorem 5.2.1 are absolutely KH-integrable on $[a, b]$ since their primitives are absolutely continuous. The same comment also applies to the corollaries.

REMARK 5.2.5   It is easy to see that if a sequence of KH-integrable functions $\{f_n\}$ is dominated, i.e. $|f_n(x)| \leq g(x)$ for almost all $x \in [a, b]$ and for all n, then the primitives F_n of f_n are uniformly absolutely continuous on $[a, b]$. A partial converse was given by Corollary 5.2.3. However, in general, the converse is not true, as seen from the following example.

EXAMPLE 5.2.6   Let $f_n(x) = n$ when $x \in [\frac{1}{n+1}, \frac{1}{n}]$ and 0 otherwise. It is easy to see that any function which dominates the sequence $\{f_n\}$ is not KH-integrable on $[0, 1]$. The primitive F_n of f_n is given by $F_n(x) = 0$ when $x \in [0, \frac{1}{n+1}]$, $n(x - \frac{1}{n+1})$ when $x \in [\frac{1}{n+1}, \frac{1}{n}]$ and $F_n(x) = \frac{1}{n+1}$ when $x \in [\frac{1}{n}, 1]$. Again, it is easy to see that F_n are uniformly absolutely continuous on $[0, 1]$. Note that the sequence $\{f_n\}$ has a subsequence which is dominated by a KH-integrable function on $[0, 1]$.

5.3 *AC** and *VB** on a set

The key theorem in this section is Theorem 5.3.13 which is an extension of Theorem 3.4.1. The theorem relates the two properties AC^* and VB^* which respectively extend absolute continuity and bounded variation on an interval to those on a set.

DEFINITION 5.3.1   *Let X be a subset of $[a, b]$. A function F defined on $[a, b]$ is said to be $AC^*(X)$ if for every $\varepsilon > 0$ there exists $\eta > 0$ such that for any partial division π of $[a, b]$ with u or $v \in X$*

$$\sum_\pi |v - u| < \eta \text{ implies } \sum_\pi |F(u, v)| < \varepsilon.$$

Note that in the above definition we require only one endpoint of $[u, v]$ to lie in X. When $X = [a, b]$, the above concept coincides with the usual definition of absolute continuity. We recall that the oscillation ω of F on E is defined to be

$$\omega(F;E) = \sup\{|F(x) - F(y)|;\ x,y \in E\}.$$

LEMMA 5.3.2   *Let F be continuous on $[a,b]$ and X a closed subset of $[a,b]$. Then F is $AC^*(X)$ if and only if for every $\varepsilon > 0$ there exists $\eta > 0$ such that for any partial division π of $[a,b]$ with u and $v \in X$*

$$\sum_\pi |v - u| < \eta \text{ implies } \sum_\pi \omega(F;[u,v]) < \varepsilon.$$

Proof   Suppose F is $AC^*(X)$. Since F is continuous on $[a,b]$, we have

$$\omega(F;[u,v]) = |F(s) - F(t)|$$

for some $s,t \in [u,v]$. Then

$$\omega(F;[u,v]) \le |F(s) - F(u)| + |F(t) - F(u)|$$

or, alternatively,

$$\omega(F;[u,v]) \le |F(v) - F(s)| + |F(v) - F(t)|.$$

Hence the condition follows from the definition of $AC^*(X)$.

Conversely, suppose the condition holds. Since X is closed, we write $(a,b) - X = \bigcup_{k=1}^\infty (a_k, b_k)$. Choose an integer N such that $\sum_{k=N+1}^\infty (b_k - a_k) < \eta$. In view of the condition, we have

$$\sum_{k=N+1}^\infty \omega(F;[a_k,b_k]) \le \varepsilon. \tag{5.3}$$

Since F is continuous on $[a,b]$, it is uniformly continuous there, i.e. there exists $\eta_1 > 0$ such that whenever $|x - y| < \eta_1$ we have

$$|F(x) - F(y)| < \frac{\varepsilon}{N}. \tag{5.4}$$

Finally, choose $\eta_2 = \text{Min}(\eta, \eta_1)$.

Now take any partial division π with u or $v \in X$ and $\sum_\pi |v - u| < \eta_2$. Sort the intervals $[u,v]$ in π into three classes π_1, π_2 and π_3 as follows:

(i)   both $u,v \in X$,

(ii) $u \in X$ and $v \in (a_k, b_k)$ for $k = 1,2,\ldots,N$, or $v \in X$ and $u \in (a_k, b_k)$ for $k = 1,2,\ldots,N$.

(iii) $u \in X$ and $v \in (a_k, b_k)$ for $k > N$, or $v \in X$ and $u \in (a_k, b_k)$ for $k > N$.

In cases (ii) and (iii), we may always assume $u = a_k$ or $v = b_k$. Otherwise we partition $[u, v]$ into $[u, a_k] \cup [a_k, v]$ or $[u, b_k] \cup [b_k, v]$ and have additional intervals $[u, a_k], [b_k, v]$ with both endpoints in X falling under (i).

Applying the condition, (5.4) and (5.3) respectively, we obtain

$$
\begin{aligned}
\sum_\pi |F(u, v)| &\leq \sum_{\pi_1} |F(u, v)| \\
&\quad + \sum_{\pi_2} |F(u, v)| + \sum_{\pi_3} |F(u, v)| \\
&< \varepsilon + \frac{\varepsilon}{N} \cdot 2N + 2\varepsilon = 5\varepsilon.
\end{aligned}
$$

The last two terms involve a factor 2 because for each k we may have two intervals from π both intersecting $[a_k, b_k]$. Hence we have proved that F is $AC^*(X)$. ●

LEMMA 5.3.3   *Let F be continuous on $[a, b]$ and $X \subset [a, b]$. If F is $AC^*(X)$ then F is $AC^*(\overline{X})$ where $\overline{X}$ is the closure of X.*

Proof   Suppose F is $AC^*(X)$. Then for every $\varepsilon > 0$ there exists $\eta > 0$ such that for any partial division π with u or $v \in X$

$$
\sum_\pi |v - u| < \eta \quad \text{implies} \quad \sum_\pi |F(u, v)| < \varepsilon.
$$

Now take any partial division π with u or $v \in \overline{X}$ and $\sum_\pi |v - u| < \eta$.

Consider a typical interval $[u, v]$ in π and suppose $u \in \overline{X}$. Since F is continuous at u, there is $t \in X$ such that

$$
|F(t) - F(u)| < \varepsilon/N
$$

where N is the number of intervals in π. Denote $[t, v]$ by $[u', v']$ and $[u, t]$ or $[t, u]$ by $[u'', v'']$. It is possible that $t = u$, in which case $[u', v'] = [u, v]$ and $F(u'', v'') = 0$. Similarly, suppose $v \in \overline{X}$; we obtain $[u', v']$ and $[u'', v'']$ such that one of u' and v' belongs to X and

$$
|F(u'') - F(v'')| < \varepsilon/N.
$$

Since u or $v \in \overline{X}$, we can choose $\pi' = \{[u', v']\}$ such that $\sum_{\pi'} |v' - u'| < \eta$. Therefore we can apply $AC^*(X)$ and obtain

$$
\begin{aligned}
\sum_\pi |F(u, v)| &\leq \sum_{\pi'} |F(u', v')| + \sum_{\pi''} |F(u'', v'')| \\
&< \varepsilon + \frac{\varepsilon}{N} \cdot N = 2\varepsilon.
\end{aligned}
$$

Hence F is $AC^*(\overline{X})$.   ●

DEFINITION 5.3.4   *Let X be a subset of $[a,b]$. A function F defined on $[a,b]$ is said to be $AC(X)$ if for every $\varepsilon > 0$ there exists $\eta > 0$ such that for any partial division π of $[a,b]$ with u and $v \in X$*

$$\sum_\pi |v - u| < \eta \text{ implies } \sum_\pi |F(u,v)| < \varepsilon$$

where $F(u,v) = F(v) - F(u)$.

REMARK 5.3.5   When $X = [a,b]$, the two concepts $AC^*(X)$ and $AC(X)$ coincide. In general, $AC^*(X)$ implies $AC(X)$ though not vice versa. To proceed from $AC(X)$ to $AC^*(X)$, we either relax the condition u and $v \in X$ in Definition 5.3.4 to at least one endpoint belonging to X, or impose a stronger requirement on F, namely, the oscillation of F on $[u,v]$ in place of the difference $F(v) - F(u)$ as in Lemma 5.3.2. Both approaches lead to the same definition of $AC^*(X)$ assuming that certain conditions hold, as we can see from Lemma 5.3.2. Since the approach using one endpoint and the difference is easier when proving theorems, we use it as the definition and regard the approach using the oscillation and two endpoints as a theorem. Although the definitions of $AC(X)$ and $AC^*(X)$ look similar the concepts are very different. Whether a function F is $AC(X)$ is determined solely by the behaviour of F on X whereas whether $AC^*(X)$ is influenced† by values of F on the smallest interval containing X. For example, if $X = \mathbb{Q} \cap [0,1]$ then the characteristic function of the rationals is $AC(X)$ but not $AC^*(X)$.

The relationship between AC and AC^* is also the topic of the next lemma.

LEMMA 5.3.6   *Let F be continuous on $[a,b]$ and X a closed subset of $[a,b]$ with $(a,b) - X = \bigcup_{k=1}^\infty (a_k,b_k)$. Then F is $AC^*(X)$ if and only if F is $AC(X)$ and*

$$\sum_{k=1}^\infty \omega(F;[a_k,b_k]) < +\infty$$

where ω denotes the oscillation of F on $[a_k,b_k]$.

The proof mimics that of Lemma 5.3.2.

† But not necessarily determined.

DEFINITION 5.3.7   *A function F defined on $[a, b]$ is said to be $VB^*(X)$ if*

$$\sup_{\pi} \sum |F(u, v)| < +\infty$$

where the supremum is taken over all partial divisions π of $[a, b]$ with u or $v \in X$.

LEMMA 5.3.8   *Let F be bounded on $[a, b]$ and $X \subset [a, b]$. Then F is $VB^*(X)$ if and only if*

$$\sup_{\pi} \sum \omega(F; [u, v]) < +\infty$$

where the supremum is taken over all partial divisions π of $[a, b]$ with u and $v \in X$.

Proof   Suppose F is $VB^*(X)$. Then there is $M > 0$ such that

$$\sum_{\pi} |F(u, v)| \leq M$$

whenever π is a partial division of $[a, b]$ with u or $v \in X$. Take π to be any partial division π of $[a, b]$ with u and $v \in X$. Let $\varepsilon > 0$ and N be the number of intervals in π. Then for each $[u, v]$ in π there exist $s, t \in [u, v]$ such that

$$
\begin{aligned}
\omega(F; [u, v]) &\leq |F(s) - F(t)| + \frac{\varepsilon}{N} \\
&\leq |F(s) - F(u)| + |F(t) - F(u)| + \frac{\varepsilon}{N}.
\end{aligned}
$$

Summing over π, we have

$$\sum_{\pi} \omega(F; [u, v]) \leq 2M + \varepsilon.$$

Hence the condition holds.

Conversely, suppose the condition holds. Then there is $M > 0$ such that

$$\sum_{\pi} \omega(F; [u, v]) \leq M$$

for any partial division of $[a, b]$ with u and $v \in X$. Now take any partial division π of $[a, b]$ with u or $v \in X$. Sort the intervals into two classes, one with the left endpoint in X and another with the right endpoint in X. Note that by assumption F is bounded and therefore $|F(x)| \leq M_1$

for all x and for some M_1. Denote the above left endpoints by $a_1 < a_2 < \ldots < a_n$ and the right endpoints by $b_1 < b_2 < \ldots < b_m$. Thus we have

$$\sum_{\pi} |F(u,v)| \leq \sum_{i=1}^{n-1} \omega(F; [a_i, a_{i+1}])$$
$$+ \sum_{i=1}^{m-1} \omega(F; [b_i, b_{i+1}]) + 4M_1,$$

where $4M_1$ on the right side is to cover the last interval in the first class and the first interval in the second class, which were possibly not included in the previous two sums. Consequently,

$$\sum_{\pi} |F(u,v)| \leq 2M + 4M_1$$

whenever π is a partial division of $[a,b]$ with u or $v \in X$. Hence F is $VB^*(X)$.   •

LEMMA 5.3.9   Let F be bounded on $[a,b]$ and $X \subset [a,b]$. If F is $VB^*(X)$ then F is $VB^*(\overline{X})$ where $\overline{X}$ is the closure of X.

Proof   Suppose F is $VB^*(X)$. Then there is $M > 0$ such that for any partial division π of $[a,b]$ with u or $v \in X$

$$\sum_{\pi} |F(u,v)| \leq M.$$

Now take a partial division π of $[a,b]$ with u or $v \in \overline{X}$. We can insert additional points of X into π to form a new division π_1 of $[a,b]$ such that each interval in π_1 has at least one endpoint belonging to X except perhaps the first or the last interval. Since F is bounded on $[a,b]$, we have $|F(x)| \leq M_1$ for all x and for some M_1. Then

$$\sum_{\pi} |F(u,v)| \leq \sum_{\pi_1} |F(u,v)| \leq M + 4M_1$$

and F is $VB^*(\overline{X})$.   •

REMARK 5.3.10   Note that in Lemmas 5.3.8 and 5.3.9 we assume F to be bounded whereas in the earlier lemmas involving AC^* we assume continuity. Similarly, we can define $VB(X)$ and prove a result like Lemma 5.3.6 with AC^* or AC replaced by VB^* or VB, respectively, and with F assumed to be bounded instead of continuous.

THEOREM 5.3.11   *Let F be bounded on $[a, b]$ and $X \subset [a, b]$ be closed. Then F is $AC^*(X)$ if and only if F is $AC(X)$, $VB^*(X)$ and continuous on X.*

Proof   Suppose F is $AC^*(X)$. It is easy to see that F is $AC(X)$ and continuous on X. We shall show that F is $VB^*(X)$.

Given $\varepsilon > 0$ there is $\eta > 0$ such that for any partial division π of $[a, b]$ with u or $v \in X$

$$\sum_\pi |v - u| < \eta \quad \text{implies} \quad \sum_\pi |F(u, v)| < 1.$$

Since X is closed, we write $(a, b) - X = \bigcup_{k=1}^\infty (a_k, b_k)$. Choose an integer N such that $\sum_{k=N+1}^\infty (b_k - a_k) < \eta$. Since F is bounded on $[a, b]$, there is $M > 0$ such that

$$\sum_{k=1}^N \omega(F; [a_k, b_k]) \le M.$$

Next, since $X \subset [a, b]$ and is therefore bounded, there is an integer n such that $n\eta \ge b - a$ and there is a division π_1 of $[a, b]$ such that an interval in π_1 is either one of $[a_k, b_k]$, $k = 1, 2, \dots, N$, or has length less than η. Now take any partial division π of $[a, b]$ with u or $v \in X$. Insert additional points into π and obtain π_2 which is finer than π_1. Then

$$\sum_\pi |F(u, v)| \le n + M.$$

That is, F is $VB^*(X)$.

The sufficiency follows from the proof of Lemma 5.3.2 or Lemma 5.3.6.

●

REMARK 5.3.12   In general, F being $AC^*(X)$ does not imply that F is $VB^*(X)$, even if X is closed. Note that the boundedness condition was used in an essential way in the proof of Theorem 5.3.11. This is also the case in the proof of Lemma 5.3.8 or Lemma 5.3.3. However, when F is defined on $[a, b]$ and $VB^*(X)$ with $X \subset [a, b]$, then F must be bounded on $[a, b]$.

THEOREM 5.3.13   *If f is KH-integrable on $[a, b]$ and its primitive is $VB^*(X)$ where $X \subset [a, b]$, then F is $AC^*(X)$.*

Proof   In view of Lemma 5.3.9, we assume that X is closed. We write

$$(a,b) - X = \bigcup_{k=1}^{\infty} (c_k, d_k),$$

and put $G(x) = F(x)$ when $x \in X$, $G(a) = F(a)$, $G(b) = F(b)$ and linearly otherwise, i.e.

$$G(x) = F(c_k) + \frac{F(d_k) - F(c_k)}{d_k - c_k}(x - c_k) \quad \text{when} \quad x \in [c_k, d_k].$$

We shall show that G is absolutely continuous on $[a,b]$. If so, G is $AC(X)$ and so is F. Then the result follows from Lemma 5.3.6 or Theorem 5.3.11.

Since f is KH-integrable on $[a,b]$, there is $\delta : [a,b] \to R_+$ such that for any δ-fine partition π of $[a,b]$ we have

$$\sum_{\pi} |f(\xi)(v - u) - F(u,v)| < \varepsilon. \tag{5.5}$$

In view of the fact that F is $VB^*(X)$, we have

$$\sum_{k=1}^{\infty} \omega(F; [c_k, d_k]) < +\infty.$$

Given $\varepsilon > 0$, we can find an integer N such that

$$\sum_{k=N+1}^{\infty} \omega(F; [c_k, d_k]) < \varepsilon. \tag{5.6}$$

Next, we shall modify δ as follows. When $\xi \in (c_k, d_k), k = 1, 2, \ldots$, we put $(\xi - \delta(\xi), \xi + \delta(\xi)) \subset (c_k, d_k)$. When $\xi \notin [c_k, d_k]$ for $k = 1, 2, \ldots, N$, put

$$(\xi - \delta(\xi), \xi + \delta(\xi)) \subset (a,b) - \bigcup_{k=1}^{N} [c_k, d_k].$$

When $\xi = c_k, d_k, k = 1, 2, \ldots, N$, put $(\xi - \delta(\xi), \xi + \delta(\xi))$ so that it contains no other points of $c_k, d_k, k = 1, 2, \ldots, N$, except ξ itself.

Further, put $g(x) = f(x)$ when $x \in X$ and $g(x) = G'(x)$ otherwise. Note that when $x \in (c_k, d_k)$ for some k, we have $g(x) = F(c_k, d_k)/(d_k - c_k)$. Now we shall prove that g is KH-integrable on $[a,b]$.

Take any δ-fine partition π of $[a,b]$. We write $\pi = \pi_1 \cup \pi_2 \cup \pi_3$ in which π_1 is a partition of the union of $[c_k, d_k], k = 1, 2, \ldots, N, \pi_2$ contains those

$(\xi, [u, v])$ with $\xi \in X$ and not in π_1, and π_3 those in π but not in π_1 or π_2. Then by 5.5 and 5.6 we obtain

$$\left| \sum_{\pi} g(\xi)(v - u) - F(a, b) \right| \leq \left| \sum_{\pi_1} g(\xi)(v - u) - G(u, v) \right|$$
$$+ \sum_{\pi_2} |f(\xi)(v - u) - F(u, v)|$$
$$+ \sum_{\pi_3} |g(\xi)(v - u)| + \sum_{\pi_3} |F(u, v)|$$

$$< \varepsilon + \varepsilon + \varepsilon = 3\varepsilon.$$

Note that $F(c_k, d_k) = G(c_k, d_k)$ for all k and the first sum $\sum_{\pi_1}$ above is 0. Hence g is KH-integrable on $[a, b]$.

It is easy to see that G is of bounded variation on $[a, b]$. Therefore, by Theorems 3.4.1 and 3.6.4 the function G is absolutely continuous on $[a, b]$. •

Recalling Definition 3.13.1 which says that function f is KH-integrable on X if $f\mathbf{1}_X$ is KH-integrable on $[a, b]$ we obtain an extension of Theorem 3.4.1.

THEOREM 5.3.14 *If f is KH-integrable on $[a, b]$ and its primitive is $VB^*(X)$ where $X \subset [a, b]$, then f is absolutely KH-integrable on X.*

5.4 *ACG** functions

In this section, we shall characterize the primitive of a KH-integrable function and give a descriptive definition of the KH-integral. As an application of the descriptive definition, we prove a convergence theorem. Other convergence theorems will be discussed in Section 5.5.

DEFINITION 5.4.1 *A function F is said to be ACG^* on X if X is the union of $X_i, i = 1, 2, \ldots$, such that F is $AC^*(X_i)$ for each i.*

EXAMPLE 5.4.2 We shall show that a function F which is differentiable everywhere in $[a, b]$ is ACG^* on $[a, b]$. Let X_n denote the set of all points $x \in [a, b]$ such that

$$|F(t) - F(x)| \leq n|t - x| \quad \text{whenever} \quad |t - x| \leq \frac{1}{n}.$$

Obviously, $[a, b]$ is the union of X_n, $n = 1, 2, \ldots$. Given $\varepsilon > 0$, choose $\eta > 0$ such that $\eta < 1/n$ and $n\eta < \varepsilon$. Then for any partial division π with u or $v \in X_n$ satisfying $\sum_\pi |v - u| < \eta$ we have

$$\sum_\pi |F(v) - F(u)| \leq n\eta < \varepsilon.$$

That is, F is $AC^*(X_n)$. Consequently, F is ACG^* on $[a, b]$.

In particular, the example of F given in Example 5.1.1 is ACG^* on $[0, 1]$ though not absolutely continuous there.

Other concepts in Section 5.3 can be extended similarly, as shown below.

DEFINITION 5.4.3   *A function F is said to be ACG on X if X is the union of $X_i, i = 1, 2, \ldots$, such that F is $AC(X_i)$ for each i.*

DEFINITION 5.4.4   *A function F is said to be VBG^* on X if X is the union of X_i, $i = 1, 2, \ldots$, such that F is $VB^*(X_i)$ for each i.*

THEOREM 5.4.5   *A function F is ACG^* on $[a, b]$ if and only if it is ACG, VBG^* and continuous on $[a, b]$.*

Proof   This follows from Theorem 5.3.11.   •

THEOREM 5.4.6   *If f is KH-integrable on $[a, b]$ and its primitive F is VBG^* on $[a, b]$, then F is ACG^* on $[a, b]$.*

Proof   This follows from Theorem 5.3.13.   •

THEOREM 5.4.7   *If f is KH-integrable on $[a, b]$, then its primitive F is VBG^* on $[a, b]$.*

Proof   By Henstock's lemma, for every $\varepsilon > 0$ there exists $\delta : [a, b] \to R_+$ such that for any δ-fine partition π

$$\sum_\pi |f(x)(v - u) - F(u, v)| < \varepsilon$$

where F is the primitive of f and $F(u, v) = F(v) - F(u)$. Let X_i denote the set of all $x \in [a, b]$ such that

$$|f(x)| \leq i \quad \text{and} \quad \delta(x) > 1/i.$$

Next, we partition X_i into a finite number of subsets X_{ik}, $k = 1, 2, \ldots, p$, such that $|X_{ik}| < 1/i$ for each k. Now take any partial partition π with all tags $x \in X_{ik}$. Denote by π_1 the two intervals in π at both ends and by π_2 all other intervals. Note that F is continuous on $[a, b]$ and therefore bounded, i.e. $|F(x)| \le M$ for all $x \in [a, b]$ and for some M. Also, each interval $[u, v]$ in π_2 satisfies $|v - u| < 1/i$ and therefore is δ-fine. Hence we have

$$
\begin{aligned}
\sum_{\pi} |F(u, v)| &\le \sum_{\pi_1} |F(u, v)| + \sum_{\pi_2} |F(u, v) - f(x)(v - u)| \\
&\quad + \sum_{\pi_2} |f(x)(v - u)| \\
&\le 4M + \varepsilon + i(b - a).
\end{aligned}
$$

It follows that F is $VB^*(X_{ik})$ and consequently VBG^* on $[a, b]$.   •

LEMMA 5.4.8 *If F is ACG^* then it is SL.*

Proof It is sufficient to show that F is SL on a set on which it is AC^*. This is very similar to the proof that F is SL if it is AC.   •

THEOREM 5.4.9   *A function f is KH-integrable on $[a, b]$ with the primitive F if and only if F is an ACG^* function such that $F'(x)$ exists and $F'(x) = f(x)$ for almost all $x \in [a, b]$.*

Proof Suppose f is KH-integrable on $[a, b]$. It follows from Theorems 5.4.7 and 5.4.6 that its primitive F is ACG^*. By Theorem 3.8.2, the derivative $F'(x)$ exists and $F'(x) = f(x)$ almost everywhere.

Conversely, suppose F is ACG^* on $[a, b]$ and $F'(x) = f(x)$ for $x \in [a, b] - E$ where $|E| = 0$. It is easy to see that if F is ACG^* on $[a, b]$ then it has the strong Luzin condition. Hence the KH-integrability of f on $[a, b]$ follows from Theorem 3.9.1.   •

DEFINITION 5.4.10   *A sequence $\{F_n\}$ of functions defined on $[a, b]$ is said to be $UAC^*(X)$ where $X \subset [a, b]$ if for every $\varepsilon > 0$ there exists $\eta > 0$, independent of n, such that for any partial division π of $[a, b]$ with u or $v \in X$*

$$
\sum_{\pi} |v - u| < \eta \text{ implies } \sum_{\pi} |F_n(u, v)| < \varepsilon
$$

for all n. Further, $\{F_n\}$ is $UACG^*$ if $[a, b] = \bigcup_{i=1}^{\infty} X_i$ such that $\{F_n\}$ is $UAC^*(X_i)$ for each i.

THEOREM 5.4.11   *If the following conditions are satisfied:*

(i) $f_n(x) \to f(x)$ almost everywhere in $[a, b]$ where each f_n is KH-integrable on $[a, b]$,

(ii) the primitives F_n of f_n are $UACG^*$,

(iii) F_n converges to a function F uniformly on $[a, b]$,

then f is KH-integrable on $[a, b]$ with the primitive F. Furthermore,

$$\int_a^b f_n(x)dx \to \int_a^b f(x)dx \quad \text{as } n \to \infty.$$

Proof   In view of (ii) and (iii), F is ACG^* on $[a, b]$. By the previous theorem, it remains to prove that $F'(x) = f(x)$ almost everywhere. Since F_n are $UACG^*$, we have $[a, b] = \cup_i X_i$ such that F_n are $UAC^*(X_i)$ for each i. We may also assume that each X_i is closed.

Write $X = X_i$. Suppose F_n are $UAC^*(X)$ and X is closed. Define $G_n(x) = F_n(x)$ when $x \in X$, $G_n(a) = F_n(a)$, $G_n(b) = F_n(b)$ and linearly otherwise (as in the proof of Theorem 5.3.13). Further, define $g_n(x) = f_n(x)$ when $x \in X$ and $g_n(x) = G_n'(x)$ when $x \notin X$. Note that G_n are uniformly absolutely continuous on $[a, b]$ and $G_n'(x) = g_n(x)$ almost everywhere. Similarly, define $G(x) = F(x)$ when $x \in X$, $G(a) = F(a)$, $G(b) = F(b)$, and linearly otherwise. Also, define $g(x) = f(x)$ when $x \in X$ and $g(x) = G'(x)$ when $x \notin X$. Obviously, $G_n(x) \to G(x)$ for all x and $g_n(x) \to g(x)$ for almost all $x \in [a, b]$.

Apply Theorem 5.2.1 and we obtain that g is KH-integrable on $[a, b]$ and $f(x) = g(x) = G'(x) = F'(x)$ for almost all $x \in X$. Hence the theorem is proved.   ●

REMARK 5.4.12   We see that the sequence $\{f_n\}$ in Example 5.1.2 satisfies the conditions in Theorem 5.4.11, and therefore the theorem applies. We shall show in the next section that condition (iii) in Theorem 5.4.11 is redundant.

5.5 Controlled convergence

First, we prove the Controlled Convergence Theorem (Theorem 5.5.2). Then we show that other convergence theorems follow. Making the Controlled Convergence Theorem a starting point is to a degree a matter of taste.

LEMMA 5.5.1   If $\{F_n\}$ is uniformly bounded and equicontinuous on $[a, b]$ then it has a uniformly convergent subsequence on $[a, b]$.

Proof Since $\{F_n\}$ is uniformly bounded on $[a,b]$, there is $M > 0$ such that $|F_n(x)| \le M$ for $x \in [a,b]$ and for all n. Consider the rational numbers in $[a,b]$. Since they are countable, we can write $\{r_1, r_2, r_3, \dots\}$. We shall use the diagonal process to select a subsequence of $\{F_n\}$ which is convergent at each r_i.

First, the sequence $\{F_n(r_1)\}$ is bounded and every bounded sequence has a convergent subsequence. Then there is a subsequence $\{F_{1n}(r_1)\}$ such that the subsequence converges. Next, consider $\{F_{1n}(r_2)\}$ which is also bounded. Again, there is a convergent subsequence $\{F_{2n}(r_2)\}$ of $\{F_{1n}(r_2)\}$, and so on. Consequently, we have

$$\{F_{1n}\}, \{F_{2n}\}, \{F_{3n}\}, \dots$$

each of which is a subsequence of the preceding sequence, and $F_{in}(r_i)$ converges as $n \to \infty$ for each i. Now choose $\{F_{nn}\}$ which is a subsequence of $\{F_n\}$, and indeed $\{F_{nn}\}_{n\ge i}$ is a subsequence of $\{F_{in}\}_{n\ge 1}$ for each i. Consequently, $\{F_{nn}(x)\}$ converges for every rational x.

Next, $\{F_n\}$ is equicontinuous on $[a,b]$, i.e., for every $\varepsilon > 0$ there exists $\eta > 0$ such that

$$|F_n(x) - F_n(y)| < \frac{\varepsilon}{3} \quad \text{whenever} \quad |x-y| < \eta$$

and for all n. Take a division of $[a,b]$ as follows:

$$a = x_0 < x_1 < x_2 < \cdots < x_p < x_{p+1} = b$$

such that $x_1, x_2, \dots, x_p$ are rational and $x_i - x_{i-1} < \eta$ for $i = 1, \dots, p+1$. By what we have previously proved, there is an integer N such that

$$|F_{nn}(x_i) - F_{mm}(x_i)| < \frac{\varepsilon}{3} \quad \text{whenever} \quad n, m \ge N$$

and for $i = 1, 2, \dots, p$. Take any $x \in [a,b]$ and $|x - x_i| < \eta$ for some i. Hence for $n, m \ge N$

$$
\begin{aligned}
|F_{nn}(x) - F_{mm}(x)| &\le |F_{nn}(x) - F_{nn}(x_i)| \\
&\quad + |F_{nn}(x_i) - F_{mm}(x_i)| + |F_{mm}(x_i) - F_{mm}(x)| \\
&< \frac{\varepsilon}{3} + \frac{\varepsilon}{3} + \frac{\varepsilon}{3} = \varepsilon.
\end{aligned}
$$

That is, $\{F_{nn}\}$ converges uniformly on $[a,b]$. •

We remark that Lemma 5.5.1 is known as Ascoli's theorem. The process of selecting the sequence $\{F_{nn}\}$ is often referred to as a diagonal process.

THEOREM 5.5.2 (Controlled Convergence Theorem) *If the following conditions are satisfied:*

(i) $f_n(x) \to f(x)$ *almost everywhere in* $[a, b]$ *where each* f_n *is KH-integrable on* $[a, b]$,

(ii) *the primitives* F_n *of* f_n *are* $UACG^*$,

then f *is KH-integrable on* $[a, b]$ *and*

$$\int_a^b f_n(x)dx \to \int_a^b f(x)dx \quad \text{as} \quad n \to \infty.$$

Proof By the definition of $UACG^*$, the family $\{F_n; \ n = 1, 2, \dots\}$ is equicontinuous on $[a, b]$. That is, for every $\varepsilon > 0$ there exists $\eta > 0$ such that

$$|F_n(x) - F_n(y)| < \varepsilon \quad \text{whenever} \quad |x - y| < \eta.$$

and for all n. We may assume $F_n(a) = 0$ for all n. Take a division of $[a, b]$

$$a = x_0 < x_1 < \cdots < x_m = b$$

such that $x_i - x_{i-1} < \eta$ for $i = 1, 2, \dots, m$. For any $x \in [a, b]$ and any n we have $|x - x_i| < \eta$ for some i. Then

$$\begin{aligned}
|F_n(x)| &\leq |F_n(x) - F_n(x_i)| + \sum_{k=1}^{i} |F_n(x_k) - F_n(x_{k-1})| \\
&< \varepsilon + i\varepsilon \leq (m+1)\varepsilon.
\end{aligned}$$

Therefore $\{F_n; \ n = 1, 2, \dots\}$ is uniformly bounded on $[a, b]$.

Now we apply Lemma 5.5.1 and obtain a subsequence $\{F_{n(i)}\}$ of $\{F_n\}$ such that the subsequence converges uniformly on $[a, b]$. Hence by Theorem 5.4.11 the function f is KH-integrable on $[a, b]$ and

$$\int_a^b f_{n(i)}(x)dx \to \int_a^b f(x)dx \quad \text{as} \quad n(i) \to \infty.$$

Since every subsequence of $\{f_n\}$ has a sub-subsequence satisfying the above property, the sequence itself has the property.   •

The validity of the next corollary was established as a part of Theorem 2.9.3. This time the proof is easy.

COROLLARY 5.5.3   *If* f *is KH-integrable on* $[x, b]$ *for every* $x \in (a, b)$ *and*

$$\lim_{x \to a} \int_x^b f(t)dt = A \quad \text{exists },$$

then f is KH-integrable to A on $[a, b]$.

Proof   Put $f_n(x) = f(x)$ when $x \in [a + \frac{b-a}{n+1}, b]$ and 0 otherwise, and apply Theorem 5.5.2.   •

COROLLARY 5.5.4   *Let X be a closed set and the intervals (a_k, b_k) pairwise disjoint with $(a, b) - X = \bigcup_{k=1}^{\infty}(a_k, b_k)$. If f is KH-integrable on X and for F_k, the primitives of f on $[a_k, b_k]$, we have*

$$\sum_{k=1}^{\infty} \omega(F_k; [a_k, b_k]) < +\infty$$

then f is KH-integrable on $[a, b]$ and

$$\int_a^b f(x)dx = \int_X f(x)dx + \sum_{k=1}^{\infty} \int_{a_k}^{b_k} f(x)dx.$$

Proof   Put $f_n(x) = f(x)$ when $x \in X$ or when $x \in [a_k, b_k]$ for $k = 1, 2, \dots, n$, and 0 otherwise. Again, we apply Theorem 5.5.2 and obtain the result.   •

THEOREM 5.5.5   *If the following conditions are satisfied:*

(i)   *$f_n(x) \to f(x)$ almost everywhere in $[a, b]$ where each f_n is KH-integrable on $[a, b]$,*

(ii)   *there exist closed sets X_i with $\bigcup_{i=1}^{\infty} X_i = [a, b]$ and functions G_i, H_i being $VB^*(X_i)$ for $i = 1, 2, \dots$ such that whenever u or $v \in X_i$ we have*

$$G_i(u, v) \leq F_n(u, v) \leq H_i(u, v)$$

for all n, where F_n are the primitives of f_n on $[a, b]$,

(iii)   *for each i we have $G_i'(x) \leq f_n(x) \leq H_i'(x)$ for almost all x in X_i and for all n,*

(iv)   *F_n converges uniformly on $[a, b]$,*

then f is KH-integrable on $[a, b]$ and

$$\int_a^b f_n(x)dx \to \int_a^b f(x)dx \text{ as } n \to \infty.$$

Proof   We shall prove that F_n are $UACG^*$ on $[a, b]$. If so, then the result follows from the Controlled Convergence Theorem (Theorem 5.5.2).

Write $X = X_i$. Following the proof of Theorem 5.3.11, we can show that F_n are $UAC^*(X)$ if and only if F_n are $UAC(X)$, $UVB^*(X)$ and equicontinuous on X. In view of (ii), F_n are $UVB^*(X)$. The equicontinuity of $\{F_n\}$ follows from (iv). Therefore it remains to show that F_n are $UAC(X)$. Let $(a,b) - X = \bigcup_{k=1}^{\infty}(a_k,b_k)$.

Since F_n are $UVB^*(X)$, we obtain that

$$\sum_{k=1}^{\infty}\omega(F_n;[a_k,b_k]) \quad \text{converges uniformly in} \quad n$$

Applying Corollary 5.5.4, we obtain for any $[u,v] \subset [a,b]$ with $u,v \in X$

$$\int_u^v f_n(x)dx = \int_{[u,v]\cap X} f_n(x)dx + \sum_{k=1}^{\infty}\int_{[u,v]\cap[a_k,b_k]} f_n(x)dx.$$

In view of (iii), for every $\varepsilon > 0$ there exists $\eta_1 > 0$, independent of n, such that whenever $E \subset X$ and $|E| < \eta_1$ we have

$$\int_E |f_n(x)|dx < \varepsilon.$$

Choose an integer N such that

$$\sum_{k=N+1}^{\infty}\omega(F;[a_k,b_k]) < \varepsilon.$$

Put $\eta \le \eta_1$ and $\eta < \text{Min}\{b_k - a_k; k=1,2,\ldots,N\}$. Then for any partial division π of $[a,b]$ with $u,v \in X$ and $\sum_\pi |v-u| < \eta$ we have

$$E = \bigcup_\pi [u,v] \quad \text{and} \quad |E \cap X| \le |E| < \eta.$$

Consequently,

$$\sum_\pi |F_n(u,v)| \le \int_{E\cap X} |f_n(x)|dx$$
$$+ \sum_{k=N+1}^{\infty}\omega(F;[a_k,b_k])$$
$$< 2\varepsilon.$$

Hence F_n are $UAC(X)$ and the proof is complete. $\bullet$

THEOREM 5.5.6   *If the following conditions are satisfied:*

(i) $f_n(x) \to f(x)$ *almost everywhere in* $[a,b]$ *where each* f_n *is KH-integrable on* $[a,b]$,

(ii) *there exist closed sets X_i such that $\bigcup_{i=1}^{\infty} X_i = [a,b]$ and for every i and $\varepsilon > 0$ there is an integer N such that for any partial division π of $[a,b]$ with u or $v \in X_i$*

$$\sum_{\pi} |F_n(u,v) - F_m(u,v)| < \varepsilon \quad \text{whenever} \quad n, m \geq N,$$

(iii) *F_n converges uniformly on $[a,b]$,*

then f is KH-integrable on $[a,b]$ and

$$\int_a^b f_n(x)dx \to \int_a^b f(x)dx \quad \text{as} \quad n \to \infty.$$

Proof   It suffices to show that the conditions in Theorem 5.5.5 hold. Now suppose (ii) of the present theorem holds. We may assume that F_N is $VB^*(X_i)$. If not, we can write $X_i = \cup_{k=1}^{\infty} Y_k$ where each Y_k is closed and then consider Y_k in place of X_i. Take any partial division π of $[a,b]$ with u or $v \in X_i$ and we have

$$\sum_{\pi} |F_n(u,v)| \leq \sum_{\pi} |F_n(u,v) - F_N(u,v)|$$
$$+ \sum_{\pi} |F_N(u,v)|$$
$$\leq \varepsilon + M$$

for some $M > 0$ and for all $n \geq N$. Hence $F_n, n = N, N+1, \ldots$, and consequently $F_n, n = 1, 2, \ldots$, are $UVB^*(X_i)$ for each i. Therefore a subsequence of $\{F_n\}$ satisfies condition (ii) of Theorem 5.5.5. For convenience, we may assume that the sequence $\{F_n\}$ itself satisfies the condition.

Next, put $G_n(x) = F_n(x)$ when $x \in X_i$, $G_n(a) = F_n(a)$, $G_n(b) = F_n(b)$, and linearly otherwise (as in the proof of Theorem 5.3.13). Write $g_n(x) = G_n'(x)$ almost everywhere. Then it follows from (ii) and Theorem 3.4.1 that

$$\int_a^b |g_n(x) - g_m(x)|dx \leq \varepsilon \quad \text{whenever} \quad n, m \geq N.$$

There is a subsequence $\{g_{n_j}\}$ of $\{g_n\}$ such that

$$h(x) = \sum_{j=1}^{\infty} |g_{n_{j+1}}(x) - g_{n_j}(x)|$$

is KH-integrable on $[a,b]$. Hence

$$g_{n_1}(x) - h(x) \leq g_{n_j}(x) \leq g_{n_1}(x) + h(x)$$

for almost all $x \in [a, b]$, or

$$g_{n_1}(x) - h(x) \leq f_{n_j}(x) \leq g_{n_1}(x) + h(x)$$

for almost all $x \in X_i$ and for all n_j. That is, condition (iii) of Theorem 5.5.5 holds for the subsequence $\{f_{n_j}\}$.

Now apply Theorem 5.5.5 and obtain that f is KH-integrable on $[a, b]$ and

$$\int_a^b f_{n_j}(x)dx \rightarrow \int_a^b f(x)dx \quad \text{as} \quad n_j \rightarrow \infty.$$

Since every subsequence of $\{f_n\}$ has a sub-subsequence satisfying the above property, the sequence $\{f_n\}$ itself satisfies the property. Hence the theorem is proved.   •

THEOREM 5.5.7   *If conditions (ii) and (iii) of Theorem 5.5.6 hold, then there is a subsequence $\{g_n\}$ of $\{f_n\}$ such that $g_n(x) \rightarrow f(x)$ almost everywhere in $[a, b]$. Furthermore, f is KH-integrable on $[a, b]$ and*

$$\int_a^b f_n(x)dx \rightarrow \int_a^b f(x)dx \quad \text{as} \quad n \rightarrow \infty.$$

Proof   Following the proof of Theorem 5.5.6, we obtain for each i that there is a subsequence $\{f_{in}\}$ of $\{f_n\}$ such that

$$f_{in}(x) \rightarrow f(x) \quad \text{as} \quad n \rightarrow \infty$$

for almost all $x \in X_i$. However, we may choose $\{f_{1n}\}$ first, then choose $\{f_{2n}\}$ as a subsequence of $\{f_{1n}\}$, and so on. More precisely, we obtain

$$\{f_{1n}\}, \{f_{2n}\}, \cdots$$

each of which is a subsequence of the preceding sequence. Finally, we choose $\{f_{nn}\}$ which is the required sequence since

$$f_{nn}(x) \rightarrow f(x) \quad \text{as} \quad n \rightarrow \infty$$

for almost all x in $[a, b]$. The rest is a consequence of Theorem 5.5.5.   •

We remark that Corollary 5.5.3 is known as the Cauchy extension of the KH-integral, Corollary 5.5.4 is the Harnack extension. We call Theorem 5.5.5 the generalized dominated convergence theorem and Theorem 5.5.6 the generalized mean convergence theorem.

In fact, in condition (ii) of Theorem 5.5.5 it is enough to consider only δ-fine partial partition π of E with tags in X_i.

In the classical approach, one proves first the Cauchy extension and the Harnack extension. Then one uses these to prove the controlled convergence theorem by means of the Baire category theorem.† However, here we have deduced the two extensions, perhaps in a more elementary fashion, as corollaries of the Controlled Convergence Theorem.

EXAMPLE 5.5.8   Let $F_n(x) = \sin 2n\pi x$ when $0 \leq x \leq 1/n$ and 0 elsewhere and $f_n(x) = F'_n(x)$. Then the sequence $\{f_n\}$ satisfies the conditions in Theorem 5.5.5 but not those in Theorem 4.4.11 with $f(x) = 0$ for all x. Note that $F_n(x) \to 0$ as $n \to \infty$ for every $x \in [0,1]$ but not uniformly on $[0,1]$. This example is due to Liao Kecheng.

EXAMPLE 5.5.9   Consider the sequence $\{f_n\}$ in Example 5.1.1, i.e. $f_n(x) = F'(x)$ for $1/n \leq x \leq 1$ and 0 elsewhere in $[0,1]$, where $F(0) = 0$ and

$$F(x) = x^2 \sin x^{-2} \quad \text{when} \quad x \neq 0.$$

Define $H(x) = F(x)$ when $F'(x) \geq 0$ and 0 elsewhere in $[0,1]$, and also $G(x) = F(x)$ when $F'(x) \leq 0$ and 0 elsewhere in $[0,1]$. Then conditions (i), (iii) and (iv) of Theorem 5.5.5 are satisfied with $H_i = H$ and $G_i = G$ for all i but not (ii). In other words, we must employ a countable number of functions H_i and G_i in (ii) if we wish to apply Theorem 5.5.5 in this case.

5.6 Exercise

EXERCISE 5.1 *Show that if F is absolutely continuous on $[a,b]$ then F is $AC^*(X)$ for every closed set $X \subset [a,b]$. Give an example of a function F which is $AC(X)$ but not $AC^*(X)$ where X is a closed subset of $[a,b]$.*

EXERCISE 5.2 *Show that F is absolutely continuous on $[a,b]$ if and only if for every $\varepsilon > 0$ there exists $\eta > 0$ such that for every sequence $\{[a_i, b_i]\}$ of non-overlapping intervals with $\sum_{i=1}^{\infty} |b_i - a_i| < \eta$ we have $\sum_{i=1}^{\infty} |F(a_i, b_i)| < \varepsilon$.*

EXERCISE 5.3 *Is the continuity condition on F necessary in Lemma 5.3.3? Similarly, is the boundedness condition on F necessary in Lemma 5.3.9?*

† which can be found in [4] p. 51.

EXERCISE 5.4 ⓘ*Suppose for every positive ϵ there exist functions M_P and m_P which satisfy the following conditions:*
(a) M_P is ACG^* on $[a,b]$ and $f(x) \leq \underline{D}M_P(x)$ for almost all $x \in [a,b]$,
(a) m_P is ACG^* on $[a,b]$ and $f(x) \geq \overline{D}m_P(x)$ for almost all $x \in [a,b]$,
(c) $M_P(b) - M_P(a) - [m_P(b) - m_P(a)] < \epsilon$.
Show that f is Perron integrable on $[a,b]$.

EXERCISE 5.5 *It is known that each of the following conditions implies the next one: (a) f is absolutely continuous on $[a,b]$; (b) f is ACG^* on $[a,b]$; (c) f is SL on $[a,b]$; and (d) f satisfies the Luzin condition (condition N). Show that each converse is not true.*

EXERCISE 5.6 *State and prove a version of the controlled convergence theorem for an infinite interval $[A, B]$.*

EXERCISE 5.7 ⓘⓘ *A function f is KH-integrable on $[a,b]$ if and only if there is a sequence $\{\varphi_n\}$ of step functions such that $\varphi_n \to f$ almost everywhere and the primitives Φ_n of φ_n are $UACG^*$. This is known as the Riesz-type definition of the KH-integral. [Hint: Suppose the primitive F of f is $AC^*(X_i)$ for each i with X_i closed and their union $[a,b]$. Put $F_n(x) = F(x)$ when $x \in X_1 \cup \cdots \cup X_n$ and linearly otherwise (as in the proof of Theorem 5.3.13), and $f_n(x) = F'_n(x)$ almost everywhere. Show that $\{f_n\}$ or a subsequence of it satisfies the conditions. Finally choose step functions φ_n such that $\int_a^b |f_n - \varphi_n| < 2^{-n}$ and show that $\{\varphi_n\}$ satisfies the conditions.]*

EXERCISE 5.8 *Show that if the conditions in Theorem 5.5.2 are satisfied then so is condition (ii) in Theorem 5.5.6. In words, it says roughly that controlled convergence implies generalized mean convergence. [Hint: Use the linearization technique in the proof of Theorem 5.4.11 then apply Theorem 5.2.1.]*

EXERCISE 5.9 *Prove Exercise 4.17 without the condition that f is non-negative. [Hint: Following the hint of Exercise 5.7 we construct a sequence $\{X_n\}$ of closed sets such that $f_n = f1_{X_n}$ is absolutely KH-integrable with primitive F_n and the primitives F_n are $UACG^*$. Then F_n satisfies condition (ii) of Theorem 5.5.6 and there is an H_1-integrable function g_n such that $f_n = g_n$ almost everywhere. Define δ as in the hint of Exercise 4.16 and apply the generalized mean convergence theorem in place of $I - F_N(a,b) < \varepsilon$.]*

EXERCISE 5.10 If f is KH-integrable on $[a, b]$ then there is a sequence $\{X_n\}$ of closed subsets of $[a, b]$ such that $X_n \subset X_{n+1}$ for all n, the complement $[a, b] - \bigcup_{n=1}^{\infty} X_n$ is of measure zero, f is absolutely integrable on each X_n, and

$$\lim_{n \to \infty} \int_{X_n} f = \int_a^b f.$$

[Hint: Consider $a_n = \int_{A_n} f$ and $b_n = \int_{B_n} f$ where $A_n = \{x : n - 1 \leq f(x) < n\}$ and $B_n = \{x : -n \leq f(x) < -n + 1\}$ for $n = 1, 2, \cdots$. If the integral of f is I, then we can arrange a_n and b_n in such a way that their sum is conditionally convergent to I.]

EXERCISE 5.11 ⓧⓧ Let f be KH-integrable on each measurable set X_n with $\bigcup_{n=1}^{\infty} X_n = [a, b]$ and

$$\lim_{n \to \infty} \int_{X_n} f = I.$$

Then f is KH-integrable to I on $[a, b]$ if and only if the following (LG) condition holds: for every positive ϵ there is a positive integer N such that for any $n \geq N$ there is $\delta_n : [a, b] \to \mathbb{R}_+$ satisfying the condition that for any δ_n-fine partition π we have

$$\left| \sum_{\pi^*} f(\xi)(v - u) \right| < \epsilon,$$

where π^* is the part of π having tags not in X_n. [Hint: Write $I_n = \int_{X_n} f$. Then the (LG) condition follows from the inequality

$$\left| \sum_{\pi^*} f \right| \leq \left| \sum_{\pi} f - I \right| + |I - I_n| + \left| I_n - \sum_{\pi - \pi^*} f \right|.$$

A re-arrangement of the inequality gives the converse.]

EXERCISE 5.12 If the following conditions are satisfied:
(i) $f_n \to f$ almost everywhere and each f_n is KH-integrable on $[a, b]$,
(ii) the primitives F_n of f_n satisfy the conditions that $[a, b]$ is the union of closed sets X_i, $i = 1, 2, \ldots$, and that for every i and for every $\epsilon > 0$ there exists $\eta > 0$ (depending on i and ϵ but independent of n) such that for any partial division π of $[a, b]$ with u and $v \in X_i$

$$\sum_{\pi} |v - u| < \eta \text{ implies } \sum_{\pi} \omega(F_n; [u, v]) < \epsilon,$$

(iii) $F_n(x)$ converges to a continuous function $F(x)$ for every $x \in [a, b]$, then f is KH-integrable on $[a, b]$ with the primitive F. Furthermore,

$$\int_a^b f_n \to \int_a^b f \text{ as } n \to \infty.$$

[Hint: Follow the proof of Theorem 5.4.11.]

EXERCISE 5.13 If condition (iii) in Exercise 5.12 is replaced by (iv) $F_n(x)$ converges to $F(x)$ nearly everywhere, i.e. everywhere except for a countable number of points, where the limit function F is continuous nearly eveywhere, and the series

$$\sum_{x \in (a,b)} |F(x+) - F(x-)| + |F(a+) - F(a)| + |F(b) - F(b-)|$$

is finite, then f is KH-integrable on $[a, b]$ and for nearly all $x \in [a, b]$

$$\int_a^x f_n \to \int_a^x f + \sum_{y < x} \{F(y+) - F(y-)\}.$$

⑦Exercises 5.14 to 5.18 solve the multiplier problem for the KH-integral. A multiplier is defined in Exercise 5.18.

EXERCISE 5.14 ⚠ Let $a_n > 0$ and $s_n = a_1 + \cdots + a_n$ for each n. If $\sum_{n=1}^{\infty} a_n$ is divergent, then so is $\sum_{n=1}^{\infty} a_n/s_n$ (Abel and Dini 1867). [Hint: $a_{n+1}/s_{n+1} + \cdots + a_{n+k}/s_{n+k} \geq (a_{n+1} + \cdots + a_{n+k})/s_{n+k} = 1 - s_n/s_{n+k}$ and $s_n/s_{n+k} < 1/2$ for all n and for some large k depending on n.]

EXERCISE 5.15 ⚠ Show that if g is bounded but not of bounded variation on $[a, b]$ then there exist a point $c \in [a, b]$ and an increasing (or decreasing) sequence $\{x_n\}$ with the limit c such that $\sum_{n=1}^{\infty} (M_n - m_n)$ diverges, where

$$M_n = \sup \{g(x); x_n \leq x \leq x_{n+1}\} \text{ and } m_n = \inf \{g(x); x_n \leq x \leq x_{n+1}\}.$$

EXERCISE 5.16 ⚠ Suppose g is bounded but not of bounded variation on $[a, b]$ and $\{x_n\}$, M_n and m_n are defined as in the previous exercise. Let $f(x) = p_n$ when $x \in X_n$, $f(x) = -p_n$ when $x \in Y_n$ for $n = 1, 2, \ldots$ and $f(x) = 0$ elsewhere, where X_n, Y_n, and p_n satisfy the following conditions for each n:

$$X_n, \ Y_n \subset [x_n, x_{n+1}] \text{ and } m(X_n) = m(Y_n) = \delta_n > 0;$$

$$g(x) \geq \frac{3}{4}M_n + \frac{1}{4}m_n \text{ when } x \in X_n, \ g(x) \leq \frac{1}{4}M_n + \frac{3}{4}m_n \text{ when } x \in Y_n;$$

$$p_n = [\delta_n \sum_{i=1}^{n}(M_i - m_i)]^{-1}.$$

Show that $|\int_a^u f| \leq 2p_n\delta_n$ when $x_n \leq u < x_{n+1}$, and

$$\int_{x_n}^{x_{n+1}} fg \geq \frac{1}{2}(M_n - m_n)/\sum_{i=1}^{n}(M_n - m_n).$$

EXERCISE 5.17 ① *Show that if fg is KH-integrable on $[a,b]$ for all absolutely KH-integrable functions f, then g is bounded almost everywhere.*

EXERCISE 5.18 ① *Show that if fg is KH-integrable on $[a,b]$ for all KH-integrable functions f, then g is almost everywhere equal to a function of bounded variation on $[a,b]$. The above function g is sometimes called a multiplier. [Hint: Suppose g is bounded and not of bounded variation. Construct f as in the previous exercises and show that f is KH-integrable but fg is not.]*

6

Integration in Several Dimensions

6.1 Introduction

In practice there is a need for integration in n dimensions. Since we live in a three-dimensional world there are more applications of two- and three-dimensional integrals than of one-dimensional ones. In generalizing results from Chapters 2 and 3 we want to create a theory which also covers integration over infinite intervals and in doing so we have a choice between analogues of Definitions 2.9.1 and 2.10.1. We employ the latter. Most of the theory is very similar and we shall not repeat almost identical arguments, leaving it to the reader to make the necessary modifications, if any are needed. Naturally there are differences; the first one concerns the existence of δ-fine partitions for possibly infinite intervals, which we prove in Section 6.2. The definition of the KH-integral in n dimensions and the immediate consequences of the definition are dealt with in Section 6.3. Theorems which are easily generalized to n dimensions are collected in Section 6.4 whereas theorems which need adjustment either in the formulation or in the proof are dealt with in Section 6.5. The main difference between integration in one and in several dimensions is the absence of a theorem like the Fundamental Theorem which allows direct systematic evaluation of integrals; in several dimensions we rely on successive repeated one-dimensional integrations. This topic is covered by the renowned theorems of Fubini and Tonelli in Section 6.6. There is an analogue of the Fundamental Theorem, namely the theorem in Exercise 6.6, but it offers little help for evaluation of integrals. Non-absolute convergence is a far more complex phenomenon in several dimensions than in one and absolute integrability becomes even more important. This comes to the foreground in Section 6.7 dealing with change of variables.

6.1.1 Sets in $\mathbb{R}^n$

We start by reminding ourselves of some facts and notation from elementary set theory. If A_1, A_2 ... , A_n are sets of whatever kind then $A_1 \times A_2 \times \cdots \times A_n$ is the set of n-tuples $(a_1, a_2, \dots , a_n)$ with $a_i \in A_i$ for $i = 1, 2, \dots , n$. The set $A_1 \times A_2 \times \cdots \times A_n$ is called the cartesian product of the sets A_1, $A_2, \dots , A_n$. If $\emptyset \neq B_1 \times B_2 \times \cdots \times B_n \subset A_1 \times A_2 \times \cdots \times A_n$ then $B_i \subset A_i$ for $i = 1, \dots , n$. It follows that the factors A_i are uniquely determined by the set† $A_1 \times A_2 \times \cdots \times A_n$. It is convenient to invent some shorthand writing: instead of $A_1 \times A_2 \cdots \times A_n$ we shall write $A_1 \times \blacksquare$. We shall also abbreviate other expressions similarly,

$$B_1 \subset A_1, \ \blacksquare$$

standing for n inclusions or $a^1 + \blacksquare$ for $\sum_{k=1}^{n} a^k$, the symbol $\blacksquare$ indicating summation from 1 to n. The cartesian product is obviously not commutative but it is not even associative. All the three sets

$$A \times B \times C, \ (A \times B) \times C, \ A \times (B \times C) \tag{6.1}$$

are, generally speaking‡, distinct. The first set is a set of triplets, the second is a set of couples in which the first element is again a couple. Obviously the distinction between the sets in (6.1) is only formal. We shall ignore it and informally§ regard all three sets (and similar sets with more factors) as equal. The cartesian product of n equal sets A is denoted A^n and we shall be dealing with $\mathbb{R}^n$ and $\overline{\mathbb{R}}^n$. A typical element $x \in \overline{\mathbb{R}}^n$ will be denoted by $(x^1, x^2, \dots , x^n)$ or $(x^1, \blacksquare)$ with components indicated by superscripts rather than subscripts, saving these for distinguishing between individual elements in $\overline{\mathbb{R}}^n$. A point in $\overline{\mathbb{R}}^n$ is finite if it belongs to $\mathbb{R}^n$, and infinite if one of its components is ∞ or $-\infty$.

If f is a function defined on a subset of $\overline{\mathbb{R}}^n$ then we shall use the notation $f(x)$ and $f(x^1, \dots , x^n)$ interchangeably. By $f(c^1, \dots , c^k, \dots)$ we denote the function $(x^{k+1}, \dots , x^n) \mapsto f(c^1 \dots , c^k, x^{k+1}, \dots , x^n)$.

An interval $J \subset \overline{\mathbb{R}}^n$ is the cartesian product of n intervals in $\overline{\mathbb{R}}$, i.e.

$$J = J_1 \times \blacksquare \tag{6.2}$$

with one-dimensional intervals $J_1 \subset \overline{\mathbb{R}}$, $\blacksquare$. The interval J is degenerate if one of J_k is degenerate. The interval J is open if all J_k are open. An open interval can be degenerate only if it is empty. J is closed if all J_k are closed. As in one dimension, the closure of an interval is the

---

† As long as this set is not empty.
‡ The sets are equal if e.g. $A = \emptyset$.
§ This process can be made formally logically rigorous.

smallest closed interval containing it, and the interior of J is the largest open interval contained in J. The closure $\bar{I}$ of an interval $I \subset \mathbb{R}$ with $(a, b) \subset I \subset [a, b]$ is $[a, b]$, the interior I° of I is (a, b). The closure or the interior of J is $\bar{J}_1 \times \bullet$ or $J_1^\circ \times \bullet$, respectively. If $A, B \in \overline{\mathbb{R}}^n$ then we also denote

$$[A^1, B^1] \times \bullet = [A, B].$$

Two intervals are non-overlapping if their interiors are disjoint; intervals of a family are non-overlapping if they are pairwise non-overlapping. We shall denote by $\mathbf{b}J$ the set $(\bar{J} \setminus J^\circ) \cap \mathbb{R}^n$ and call it the boundary of J. A set of the form $H_c = \{x^i = c \, ; \, x \in \mathbb{R}^n\}$ is called a hyperplane. The boundary of J is the union of intersections of J with H_c where c is an endpoint of one of the J_k. Obviously there are at most $2n$ such sets and they are called sides of J. For the interval J the numbers $|J_1|, \bullet$ are the edge-lengths. An interval which has all its edge-lengths positive, finite and equal is called a cube.† If $a \in \mathbb{R}^n$ and $h > 0$ then by $C(a, h)$ we denote the (open) cube $(a^1 - h, a^1 + h) \times \bullet$. For elements of $\mathbb{R}^n$ we introduce two norms, the maximum norm $|x|_u = \text{Max}(|x^1|, \bullet)$ and the Euclidean norm $|x|_2 = \sqrt{(x^1)^2 + \bullet}$. For most of what we do either norm could be used but on occasion one or the other can be more convenient. For example $C(a, h) = \{x; \, |x - a|_u < h\}$. The the content $|J|$ of the interval J is by definition the product of its edge-lengths, i.e $|J| = |J_1||J_2| \ldots |J_n|$. In one dimension the content of an interval is its length, in two dimensions it is its area. It is utterly obvious that length is additive, i.e. if an interval is a union of non-overlapping subintervals then its length is the sum of lengths of the subintervals. In several dimensions content is also additive but the fact is less obvious, see Figure 6.1. It is certainly unwise to rely on intuition in dimension more than three but even in two dimensions we feel that the additivity of content requires proof, which we supply in the next section.

6.2 Divisions, partitions

The concepts of a division and a partition in $\overline{\mathbb{R}}^n$ are similar to those in $\overline{\mathbb{R}}$. A family of intervals

$$\{K_i \, ; \, i = 1, \ldots, r\} \tag{6.3}$$

---

† In two dimensions a cube is a square.

is a *partial division* of a closed interval J if K_i are closed, non-overlapping and $K_i \subset J$ for $i = 1, \ldots, r$. If moreover

$$\bigcup_1^r K_i = J \tag{6.4}$$

then $\{K_i;\ i = 1, \ldots, r\}$ is a *division* of J. A *(sub)partition* of J or a *(partial) tagged division* of J is a set of couples $\{(x_i,\ K_i)\,;\ i = 1, \ldots, r\}$ such that the intervals K_i form a (partial) division of J and $x_i \in K_i$. If there is no danger of misunderstanding we shall abbreviate the notation for a partition or subpartition to $\{(x, K)\}$. Phrases like tagged in S or tagged partial division on S have the same meaninig as in Section 3.2. If $\pi \equiv \{(x, K)\}$ is a partition of J then by π° we shall denote the part of π tagged in J°. More precisely

$$\pi^\circ = \{(x, K); x \in J^\circ,\ (x, K) \in \pi\}. \tag{6.5}$$

A *brick layer division* or *bl-division* for short is so called because the intervals of the division lie in layers like bricks in a wall. Not every division is a bl-division, see Figure 6.1. If $\{([u_i^2, v_i^2],\ i = 1, \ldots, s)\}$ is a

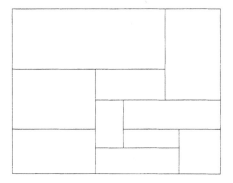

Fig. 6.1. A division in $\mathbb{R}^2$

division of J_2 and $\{[u_{ik}^1, v_{ik}^1],\ k = 1, \ldots, r_i\}$ is for each i a division of J_1 then the set of $r_1 + \cdots + r_s$ intervals

$$[u_{ik}^1, v_{ik}^1] \times [u_i^2, v_i^2] \quad \text{for} \quad i = 1, \ldots, s,\ k = 1, \ldots, r_i,$$

form a bl-division of J. In n dimensions the situation is similar, the rôle of J_2 being played by J_n and of J_1 by $J_1 \times \cdots \times J_{n-1}$. The bl-divisions just described have the layers mounting in the n-th direction, obviously there are bl-divisions in other directions as well. Other authors call a bl-division a *compound division*. A partition is a bl-partition if its intervals

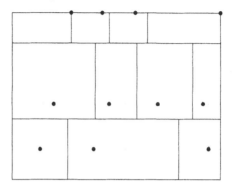

Fig. 6.2. bl-partition

form a bl-division *and* tags in each layer are at the same 'height'. See Figure 6.2. More precisely, for the partition

$$\left\{ \left((x_{ik}^1, x_{ik}^2), \ [u_{ik}^1, \ v_{ik}^1] \times [u_i^2, \ v_i^2] \right) \right\}$$

to be a bl-partition it is necessary that $x_{ik}^2 = y_i$ for all $k = 1, \dots, r_i$ and some y_i with $i = 1, \dots s$. Let $J = J_1 \times \bullet$ and $\{j_{k i_k}; i_k = 1, \dots, p_k\}$ be a division of J_k with $1 \le k \le n$. The set of $p_1 \dots p_n$ intervals

$$j_{1k_1} \times \bullet \tag{6.6}$$

is a special division called a net of J. A net is a bl-division in every direction. See Figure 6.3. Now we prove the following lemma, which is

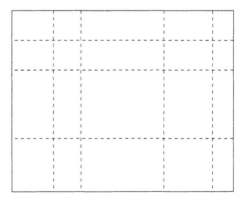

Fig. 6.3. A net

obvious in $\mathbb{R}$. See Figure 6.4.

LEMMA 6.2.1 *If $K_1, \ldots, K_r, J$ are closed intervals with $K_i \subset J \subset \overline{\mathbb{R}}^n$ then there exists a net (6.6) of J such that the intervals $j_{1k_1} \times \bullet$ which overlap with K_i form a net of K_i.*

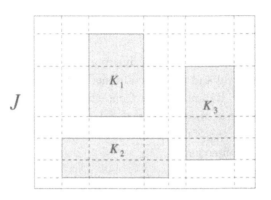

Fig. 6.4. Simultaneous net

COROLLARY 6.2.2 *There exists a division D of J such that the intervals K_i for $i = 1, \ldots, r$ are subintervals of D.*

Proof of Lemma 6.2.1 Let $K_i = k_{1i} \times \bullet$, $J = J_1 \bullet$ and

$$x_0^1 < x_1^1 < \ldots < x_{p_1}^1,$$

$$\bullet$$

be the endpoints of $k_{1i} \bullet$ and $J_1 \bullet$.† Setting $j_{1i_1} = [x_{i_1}^1, x_{i_1+1}^1] \times \bullet$ for $i_1 = 1, \ldots, p_i$ makes the net (6.6) the required one. $\qquad \bullet$

Adjusting the intervals a little, e.g. taking $j_{rs} = [x_s^r, x_{s+1}^r]$ if x_{s+1}^r is not an endpoint of J_s, we obtain

LEMMA 6.2.3 *Let K_i be as in Lemma 6.2.1. There exist disjoint intervals $\tilde{j}_k$ with $k = 1, \ldots, p$ and $p = p_1 p_2 \ldots p_n$ such that*

$$\bigcup_{k=1}^{p} \tilde{j}_k = J$$

and each $\tilde{j}_k$ is a subset of some K_i.

An *interval function* F *on* J is a mapping which associates with every closed interval $K \subset J$ a real number $\mathsf{F}(K)$. An interval function on $\mathbb{R}^n$

---

† Writing each endpoint only once.

(or on $\overline{\mathbb{R}}^n$) is called simply an interval function. An interval function F on J is called *additive* if, for every pair of closed bounded and non-overlapping intervals K, $L \subset J$ such that $K \cup L$ is again an interval, the equation

$$\mathsf{F}(K) + \mathsf{F}(L) = \mathsf{F}(K \cup L) \qquad (6.7)$$

is satisfied. Examples of additive functions of intervals in $\mathbb{R}$ are the length of the interval and the interval function F defined by $\mathsf{F}([\alpha, \beta]) = \int_\alpha^\beta f$ provided f is KH-integrable. Content is an additive function, indeed if $K = K_1 \times \bullet$, $L = L_1 \times \bullet$ and $I = K \cup L$ is an interval then (Exercise 6.1) $I_p = K_p \cup L_p$ for some p and $K_i = L_i$ for $1 \le i \ne p \le n$. See Figure 6.5. Consequently

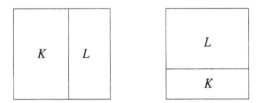

Fig. 6.5. Union of intervals again an interval

$$
\begin{aligned}
|I| &= |K_1||K_2|\dots|K_p \cup L_p|\dots|K_n| \\
&= |K_1|\dots|K_p|\dots|K_n| + |K_1|\dots|L_p|\dots|K_n| \\
&= |K| + |L|.
\end{aligned}
$$

The next lemma is fairly plausible but we need Lemma 6.2.1 to prove it.

LEMMA 6.2.4 *If* F *is an additive interval function on a closed bounded interval* J *and* K_i, $i = 1, \dots, r$ *form a division of* $J \subset \mathbb{R}^n$ *then*

$$\mathsf{F}(J) = \sum_{i=1}^r \mathsf{F}(K_i). \qquad (6.8)$$

Proof Let (6.6) be a net of J. We prove first that

$$\mathsf{F}(J) = \sum_1^{p_1} \sum_2^{p_2} \dots \sum_n^{p_n} \mathsf{F}(j_{1i_1} \times \bullet). \qquad (6.9)$$

Equation (6.7) is easily extended from two to any number of intervals as long as the intervals form a 'layer' and then

$$\mathsf{F}\left(\bigcup_{k=1}^{p_1} j_{1k} \times J_2 \times \, \blacksquare \, \right) = \sum_{k=1}^{p_1} \mathsf{F}(j_{1k} \times J_2 \times \, \blacksquare \,). \qquad (6.10)$$

For a fixed t the intervals

$$j_{1t} \times j_{2k} \times J_3 \times \, \blacksquare \quad k = 1, \dots, p_2$$

constitute a net of $j_{1t} \times J_2 \times \, \blacksquare$ and hence

$$\mathsf{F}(j_{1t} \times J_2 \times \, \blacksquare \,) = \sum_{k=1}^{p_2} \mathsf{F}(j_{1t} \times j_{2k} \times J_3 \times \, \blacksquare \,).$$

Substituting this into (6.10) and continuing in this fashion we obtain (6.9). The rest of the proof is easy: we choose intervals (6.6) according to Lemma 6.2.1 and then the sum $\sum \mathsf{F}(j_{1i_1} \times \, \blacksquare \,)$ extended over all those intervals which overlap with K_i equals $\mathsf{F}(K_i)$. Summation over i now leads to (6.8). $\qquad\qquad\qquad\qquad\qquad\qquad\qquad\bullet$

COROLLARY 6.2.5

$$|J| = \sum_{i=1}^{r} |K_i|.$$

A *gauge* δ on a set $S \subset \overline{\mathbb{R}}^n$ is a mapping from S into a set of open n-dimensional intervals such that $x \in \delta(x)$ for every $x \in S$ and $\delta(x)$ is a bounded interval if $x \in \mathbb{R}^n$. A tagged division $\pi \equiv \{(x, K)\}$ of J is δ-fine, in symbols $\pi \ll \delta$, if δ is a gauge on J and $K \subset \delta(x)$. If J is a bounded closed interval then for a gauge δ on J there is a δ-fine partition of J. This n-dimensional version of Cousin's lemma can be as easily proved as in one dimension by a similar bisection argument, the difference being that at each step the interval is divided into 2^n subintervals. In contrast to one dimension where there were only two infinite elements the transition to unbounded intervals is not immediate.

LEMMA 6.2.6 (The Cousin lemma in $\mathbb{R}^n$) *If δ is a gauge on an interval $I \subset \overline{\mathbb{R}}^n$ then there exists a δ-fine partition of I.*

Proof We have just seen that the lemma holds if J is a compact interval. Let $Q = [-1, 1] \times \bullet$ and $h : \overline{\mathbb{R}}^n \mapsto Q$ where

$$
\begin{aligned}
h^i(x) &= \frac{x^i}{1 + |x^i|} \quad \text{if} \quad x^i \in \mathbb{R}, \\
h^i(\infty) &= 1, \\
h^i(-\infty) &= -1.
\end{aligned}
$$

The mapping h has the following properties

  (i) it is one-to-one;

  (ii) it preserves cartesian products;

  (iii) h and its inverse g map open intervals onto open intervals and closed intervals onto closed intervals.

If $J \subset \overline{\mathbb{R}}^n$ and δ is a gauge on J then there exists an $(h \circ \delta)$-fine partition of $h(J)$ because $h(J)$ is a bounded closed interval. If this partition is $\{(y, L)\}$ then $\{(g(y), g(L))\}$ is δ-fine partition of J.   •

6.3 The definition

Riemann sums are defined similarly as in $\mathbb{R}$. If f is defined on a closed interval J and $\pi \equiv \{(\xi_i, K_i); i = 1, \dots, r\}$ is a partition of J then

$$
\sum_\pi f = \sum_{i=1}^r f(\xi_i)|K_i|
$$

and it is understood that the term $f(\xi_j)|K_j|$ is not included in the sum if K_j is an unbounded interval. This is the same as assuming f to be zero on $\overline{\mathbb{R}}^n \setminus \mathbb{R}$ and, as in (2.111), interpreting $f(\xi_j)|K_j| = 0.\infty = 0$. The definition of the KH-integral in $\overline{\mathbb{R}}^n$ is an almost verbatim repetition of Definition 2.10.1.

DEFINITION 6.3.1 *A number $I \in \mathbb{R}$ is the Kurzweil–Henstock integral (or integral) of f over an interval $J \subset \overline{\mathbb{R}}^n$ (or on J) if for every positive ε there is a gauge δ such that for every δ-fine partition π of J*

$$
|\sum_\pi f - I| < \varepsilon. \tag{6.11}
$$

The function f is then called KH-integrable.

Clearly I is uniquely determined; the proof is the same as in Theorem 2.4.6. Any of the following symbols will be used for the integral:

$$I = \int_K f = \int \cdots \int_K f \tag{6.12}$$

$$= \int_K f(x)\, dx = \int \cdots \int_K f(x)\, dx \tag{6.13}$$

$$= \int_K f(x^1,\dots,x^n)\, dx^1 \dots dx^n = \int \cdots \int_K f(x^1,\dots,x^n)\, dx^1 \dots dx^n. \tag{6.14}$$

Similarly as in one dimension the notation in (6.12) is logically consistent (we integrate a function and not function values) whereas (6.14) is more convenient in concrete situations where the function values are given by some algebraic formula. Occasionally we may place the integration domain below the integral sign thus:

$$\int_K f(x)\, dx \quad \text{or} \quad \int \cdots \int_K f(x^1,\dots,x^n)\, dx^1 \dots dx^n$$

We may also use the prefix KH with any of the symbols in (6.12)—(6.14). Keeping to established notation we write

$$\int_{\mathbb{R}^n} f \quad \text{rather than} \quad \int_{\underline{\mathbb{R}}^n} f$$

if $J = \overline{\mathbb{R}}^n$. In two or three dimensions we write $\iint$ or $\iiint$, respectively, instead of $\int \cdots \int$. We call any of the integrals in (6.12)—(6.14) a multiple integral if we want to emphasize that the integration is over an n-dimensional interval. If $n = 2$ or $n = 3$ we use the term double integral or triple integral, respectively. If $J = [a^1, b^1] \times \blacksquare$ then it is customary to write

$$\int_J f = \int_{a^1}^{b^1} \cdots \int_{a^n}^{b^n} f(x^1,\dots,x^n)\, dx^1 \dots dx^n,$$

$$\int_J f = \int_a^b f,$$

with $[a, b] = [a_1, b_1] \times \blacksquare$. If f is KH-integrable on some interval J then on occasions we set for brevity

$$\mathsf{F}(K) = \int_K f$$

for any interval $K \subset J$.

EXAMPLE 6.3.2 If $J \subset \mathbb{R}^n$ is a closed bounded interval and $f(x) = c$ on J then

$$\int_J f = c|J|.$$

This is very plausible and we are even spared the need to choose δ since for any partition $\pi \equiv \{(x, K)\}$ of J we have

$$\sum_\pi f = c \sum_\pi |K| = c|J|. \qquad \bullet$$

However, it was necessary to employ Corollary 6.2.5 (and through it Lemma 6.2.1) on the additivity of content!

EXAMPLE 6.3.3 If $\mathbb{R}^n \supset S = \{c_1, c_2, c_3, \dots\}$ with $c_i \neq c_j$ for $i \neq j$ and $f(x) = 0$ for $x \notin S$ then f is KH-integrable on $\overline{\mathbb{R}}^n$ and $\int_{\mathbb{R}^n} f = 0$. Choose h_j so that

$$(2h_j)^n < \frac{\varepsilon}{2^{j+n}(|f(c_j)| + 1)}.$$

Define†

$$\delta(c_j) = C(c_j, h_j) \quad \text{and} \quad \delta(x) = C(x, 1) \text{ for } x \in \mathbb{R}^n \setminus S.$$

Let $\pi \equiv \{(y_i, K_i); i = 1, 2, \dots, n\}$ be a δ-fine partition of J. Clearly

$$\sum_\pi f = \sum_{y_i \in S} f(y_i)|K_i|.$$

Since the same c_j can tag 2^n distinct intervals and because of the choice of δ we have

$$\left| \sum_{y_i \in S} f(y_i)|K_i| \right| < 2^n \sum_{j=1}^\infty |f(c_j)| \frac{\varepsilon}{2^{j+n}(|f(c_j)| + 1)} < \varepsilon.$$

This proves f is KH-integrable and $\int_{\mathbb{R}^n} f = 0$. $\qquad \bullet$

Similarly as in one dimension a set S is called *negligible* or a *null set* if for any f the function‡ $f\mathbf{1}_S$ is KH-integrable and

$$\int_{\mathbb{R}^n} f\mathbf{1}_S = 0.$$

By the previous example a countable set is negligible. We rephrase Theorem 2.5.5 as

† It does not matter how δ is defined on $\overline{\mathbb{R}}^n \setminus \mathbb{R}^n$.
‡ By our convention from Section 3.11 $f\mathbf{1}_S$ is defined everywhere even if f is not.

THEOREM 6.3.4 *A set S is negligible if and only if*

$$\int_{\mathbb{R}^n} 1_S = 0.$$

The proof of Theorem 2.5.5 can easily be adapted to our more general situation. It follows easily that a subset of a negligible set is negligible and a countable union of null sets is null. In $\mathbb{R}$ it requires some effort to produce an uncountable null set (see Sections A.1 and 2.11). In $\mathbb{R}^n$ such sets abound. A useful example is the following.

EXAMPLE 6.3.5 (Piece of hyperplane negligible) We prove that the set $S = [a^1, b^1] \times \blacksquare$ with $a^1 \leq b^1 \blacksquare$ and $a^t = b^t$ for some $t \in \mathbb{N}$, $1 \leq t \leq n$ is negligible. For $x \in S$ let $\boldsymbol{\delta}(x) = C(x, \varepsilon)$, for $x \in \overline{\mathbb{R}}^n \setminus \mathbb{R}^n$ let $\boldsymbol{\delta}(x) = \overline{\mathbb{R}}^n$ and for any other x let $\boldsymbol{\delta}(x) = C(x, 1)$. If $\{(x_i, K_i); i = 1, \ldots, r\}$ is a $\boldsymbol{\delta}$-fine partition of $\overline{\mathbb{R}}^n$ then

$$\bigcup_{x_i \in S} K_i \subset [a^1 - \varepsilon, b^1 + \varepsilon] \times \blacksquare$$

and therefore

$$\sum_{i=1}^r 1_S |K_i| = \sum_{x_i \in S} |K_i| \leq \prod_{i=1}^r (b^i - a^i + 2\varepsilon) = P(\varepsilon).$$

Since $a^t = b^t$ the product $P(\varepsilon) \to 0$ as $\varepsilon \to 0$. •

It follows that for an interval $J \subset \overline{\mathbb{R}}^n$ the set $\mathbf{b}J$ is negligible.

EXAMPLE 6.3.6 In this example $n = 2$ and $k \in \mathbb{N}$. We denote by J or J_k the interval $[0, 1] \times [0, 1]$ or $[\frac{1}{k+1}, \frac{1}{k}] \times [\frac{1}{k+1}, \frac{1}{k}]$, respectively. Let $f(x) = (-1)^k k(k + 1)^2$ for $x \in J_k^\circ$ for all $k \in \mathbb{N}$. Otherwise let $f(x) = 0$. We wish to show that f is KH-integrable† and

$$\iint_J f = \sum_{k=1}^\infty \frac{(-1)^k}{k}.$$

For a partition π_k of J_k let π_k° be the part of π_k tagged in J_k°. Since

$$\int_{J_k} f = \frac{(-1)^k}{k}$$

---

† Anyone even briefly acquainted with Lebesgue theory sees immediately that f is not Lebesgue integrable.

6 Integration in Several Dimensions

and $f = 0$ on $\mathbf{b}J_k$, for every positive ε there is gauge δ_k such that for any δ-fine partition π_k we have

$$|\sum_{\pi_k^\circ} f - \frac{(-1)^k}{k}| < \frac{\varepsilon}{2^{k+2}}. \tag{6.15}$$

We now define δ in five steps.

$$\delta(0,0) = C((0,0), \frac{\varepsilon}{4}),$$

$$\delta(1,1) = \delta_1(1,1),$$

$$\delta(x_k) = \delta_k(x_k) \cap \delta_{k+1}(x_k) \quad \text{where} \quad x_k = \left(\frac{1}{k+1}, \frac{1}{k+1}\right),$$

$$\delta(x) = \delta_k(x) \quad \text{for} \quad x \in J_k,\ x \neq x_k, x_{k+1},$$

$$\delta(x) \subset J \setminus \bigcup_1^\infty J_k \quad \text{for all remaining } x.$$

Let $\pi \ll \delta$ and let $N-1$ be the first k for which J_k intersects the interval of π tagged by $(0,0)$. Then $4 < N\varepsilon$ and

$$\sum_{\pi_N^\circ} |f| = |\sum_{\pi_N^\circ} f| \leq |\sum_{\pi_N^\circ} f - \frac{(-1)^N}{N}| + \frac{1}{N} \leq \frac{\varepsilon}{2^{N+2}} + \frac{1}{N} \leq \frac{\varepsilon}{2}.$$

Finally we have

$$\left|\sum_\pi f - \sum_{k=1}^\infty \frac{(-1)^k}{k}\right|$$

$$\left|\sum_{k=1}^{N-1} \left(\sum_{\pi_k^\circ} f - \frac{(-1)^k}{k}\right)\right| + \left|\sum_{k=N}^\infty \frac{(-1)^k}{k}\right| + \sum_{\pi_n^\circ} |f|$$

$$< \frac{\varepsilon}{4} + \frac{1}{N} + \frac{\varepsilon}{2} \leq \varepsilon. \quad \bullet$$

Contrast this example with Exercise 6.2 where a similarly defined f is not KH-integrable.

Similarly as with previous definitions of an integral we have

THEOREM 6.3.7 (Bolzano–Cauchy condition) *A function f is KH-integrable on an interval $J \subset \overline{\mathbb{R}}^n$ if and only if for every positive ε there is a gauge δ on J such that*

$$\left|\sum_{\pi_1} f - \sum_{\pi_2} f\right| < \varepsilon \tag{6.16}$$

whenever $\pi_1 \ll \delta$ and $\pi_2 \ll \delta$.

Proof Necessity is obvious. The proof of sufficiency is a verbatim repetition of the proof of Theorem 3.12.2 except M-integrable, M-partition and M-integral become KH-integrable, partition and KH-integral. •

6.4 Basic theorems

The theorems from Section 2.5 are easily adapted to n dimensions. All that is needed is to replace the interval $[a, b]$ by an n-dimensional interval J and $b - a$ by $|J|$, the content of J. Exceptions are statements involving monotonicity of functions, the intermediate value property and Theorem 2.5.29 on the continuity of the indefinite integral, the analogue of which we discuss in the next section. If S is a set and f is defined on S then we define

$$\int_S f \stackrel{\text{def}}{=} \int_{\mathbb{R}^n} 1_S f \qquad (6.17)$$

if the latter integral exists. Since for any interval J the boundary is negligible† it can easily be shown that Definition (6.17) is consistent with Definition 6.3.1 if S is a non-degenerate interval. A consequence of (6.17) is that $\int_J f = 0$ if J is degenerate. However, we shall not attempt to extend the definition of the n-dimensional KH-integral as we did in one dimension in Definition 2.5.28. Sometimes we replace S with its defining relation, e.g. if

$$S = \left\{ (x, y);\ x^2 + y^2 \leq 1 \right\}$$

then we write

$$\int_S f = \iint_{x^2+y^2 \leq 1} f(x,y)\,dxdy.$$

We now consider theorems on integrability on the union of intervals and on subintervals a little more carefully. Let K and L be intervals such that $K \cup L$ is again an interval. If f is KH-integrable on K and L then f is KH-integrable on $K \cup L$. Moreover

$$\int_K f + \int_L f = \int_{K \cup L} f. \qquad (6.18)$$

This can be proved similarly as in Theorem 2.5.12, but we prefer to state and prove a more general result.

† Or empty.

THEOREM 6.4.1   *If*

(i) *J is an interval in $\overline{\mathbb{R}}^n$;*

(ii) *$\{K_i; \, i = 1, \ldots, r\}$ is a division of J;*

(iii) *f is KH-integrable on each K_i*

then f is KH-integrable on J and

$$\int_J f = \sum_{i=1}^r \int_{K_i} f. \qquad (6.19)$$

Proof Since the values of a function on a negligible set do not affect either the existence or the value of the integral, we redefine f on $\bigcup_1^r \mathbf{b}K_i$ to be zero. Given $\varepsilon > 0$ we find a gauge $\boldsymbol{\delta}_i$ on K_i such that

$$\left| \sum_{\pi_i^\circ} f - \int_{K_i} f \right| < \frac{\varepsilon}{r}, \qquad (6.20)$$

for every $\pi_i \ll \boldsymbol{\delta}_i$. If $x \in K_i^\circ$ then we define $\boldsymbol{\delta}(x) = \boldsymbol{\delta}_i(x) \cap K_i^\circ$. If x belongs to several $\mathbf{b}K_i$, say† $x \in \mathbf{b}K_{i_1} \cap \cdots \cap \mathbf{b}K_{i_p}$ and $x \notin K_j$ with $j \neq i_k$ then let $\boldsymbol{\delta}(x) = \boldsymbol{\delta}_{i_1}(x) \cap \cdots \cap \boldsymbol{\delta}_{i_p}(x)$. Consider now $\pi \equiv \{(y, L)\}$, a $\boldsymbol{\delta}$-fine partition of J. The set

$$\{(x, L \cap K_i); x \in K_i, (x, L) \in \pi\}$$

forms a $\boldsymbol{\delta}_i$-fine partition of K_i and we have by (6.20)

$$\sum_\pi f \; = \; \sum_{i=1}^r \sum_{\pi_i} f = \sum_{i=1}^r \sum_{\pi_i^\circ} f$$

$$\left| \sum_\pi f - \sum_{i=1}^r \int_{K_i} f \right| \; \leq \; \sum_{i=1}^r \left| \sum_{\pi_i^\circ} f - \int_{K_i} f \right| < r\frac{\varepsilon}{r}.$$

This proves f KH-integrable and equation (6.19).   •

THEOREM 6.4.2 (Integrability on subintervals)   *If J, L are intervals, $L \subset J$ and f is KH-integrable on J then f is KH-integrable on L.*

Proof By Theorem 6.3.7 for every positive ε there exists a gauge $\boldsymbol{\delta}$ such that

$$\left| \sum_{\Pi'} f - \sum_{\Pi''} f \right| < \varepsilon$$

---

† $p = 1$ is not excluded.

whenever $\Pi' \ll \delta$, $\Pi'' \ll \delta$. By Corollary 6.2.2 there are intervals $L_1, \ldots, L_p$ such that these together with L form a division of J. For each i there is a δ-fine partition of L_i, say π_i. Let π' and π'' be two δ-fine partitions of L. We define

$$\Pi' = \pi' \cup \pi_1 \cup \cdots \cup \pi_p,$$
$$\Pi'' = \pi'' \cup \pi_1 \cup \cdots \cup \pi_p,$$

and have

$$\left| \sum_{\Pi'} f - \sum_{\Pi''} f \right| = \left| \sum_{\pi'} f - \sum_{\pi''} f \right| < \varepsilon. \qquad \bullet$$

It now follows similarly as in Section 2.5 that step functions are KH-integrable on closed and bounded intervals, and by an analogue of Theorem 2.5.16 this extends to regulated functions including continuous functions. However, we still lack an effective means for evaluating n-dimensional integrals even for a very simple function like a polynomial. A partial solution to the problem of evaluation is our next topic.

6.4.1 Prelude to Fubini's theorem

We set $n = 2$ and write (x, y) instead of (x^1, x^2). The interval $J = [a, b] \times [c, d]$ will be bounded. The functions f, g, ϕ defined on I will be respectively continuous, regulated and step. For $x \in [a, b]$ the integral

$$Q(x) = \int_c^d f(x, \cdot)$$

exists and is a continuous function of x. Consequently $\int_a^b Q$ exists and we denote it by

$$\int_a^b \left(\int_c^d f(x, y) \, dy \right) dx$$

or simply

$$\int_a^b dx \int_c^d f(x, y) \, dy$$

and call it an iterated or repeated integral. $\iint_J f$ also exists and we shall prove that

$$\iint_J f = \int_a^b dx \int_c^d f(x, y) \, dy. \qquad (6.21)$$

If this has been established then by symmetry also

$$\iint_J f = \int_a^b dy \int_c^d f(x,y)\, dx.$$

Therefore

$$\int_a^b dx \int_c^d f(x,y)\, dy = \int_a^b dy \int_c^d f(x,y)\, dx.$$

The repeated integral can often be evaluated by using the Fundamental Theorem twice, e.g.

$$\int_0^{\frac{\pi}{2}} dx \int_0^1 x \cos(xy)\, dy = \int_0^{\frac{\pi}{2}} \sin x\, dx = 1.$$

It is easily checked that the other repeated integral is also 1 but the calculation is less straightforward. Generally speaking the repeated integrals are not equal, e.g.

$$\int_0^1 dx \int_0^1 \frac{x^2 - y^2}{(x^2 + y^2)^2}\, dy = \int_0^1 \frac{1}{x^2 + 1}\, dx = \frac{\pi}{4}$$

$$\int_0^1 dy \int_0^1 \frac{x^2 - y^2}{(x^2 + y^2)^2}\, dx = -\int_0^1 \frac{1}{1 + y^2}\, dy = -\frac{\pi}{4}.$$

We have just seen that the iterated integrals might or might not be equal; the other possibility that one exists and the other not can also occur. (See Exercise 6.5.) We shall prove in Section 6.6 that (6.21) holds not only for continuous f but for any KH-integrable function. Such a result is often referred to as Fubini's Theorem which Fubini proved within the framework of Lebesgue theory. We have here a modest aim of proving (6.21) with f replaced by a regulated function g. We begin by noting that $\iint 1_K = \int_a^b (\int_c^d 1_K(x,y)\, dy) dx$ if $K \subset J$ is an interval. It is also clear that if (6.21) holds with f replaced by h for every h in some family $\mathcal{H}$ then it also holds for any linear combination of functions from $\mathcal{H}$. It follows that

$$\iint_J \phi = \int_a^b dx \int_c^d \phi(x,y)\, dy \qquad (6.22)$$

for any step function ϕ. For a positive ε we find a step function ϕ such that $|g - \phi| < \varepsilon$ and have

$$\left| \iint_J g - \iint_J \phi \right| < \varepsilon |J|,$$

$$\left| \int_a^b dx \int_c^d g(x,y)\, dy - \int_a^b dx \int_c^d \phi(x,y)\, dy \right| < \varepsilon |J|.$$

This combined with (6.22) gives

$$\left| \iint_J g - \int_a^b dx \int_c^d g(x,y)dy \right| < 2\varepsilon|J|.$$

Since ε is arbitrary this proves the required equation

$$\iint_J g = \int_a^b dx \int_c^d g(x,y)\, dy. \tag{6.23}$$

EXAMPLE 6.4.3 Let $f(x,y) = x^y$ for $0 \le x \le 1$ and $0 < a \le y \le b$. The function f is continuous and the repeated integrals are equal. Let us try first

$$\iint_{\substack{0<x<1 \\ a<y<b}} x^y \, dxdy = \int_0^1 dx \int_a^b x^y \, dy = \int_0^1 \frac{x^b - x^a}{\log x}\, dx.$$

This looks hopeless so we integrate in the other order

$$\int_a^b dy \int_0^1 x^y \, dx = \int_a^b \frac{dy}{y+1} = \log \frac{b+1}{a+1}. \qquad \bullet$$

This example shows not only that one iterated integral can be evaluated easily and the other not but also demonstrate the possibility of finding a simple integral by a detour to an intelligently chosen double integral.

EXAMPLE 6.4.4 Let $D = \{(x,y); |x| \le 1, 1 - x^2 \le y \le 4(1 - x^2)\}$ and f be a function continuous on $\mathbb{R}^2$ and equal to zero outside D. Then we have

$$I = \iint_D f = \int_{-1}^1 dx \int_{1-x^2}^{4(1-x^2)} f(x,y)\, dy,$$

$$I = \int_0^1 dy \int_{-\sqrt{1-y/4}}^{-\sqrt{1-y}} f(x,y)\, dx$$

$$+ \int_1^4 dy \int_{-\sqrt{1-y/4}}^{\sqrt{1-y/4}} f(x,y)\, dx + \int_0^1 dy \int_{\sqrt{1-y}}^{\sqrt{1-y/4}} f(x,y)\, dx.$$

In this subsection we have become familiar with iterated integrals and their use in evaluation of multiple integrals. However, more general results are needed. For the function f defined by

$$f(x,y) = \frac{x}{\sqrt{x^2 + 4y}} \text{ for } 0 < x \leq 1 \text{ and } 0 < y \leq 1$$

the double and both repeated integrals exist and are equal, but f, despite its simplicity, is not regulated. Moreover we need theorems on evaluation of multiple integrals over unbounded intervals or for unbounded functions. This will be discussed in Section 6.6. See also Exercises 6.11 to 6.18.

6.5 Other theorems in $\mathbb{R}^n$

6.5.1 Negligible sets

The covering lemma (Lemma 2.11.4) holds in $\mathbb{R}^n$ as stated for $\mathbb{R}$, except that the inclusion $S \subset \mathbb{R}$ has to be replaced by $S \subset \mathbb{R}^n$ and inclusion (2.119) is to be interpreted as $K_i \subset C(x_i, \delta(x_i))$. The new covering lemma can be used to prove analogues of Theorems 2.11.1 to 2.11.3. This means that a set $S \subset \mathbb{R}^n$ is negligible if and only if it is of measure zero, i.e. can be covered by a countable family of open intervals J_k with $k \in \mathbb{N}$ such that $S \subset \bigcup_1^\infty J_k$. All that is needed to adjust Theorem 2.11.3 is to replace $\mathbb{R}$ by $\mathbb{R}^n$ and equation (2.118) by

$$\sum_{k=1}^\infty |J_k| \leq \int_{\mathbb{R}^n} 1_S + \varepsilon.$$

Changes like these are easy to recognize and easy to make. We have stated them explicitly in this first subsection but in future we leave them to the reader.

6.5.2 Henstock's lemma

Only obvious changes are needed to transfer Henstock's lemma and its proof to $\overline{\mathbb{R}}^n$. It is even more important for multiple integrals and as in one dimension it is crucial for any advancement of the theory. We reformulate two important consequences of Henstock's Lemma from Section 3.2.

If $\int_K f = 0$ for any interval $K \subset J$ then $f = 0$ a.e. on J.

If f is integrable on J then the interval function $\mathsf{F}(K) = \int_K f$ with $K \subset J$ is called the primitive of f. We may also define a corresponding point function as follows. Write $J = [a, b]$ and $x \in J$. Define $F(x) = 0$

when $x_j = a_j$ for at least one j and $F(x) = \mathsf{F}([a, x])$. If f is KH-integrable on J then its primitive point function obviously exists and Henstock's lemma can be used to prove that it is continuous. Conversely given a point function F defined on J we may define a corresponding interval function F as follows. Let $K = [\alpha, \beta]$ with $\alpha = (\alpha^1, \bullet)$ and $\beta = (\beta^1, \bullet)$. Write $\kappa = (\kappa^1, \bullet)$ where $\kappa^j = \alpha^j$ or β^j. Denote by $n(\kappa)$ the number of terms in κ for which $\kappa^j = \alpha^j$ and let

$$\mathsf{F}([\alpha, \beta]) = \sum_\kappa (-1)^{n(\kappa)} F(\kappa)$$

where the summation is over all the vertices κ. For example, when $n = 2$,

$$\mathsf{F}([\alpha, \beta]) = F(\beta^1, \beta^2) - F(\alpha^1, \beta^2) + F(\alpha^1, \alpha^2) - F(\beta^1, \alpha^2).$$

Also, when $n = 3$

$$\begin{aligned}\mathsf{F}([\alpha, \beta]) =\ & F(\beta^1, \beta^2, \beta^3) - F(\alpha^1, \beta^2, \beta^3) + F(\alpha^1, \alpha^2, \beta^3) \\ & -F(\beta^1, \alpha^2, \beta^3) + F(\beta^1, \alpha^2, \alpha^3) - F(\beta^1, \beta^2, \alpha^3) \\ & +F(\alpha^1, \beta^2, \alpha^3) - F(\alpha^1, \alpha^2, \alpha^3).\end{aligned}$$

Note that given a primitive point function, we may recover the primitive interval function as described above. They are mutually convertible.

An important consequence is this: If f is KH-integrable on J and $h > 0$ then the point function

$$F_h(x) = \int_{C(x,h)} \mathbf{1}_J f$$

is a continuous function of x on J. We shall need this observation when we consider measurabilty of KH-integrable functions.

6.5.3 Absolute integrability

The n-dimensional version of Theorem 3.4.1 reads

THEOREM 6.5.1 (Criterion of absolute integrability) *A function KH-integrable on J is absolutely integrable there if and only if*

$$V(f) = \sup\left\{\sum_D |\mathsf{F}(K)|;\ K \in D,\ D \text{ a division of } J\right\} < \infty. \quad (6.24)$$

REMARK 6.5.2 The proof can be modelled on the proof of Theorem 3.4.1 but in $\overline{\mathbb{R}}^n$ we prefer to employ Example 6.3.5.

Proof of Theorem 6.5.1 The aim is to show that

$$\int_J |f| = V(f). \tag{6.25}$$

Given $\varepsilon > 0$ there is a division $\{K_i \,;\, i = 1, \ldots, p\}$ such that

$$V(f) - \varepsilon < \sum_D |\mathsf{F}(K_i)| \le V(f). \tag{6.26}$$

We write $B = \bigcup_1^p \mathbf{b}K_i$. Since B is negligible it is possible to define a gauge δ_b in such a way that

$$\sum_{\pi_b} |f(x)||L| < \varepsilon \text{ if } \pi_b \ll \delta_b \text{ and tagged in } B, \tag{6.27}$$

$$\sum_{\pi_b} |\mathsf{F}(L)| < 2\varepsilon \text{ if } \pi_b \ll \delta_b \text{ and tagged in } B. \tag{6.28}$$

By the definition of the KH-integral and by Henstock's lemma there is a gauge δ_i such that if $\pi_i \ll \delta_i$ then

$$\sum_{\pi_i^\circ} |f(x)|L| - \mathsf{F}(L)| < \varepsilon, \tag{6.29}$$

$$\delta_i(x) \subset K_i^\circ \text{ if } x \in K_i^\circ. \tag{6.30}$$

It follows that

$$\sum_{\pi_i^\circ} |f(x)||L| - \varepsilon < \sum_{\pi_i^\circ} |\mathsf{F}(L)| < \sum_{\pi_i^\circ} |f(x)||L| + \varepsilon \tag{6.31}$$

If x belongs to several $\mathbf{b}K_i$, say† $x \in \mathbf{b}K_{i_1} \cap \ldots \cap \mathbf{b}K_{i_p}$, and $x \notin K_j$ with $j \ne i_k$ then let $\delta(x) = \delta_b(x) \cap \delta_{i_1}(x) \cap \ldots \cap \delta_{i_p}(x)$. If $x \in K_i^\circ$ then let $\delta(x) = \delta_i(x)$. Let now π be a δ-fine partition of J. Then

$$\pi_i = \{(x, L \cap K_i) \,;\, (x, L) \in \pi,\, L \cap K_i^\circ \ne \emptyset,\, x \in K_i\}$$

is a δ_i-fine partition of K_i for which

$$\pi_i^\circ = \{(x, L);\, (x, L) \in \pi,\, x \in K_i^\circ\}.$$

We have by (6.28)

$$2\varepsilon + \sum_{\pi_i^\circ} |\mathsf{F}(L)| \ge |\mathsf{F}(K_i)| \ge \sum_{\pi_i^\circ} |\mathsf{F}(L)|.$$

Using (6.31) this leads to

$$3\varepsilon + \sum_{\pi_i^\circ} |f(x)||L| > |\mathsf{F}(K_i)| > \sum_{\pi_i^\circ} |f(x)||L| - \varepsilon.$$

† $p = 1$ is not excluded.

Summing over i and taking into account (6.27) gives

$$3p\varepsilon + \sum_{\pi} |f(x)||L| > \sum_{1}^{p} |\mathsf{F}(K_i)| > \sum_{\pi} |f(x)||L| - p\varepsilon - \varepsilon.$$

Combining this with (6.26) leads to

$$(3p+1)\varepsilon + \sum_{\pi} |f(x)||L| > V(f) > \sum_{\pi} |f(x)||L| - (p+1)\varepsilon.$$

Since ε is arbitrary this shows (6.25).   •

The theorem we have just proved has the same consequences as Theorem 3.4.1, namely if f is KH-integrable and $|f| \leq g$ for some absolutely KH-integrable g then f is absolutely KH-integrable. In particular, if f_1 and f_2 are absolutely integrable then so are $\mathrm{Max}(f_1, f_2)$ and $\mathrm{Min}(f_1, f_2)$. The functions f^+ and f^- are KH-integrable if f is absolutely KH-integrable.

6.5.4 Convergence, measurability, AC

The monotone convergence theorem (Theorem 3.5.2) and its proof can be adjusted to $\overline{\mathbb{R}}^n$ but we prefer to give an alternative proof. We begin with

LEMMA 6.5.3 *For every positive ε there exists a positive function $\mathcal{E}$ and a gauge δ_e such that*

$$\sum_{\pi} \mathcal{E} < \varepsilon \tag{6.32}$$

for any partial division π of $\mathbb{R}^n$ such that $\pi \ll \delta_e$.

Proof We set

$$e(x) \;=\; \exp(-\sum_{i=1}^{n} |x^i|),$$

$$\mathsf{F}_e(K) \;=\; \int_{u^1}^{v^1} dx^1 \ldots \int_{u^n}^{v^n} e(x)\, dx^n \text{ with } K = [u^1, v^1] \times \blacksquare.$$

It is obvious that F_e is an additive interval function and if K is a bounded closed interval there exists $y \in K$ such that

$$\mathsf{F}_e(K) = e(y)|K|. \tag{6.33}$$

We also have

$$\mathsf{F}_e(\mathbb{R}^n) = \int_{-\infty}^{\infty} dx^1 \ldots \int_{-\infty}^{\infty} e(x)\, dx^n = 2^n. \qquad (6.34)$$

We choose positive $h(x)$ such that

$$|e(x) - e(y)| < \frac{1}{2}e(x) \qquad (6.35)$$

for $y \in C(x, h(x))$. If $\delta_e(x) = C(x, h(x))$ and $\pi \ll \delta_e$ then by (6.32), (6.33) and (6.35)

$$0 < \sum_\pi e(x)|K| < 2\sum_\pi \mathsf{F}_e(K) \le 2^{n+1}.$$

To obtain (6.32) it suffices to define $\mathcal{E}(x) = \varepsilon 2^{-n-1}e(x)$.   •

THEOREM 6.5.4 (Monotone convergence theorem) *If*

(i) *the sequence $\{f_k(x)\}$ is monotonic for almost all $x \in J \subset \overline{\mathbb{R}}^n$,*

(ii) *the functions f_k are KH-integrable and the sequence $\{\int_J f_k\}$ is bounded, i.e. $|\int_J f_k| < K$ for some K all $k \in \mathbb{N}$,*

(iii) $\lim_{k\to\infty} f_k = f$ *is finite a.e.*

then f is KH-integrable on J and

$$\int_J f = \lim_{k\to\infty} \int_J f_k. \qquad (6.36)$$

Proof For reasons outlined at the beginning of the proof of Theorem 3.5.2 we can and shall assume that $\{f_k\}$ is increasing, $f_k \ge 0$ and $f_k \uparrow f < \infty$ everywhere. The sequence $\{\int_A^B f_k\}$ is monotonic and bounded. Hence the limit on the right-hand side of (6.36) exists; let us denote it by L. Given ε we can find N such that

$$\int_J f_N > L - \frac{\varepsilon}{3}.$$

Next we find $k(x) \ge N$ such that, for $k \ge k(x)$,

$$f(x) - \frac{1}{3}\mathcal{E}(x) < f_k(x) \le f(x). \qquad (6.37)$$

By Henstock's lemma there is a gauge δ_n on $\overline{\mathbb{R}}^n$ such that

$$\sum_\pi \left| f_k(x_i)|K_i| - \int_{K_i} f_k \right| < \frac{\varepsilon}{3.2^k} \qquad (6.38)$$

whenever the tagged partial division $\pi \equiv \{(x_i, K_i); i = 1, \ldots, r\}$ is δ-fine. We define

$$\delta(x) = \delta_{k(x)}(x) \cap \delta_e(x) \text{ for } x \in J.$$

Let π be a δ-partition of J. The proof will be accomplished if we show that

$$\left| \sum_{\pi} f(x_i)|K_i| - L \right| < \varepsilon. \tag{6.39}$$

The way from the Riemann sum of f to L goes through the sums

$$\sum_{\pi} f_{k(x_i)}(x_i)|K_i|, \tag{6.40}$$

$$\sum_{\pi} \int_{K_i} f_{k(x_i)}. \tag{6.41}$$

It is easy to check that the first sum is close to the Riemann sum of f and the second to L. Indeed by (6.37) and (6.32)

$$\sum_{\pi} f_{k(x_i)}(x_i)|K_i| \leq \sum_{\pi} f(x_i)|K_i| \leq \sum_{\pi} f_{k(x_i)}(x_i)|K_i| + \frac{\varepsilon}{3}, \tag{6.42}$$

and

$$\sum_{\pi} \int_{K_i} f_{k(x_i)} \geq \sum_{\pi} \int_{K_i} f_N = \int_J f_N > L - \frac{\varepsilon}{3}. \tag{6.43}$$

Denoting by $\bar{N}$ the largest $k(x_i)$ we also have

$$\sum_{\pi} \int_{K_i} f_{k(x_i)} \leq \sum_{\pi} \int_{K_i} f_{\bar{N}} \leq \int_J f_{\bar{N}} \leq L. \tag{6.44}$$

It remains to estimate the difference between (6.40) and (6.41), i.e.

$$\left| \sum_{\pi} \left[f_{k(x_i)}(x_i)|K_i| - \int_{K_i} f_{k(x_i)} \right] \right|. \tag{6.45}$$

The $k(x_i)$ are not necessarily distinct; let $i_1, i_2, \ldots, i_\kappa$ be the distinct i such that $k(x_i) = \kappa$. In (6.45) we group together the terms with the same $k(x_i) = \kappa$ and estimate that sum using (6.38):

$$\left| \sum_{j=1}^{\kappa} \left[f_\kappa(x_{i_j})|K_{i_j}| - \int_{K_{i_j}} f_\kappa \right] \right| < \frac{\varepsilon}{3.2^\kappa}.$$

Consequently

$$\left| \sum_{\pi} \left[f_{k(x_i)}(x_i)|K_i| - \int_{K_i} f_{k(x_i)} \right] \right| < \sum_{1}^{\infty} \frac{\varepsilon}{3 \cdot 2^k} = \frac{\varepsilon}{3}. \qquad (6.46)$$

Collecting (6.42), (6.43), (6.44) and (6.46) gives on one hand

$$\sum_{\pi} f(x_i)|K_i| \geq \sum_{\pi} f_{k(x_i)}(x_i)|K_i|$$

$$\geq \sum_{\pi} \int_{K_i} f_{k(x_i)} - \frac{\varepsilon}{3} > L - \frac{2\varepsilon}{3},$$

and on the other

$$\sum_{\pi} f(x_i)|K_i| \leq \sum_{\pi} f_{k(x_i)}(x_i)|K_i| + \frac{\varepsilon}{3} \leq \sum_{\pi} \int_{K_i} f_{k(x_i)} + 2\frac{\varepsilon}{3} < L + \varepsilon.$$

The last chains of inequalities imply (6.39).   •

Infinite values for the KH-integral can now be admitted as in Definition 3.5.5.

No specific properties of the the real line were used in deducing the Beppo Levi theorem, Fatou's lemma and the Lebesgue dominated convergence theorem: all these theorems are pure logical consequences of the monotone convergence theorem. We can safely and with confidence use them in $\mathbb{R}^n$. Equiintegrability of a family of functions is defined the same way as in $\mathbb{R}$: there is a gauge from the definition the integral which is common to all members of the family. Theorem 3.7.5 on interchange of limit and integration for an everywhere convergent and equiintegrable sequence then remains valid. When we move on to measurabilty we need to show that a KH-integrable function is almost everywhere the limit of a sequence of continuous functions. In $\mathbb{R}$ we used Theorem 3.8.2; this theorem however is not easily translatable to $\mathbb{R}^n$. We start by observing that Austin's Lemma 3.8.1 remains valid in $\mathbb{R}^n$ if the intervals I_k are cubes and $\frac{1}{3}$ is replaced by 3^{-n} in inclusion (3.83). Naturally $\ell(S)$ now denotes the content of S. The following theorem yields the measurability of KH-integrable functions.

THEOREM 6.5.5   *If f is KH-integrable on J and*

$$F_h(x) = \frac{1}{|C(x,h)|} \int_{C(x,h)} f \mathbf{1}_J \qquad (6.47)$$

where $C(x, h) = \{y; |x - y|_u < h\}$ then

$$\lim_{h \downarrow 0} F_h(x) = f(x) \qquad (6.48)$$

for almost all $x \in J$.

The proof follows similar lines to the proof of Theorem 3.8.2, using the n-dimensional version of Austin's lemma mentioned above. The n-dimensional version of the theorem on differentiation of the indefinite integral is weaker: only the existence of a symmetric derivative is asserted. However, the measurability of f obviously follows. With this done the whole Section 3.11 can be translated to an n-dimensional version of it, including Theorem 3.11.10 and Luzin's and Egoroff's theorems. The Vitali–Carathéodory Theorem does not need reformulation but can be restated independently of the McShane integral as equivalence of statements (B) and (C). The proof does not need much adjustment, since the proof of the implication (C)$\Rightarrow$(A) can be used almost verbatim to show (C)$\Rightarrow$(B).

The analogue of Theorem 2.8.3 is valid in $\mathbb{R}^n$, see Exercise 6.4. However, the following theorem is more useful.

THEOREM 6.5.6 *If f is absolutely KH-integrable on each measurable set S_i, for $i \in \mathbb{N}$, and the sets S_i are disjoint then f is KH-integrable on $S = \bigcup_{i=1}^{\infty} S_i$ if and only if*

$$\sum_{1}^{\infty} \int_{S_n} |f| < \infty \qquad (6.49)$$

and then

$$\int_S f = \sum_{i=1}^{\infty} \int_{S_n} f \qquad (6.50)$$

Proof If f is absolutely KH-integrable on S then the partial sums of the series in (6.49) are bounded by $\int_S |f|$.

Going in the opposite direction write $E_m = \bigcup_1^m S_i$ and let $f_m = f^+$ on E_m and $f_m = 0$ otherwise. Then (6.50) with f replaced by f^+ follows by the monotone convergence theorem. Repeating the argument with f^- yields (6.50). $\bullet$

COROLLARY 6.5.7 *If the sets E_i are measurable, the function f is absolutely KH-integrable on each E_i and*

$$E_1 \subset E_2 \subset E_3 \subset \cdots$$

with $E = \bigcup_{i=1}^{\infty} E_i$ then

$$\int_E f = \lim_{m \to \infty} \int_{E_m} f \qquad (6.51)$$

if and only if the sequence $\left\{ \int_{E_i} |f| \right\}$ is bounded.

COROLLARY 6.5.8 *If the sets E_i are measurable, the function f is absolutely KH-integrable on each E_i and*

$$E_1 \supset E_2 \supset E_3 \supset \cdots$$

with $E = \bigcap_{i=1}^{\infty} E_i$ then (6.51) holds.

To prove this corollary it is sufficient to aply the previous corollary with E_i replaced by $E_1 \setminus E_i$ and note that $\bigcup_1^{\infty} (E_1 \setminus E_i) = E_1 \setminus \bigcap_1^{\infty} E_i$.

The relations of the KH-integral to the McShane integral and to the Lebesgue integral remain as described in Sections 3.12 and 3.13.

In Section 3.6 on absolute continuity we used properties of the real line, mainly for two reasons: to avoid the concept of measurability at that stage of development and to investigate the relation of AC functions to functions of bounded variation. We shall call a function of sets, say H, *absolutely continuous* on a measurable set S if for every positive ε there is a number η such that for every measurable subset $E \subset S$ with $m(E) < \eta$ the inequality

$$H(E) < \varepsilon$$

is satisfied. Measure in $\mathbb{R}^n$ depends on n, we shall therefore denote it by m_n. If f is KH-integrable on an interval J then $\mathsf{F}(E) = \int_E f$ is absolutely continuous. This is clearly so since by the Lebesgue dominated convergence theorem there is N such that†

$$\int_J |f - f^N| < \frac{\varepsilon}{2}$$

and

$$\left| \int_E f^N \right| \leq N m_n(E).$$

---

† f^N is defined in Example 3.5.7 equation (3.48).

6.6 The Fubini theorem

In this section J will be an interval in $\overline{\mathbb{R}}^n$ with $J = H \times K$, and the intervals H and K will belong to $\overline{\mathbb{R}}^l$ and $\overline{\mathbb{R}}^m$, $n = l + m$. For a point $z \in \overline{\mathbb{R}}^n$ we write $z = (x, y)$ with $x \in \overline{\mathbb{R}}^l$ and $y \in \overline{\mathbb{R}}^m$. Similarly for a function f on $\overline{\mathbb{R}}^n$ we write $f(z) = f(x, y)$ and for a function h_x defined by $h_x(y) = f(x, y)$ we use the notation $h_x = f(x, .)$ introduced in Subsection 6.1.1. The i-dimensional measure of a set $S \subset \mathbb{R}^i$ is denoted by $m_i(S)$. For this section the function f will be KH-integrable on J and

$$Q(x) = \int_K f(x, .) . \tag{6.52}$$

In Subsection 6.4.1 (Prelude to Fubini's theorem) we proved for a regulated f that

$$\iint_J f = \int_H Q, \tag{6.53}$$

or equivalently

$$\iint_J f = \int_H \left(\int_K f(x, y) \, dy \right) dx. \tag{6.54}$$

If f is merely KH-integrable then, for a particular x, $Q(x)$ need not be defined. Worse, if $N_H \subset H$ is an arbitrary set of measure zero in $\mathbb{R}^l$, i.e. $m_l(N_H) = 0$, then $m_n(N_H \times K) = 0$ and f can be changed at will on $N_H \times K$ without its integrability being affected. Consequently Q may fail to be defined on N_H. Fortunately this is the worst possible scenario. We have

LEMMA 6.6.1 *If f is KH-integrable on J and $S \subset H$ is the set of x for which the integral $Q(x)$ is well defined then we have, for $N = H \setminus S$, that $m_l(N) = 0$.*

Proof We employ the negation of the Bolzano–Cauchy condition for exploiting the non-integrability of $f(x, .)$. A point $x \in N$ if and only if there is $\varepsilon_0(x)$ such that for every γ on K there exist two partitions of K

$$\begin{aligned} \sigma &= \{(\eta_i, K_i) \, ; \, i = 1, \dots, a\}, & \sigma \ll \gamma \\ \sigma' &= \{(\eta_i', K_i') \, ; \, i = 1, \dots, a'\} & \sigma' \ll \gamma \end{aligned}$$

with the property that

$$\left| \sum_{i=1}^{\alpha} f(x, \eta_i)|K_i| - \sum_{i=1}^{\alpha'} f(x, \eta_i')|K_i'| \right| \geq \varepsilon_0(x). \qquad (6.55)$$

The partitions σ and σ' depend, of course, on x so in the sequel we write $\eta_i(x)$, $K_i(x)$, $\eta_i'(x)$ and $K_i'(x)$. Let

$$N_\mu = \{x; \, x \in H, \, \varepsilon_0(x) \geq \mu\}.$$

Obviously $N = \bigcup_{n=1}^{\infty} N_{\frac{1}{n}}$ and it suffices to show that $m_l(N_\mu) = 0$ for every $\mu > 0$. By Henstock's lemma for every positive ε there exists a gauge δ on J such that

$$\sum_\pi \left| f(z_k)|J_k| - \iint_{J_k} f \right| < \frac{\varepsilon\mu}{2}, \qquad (6.56)$$

for $\pi \ll \delta$. Write $\delta(x, y) = \delta_H(x, y) \times \delta_K(x, y)$ with $\delta_H \subset \overline{\mathbb{R}}^l$ and $\delta_K \subset \overline{\mathbb{R}}^m$. Choose $\gamma(y) = \delta_K(x, y)$ and find partitions σ and σ' such that (6.55) holds with $\varepsilon_0(x)$ replaced by μ. For $x \in N$ define

$$\Delta(x) = \bigcap_{i=1}^{\alpha} \delta_H(x, \eta_i(x)) \cap \bigcap_{i=1}^{\alpha'} \delta_H(x, \eta_i'(x)).$$

For $x \in S$ let $\Delta = \delta_H$. If $\pi_H \equiv \{(\xi, H(\xi))\}$ is a Δ-fine partition of H then

$$\sum_{\pi_H} \mathbf{1}_{N_\mu} = \sum_{\pi_H} \mathbf{1}_{N_\mu}(\xi)|H(\xi)|$$

$$\leq \mu^{-1} \sum_{\substack{\pi_H \\ \xi \in N_\mu}} \left| \sum_{i=1}^{\alpha} f(\xi, \eta_i(\xi))|H(\xi)||K_i(\xi)| - \sum_{i=1}^{\alpha'} f(\xi, \eta_i'(\xi))|H(\xi)||K_i'(\xi)| \right|.$$

$$(6.57)$$

Since

$$\bigcup_{i=1}^{\alpha} H(\xi) \times K_i(\xi) = \bigcup_{i=1}^{\alpha'} H(\xi) \times K_i'(\xi) = H(\xi) \times K,$$

we have

$$\sum_{i=1}^{\alpha} \iint_{H(\xi) \times K_i(\xi)} f = \sum_{i=1}^{\alpha'} \iint_{H(\xi) \times K_i'(\xi)} f = \iint_{H(\xi) \times K} f. \qquad (6.58)$$

The intervals $H(\xi) \times K_i(\xi)$ form a δ-fine partial division of J. By (6.56) and (6.58)

$$\sum_{\substack{\pi_H \\ \xi \in N_\mu}} \left| \sum_{i=1}^{\alpha} f(\xi, \eta_i(\xi)) |H(\xi)||K_i(\xi)| - \iint_{H(\xi) \times K} f \right| < \frac{\varepsilon \mu}{2}. \qquad (6.59)$$

Similarly

$$\sum_{\substack{\pi_H \\ \xi \in N_\mu}} \left| \sum_{i=1}^{\alpha'} f(\xi, \eta_i'(\xi)) |H(\xi)||K_i'(\xi)| - \iint_{H(\xi) \times K} f \right| < \frac{\varepsilon \mu}{2}. \qquad (6.60)$$

Combining these inequalities with(6.57) leads to

$$\left| \sum_{\pi} 1_{N_\mu} \right| < \mu^{-1}\left(\frac{\varepsilon \mu}{2} + \frac{\varepsilon \mu}{2}\right).$$

This means $\int_H 1_{N_\mu} = 0$.   •

During the proof we almost established

LEMMA 6.6.2 *If*

  (i) δ *is a gauge on J;*

  (ii) $\delta(x,y) = \delta_H(x,y) \times \delta_K(x,y)$ *with $\delta_H(x,y) \subset \overline{\mathbb{R}}^l$ and $\delta_K(x,y) \subset \overline{\mathbb{R}}^m$ for $(x,y) \in J$;*

  (iii) *the partition $\sigma(x) \equiv \{(\eta_i(x), K_i(x)); i = 1, \dots, \alpha\}$ is δ_K-fine for every $x \in H$*

then there exists Δ, a gauge on H, with the property that a bl-partition

$$\{((\xi, \eta_i(\xi)), H_\xi \times K_i(\xi); i = 1, \dots, \alpha\}$$

of J is δ-fine whenever $\pi_H \equiv \{(\xi, H_\xi)\}$ is a Δ-fine partition of H.

Proof It is sufficient to set

$$\Delta(x) = \bigcap_{i=1}^{\alpha} \delta(x, \eta_i(x)). \qquad \bullet$$

With all the necessary preparation done we have now

THEOREM 6.6.3 (The Fubini theorem) *Suppose f is integrable on $J = H \times K$. Then*

  (i) $Q(x) = \int_K f(x, .)$ *exists and is finite for almost all $x \in H$;*

  (ii) Q *is KH-integrable on H and the multiple integral of f is equal to the iterated integral as in (6.53).*

By symmetry we have

COROLLARY 6.6.4

$$\iint_J f = \int_H \left(\int_K f(x,y)\, dy \right) dx = \int_K \left(\int_H f(x,y)\, dx \right) dy. \qquad (6.61)$$

Proof of Fubini's theorem Assertion (i) was proved in lemma 6.6.1. We keep the notation for N and S from that lemma and put $M = N \times K$. This set is negligible, $\int_J \mathbf{1}_M f = 0$ and consequently, for every positive ε there is a gauge δ_0 such that

$$\sum_\pi |f| < \frac{\varepsilon}{6} \qquad (6.62)$$

whenever $\pi \ll \delta_0$ and is tagged in M. For ease of writing we set

$$\mathsf{F}(L) = \int_L f.$$

By Henstock's lemma there is a gauge δ_1 such that

$$\sum_\pi |f(z)|L| - \mathsf{F}(L)| < \frac{\varepsilon}{6} \qquad (6.63)$$

for a δ_1-fine partition $\pi \equiv \{(z, L)\}$. Let $\delta(z) = \delta_0(z) \cap \delta_1(z)$ for $z \in J$. Combining (6.62) and (6.63) shows that

$$\sum_\pi |\mathsf{F}(L)| < \frac{\varepsilon}{3}, \qquad (6.64)$$

if π is δ-fine and tagged in M. We choose, by Lemma 6.5.3, a gauge Δ_0 on H and a function $\mathcal{E} : H \mapsto \mathbb{R}_+$ such that

$$\sum_{\pi_0} \mathcal{E} < \frac{\varepsilon}{6} \qquad (6.65)$$

for every Δ_0-fine partition π_0 of H. For $x \in S$ let $\sigma(x)$ be a δ_K-fine partition for which

$$|Q(x) - \sum_{\sigma(x)} f(x, \cdot)| < \mathcal{E}. \qquad (6.66)$$

For $x \in N$ let $\sigma(x)$ be a δ_K-fine partition. Let Δ and

$$\pi \equiv \{((\xi, \eta_i(\xi)), H_\xi \times K_i(\xi)\}$$

be according to Lemma 6.6.2. If necessary we diminish Δ so that $\Delta \subset \Delta_0$ and have by (6.62)—(6.66)

$$\left| \sum_{\pi_H} Q(\xi)|H_\xi| - F(J) \right| \leq \left| \sum_{\pi_H} |H_\xi| \left[Q(\xi) - \sum_{\sigma(\xi)} f(\xi, \eta_i(\xi))|K_i(\xi)| \right] \right|$$

$$+ \left| \sum_{\pi} f(\xi, \eta_i(\xi))|H_\xi||K_i(\xi)| - F(H_\xi \times K_i(\xi)) \right|$$

$$\leq \sum_{\pi_H} \mathcal{E} + \frac{\varepsilon}{6} + \sum_{\substack{\pi \\ \xi \in N}} |f(\xi, \eta_i(\xi))|H_\xi||K_i(\xi)|| + \sum_{\substack{\pi \\ \xi \in N}} |F(H_\xi \times K_i(\xi))|$$

$$< \frac{\varepsilon}{6} + \frac{\varepsilon}{6} + \frac{\varepsilon}{6} + \frac{\varepsilon}{3}.$$ ●

We return the example from the end of Section 6.4. The function f with $f(x,y) = \dfrac{x}{\sqrt{x^2+4y}}$ is clearly measurable and bounded and hence integrable. By the Fubini theorem

$$\iint\limits_{\substack{0<x<1 \\ 0<y<1}} \frac{x}{\sqrt{x^2+4y}} \, dx dy = \int_0^1 (\sqrt{1+4y} + 2\sqrt{y}) dy = \frac{1}{6}(5\sqrt{5}+7).$$

The Fubini theorem can be also used to prove that f is not integrable on J. To do that it is sufficient to find a subinterval of J† on which the repeated integrals are not equal. There is a difficulty in applying the Fubini theorem: we need to establish the integrability of f on J. Apart from the case when f is bounded and measurable this can be a difficult matter. It is therefore important that Fubini's theorem can be used when one of the repeated integrals of the *absolute value* of f is finite. More generally we have

THEOREM 6.6.5 (The Tonelli theorem) *If*

(i) *f is measurable on J*

(ii) *there is a function g such that $|f| \leq g$ on J and either*

$$A_1 = \int_H \left(\int_K g(x,y) \, dy \right) dx < \infty$$

or

$$A_2 = \int_K \left(\int_H g(x,y) \, dx \right) dy < \infty$$

then f is KH-integrable on J and (6.53) is satisfied.

† This could, of course, be J itself.

Taking for $g = |f|$ gives

COROLLARY 6.6.6 *If f is measurable and either*

$$\int_H \left(\int_K |f(x,y)|\, dy \right) dx < \infty$$

or

$$\int_K \left(\int_H |f(x,y)|\, dx \right) dy < \infty$$

then (6.53) holds.

In particular

COROLLARY 6.6.7 *If f is measurable and **non-negative** then*

$$\iint_J f = \int_H \left(\int_K f(x,y)\, dy \right) dx = \int_K \left(\int_H f(x,y)\, dx \right) dy \quad (6.67)$$

provided that at least one of the three integrals exists and is finite.

Proof of Tonelli's theorem It is sufficient to prove integrability of f; the rest then follows from the Fubini theorem. For the sake of definiteness assume that $A_1 < \infty$. Let h_n be the characteristic function of the cube $\{(x,y); |(x,y)|_u \le n\}$ and $f_n = \text{Min}(f^+, nh_n)$. Then f_n is measurable and bounded by an integrable function nh_n. It is therefore itself integrable and by the Fubini theorem

$$\int_J f_n = \int_H \left(\int_K f_n(x,y)\, dy \right) dx \le A_1.$$

By the monotone convergence theorem f^+ is KH-integrable on J. Similarly f^- is integrable. •

EXAMPLE 6.6.8 Let $J = [0,\infty] \times [0,a]$ and $f(x,y) = e^{-kx} \cos xy$ with $k > 0$. In this case we choose $g(x,y) = e^{-kx}$ and have

$$\int_J f = \int_0^\infty dx \int_0^a e^{-kx} \cos xy\, dy = \int_0^\infty e^{-kx} \frac{\sin ax}{x}\, dx,$$

$$\int_J f = \int_0^a \frac{k}{k^2 + y^2}\, dy = \arctan \frac{a}{k}.$$

Taking the limit as $k \to \infty$ and using Example 3.7.8 gives (for $a > 0$)

$$\int_0^\infty \frac{\sin ax}{x}\, dx = \frac{\pi}{2}.$$

EXAMPLE 6.6.9 (The Laplace integral) We wish to evaluate the
Laplace integral $A = \int_0^\infty e^{-x^2}\, dx$. Let us consider

$$\int_0^\infty \int_0^\infty e^{-x^2-y^2}\, dx dy.$$

The integrand is continuous and non-negative. We use Corollary 6.6.7
and have

$$B = \int_0^\infty \int_0^\infty e^{-x^2-y^2}\, dx dy = \int_0^\infty \left(e^{-x^2}\int_0^\infty e^{-y^2}\, dy\right) dx = A^2 < \infty$$

$$B = \left(\int_0^\infty e^{-x^2}\, dx\right)\left(\int_0^\infty e^{-t^2 x^2} x\, dt\right) = \int_0^\infty \int_0^\infty e^{-(1+t^2)x^2} x\, dx dt$$

$$= \int_0^\infty \left(\int_0^\infty e^{-(1+t^2)x^2} x\, dx\right) dt = \int_0^\infty \frac{dt}{2(1+t^2)} = \frac{\pi}{4}.$$

Consequently $A = \sqrt{B} = \frac{1}{2}\sqrt{\pi}$.

REMARK 6.6.10 In several dimensions there is a much greater need
to integrate over sets that are not intervals than in one dimension. In
order to apply Fubini's theorem or Tonneli's theorem to $\int_S f = \int_{\mathbb{R}^n} f 1_S$
the measurability of S is required. In most cases in practice this is
easily done by inspection, e.g. S might be open or closed or some other
set whose measurability is obvious. The following observation is useful in
more difficult situations. Let M be measurable as a subset of $\mathbb{R}^l$.† Write
$(x,y) \in \mathbb{R}^n$ with $x \in \mathbb{R}^{n-1}$ and $y \in \mathbb{R}^1$. If f is a function measurable in
M then all the sets

$$S_1 = \{(x,y); x \in M, y < f(x)\},$$
$$S_2 = \{(x,y); x \in M, y > f(x)\},$$
$$S_3 = \{(x,y); x \in M, y = f(x)\}$$

are measurable. See also Exercise 6.10.

EXAMPLE 6.6.11 Let us calculate

$$A = \iiint_D dx dy dz$$

where

$$D = \left\{(x,y,z); z > 0, \frac{x^2}{a^2} + \frac{y^2}{b^2} + z^2 < 1.\right\}$$

† In our previous discussions the dimension of the space was fixed. It is necessary
to realize now that the concept of measurability depends on the dimension.

Integrating with respect to z first and splitting the double integral into two integrals, the first with respect to y, leads to

$$
\begin{aligned}
A &= \frac{1}{2} \int_{-a}^{a} dx \int_{-\sqrt{a^2-x^2}}^{\sqrt{a^2-x^2}} \left(1 - \frac{x^2+y^2}{a^2}\right) dy \\
&= \frac{2}{a^2}\frac{2}{3} \int_{0}^{a} \left(a^2 - x^2\right)^{\frac{3}{2}} = \frac{\pi a^2}{4}.
\end{aligned}
$$

There is another way to evaluate the integral, integrating with respect to z last and not splitting the inner double integral:

$$
A = \int_{0}^{1} \left(\iint_{D_z} dx\,dy \right) z\,dz
$$

where $D_z = \{(x,y); x^2 + y^2 \leq a^2(1 - z^2)\}$. The double integral is clearly the area of the circle D_z and consequently

$$
A = \pi a^2 \int_{0}^{1} z(1 - z^2)\,dz = \frac{\pi a^2}{4}.
$$

6.7 Change of variables

The theorem on substitution in multiple integrals is naturally more complicated than the corresponding theorems in one dimension. The proof is not very difficult but it is long and contains technical details. It is made in several steps, by induction on the dimension and by moving successively from less complicated sets and functions to more complicated ones. We have already changed variables in multiple integrals, more precisely in repeated integrals, but we changed only one variable at a time. This method is also used in the general case but the substitution must be represented as a composition of two substitutions each of which leaves all but one variable unchanged.

A new aspect in the multidimensional case is that the change of variable is often made not to simplify the integrated function but the domain of integration. To obtain some ideas of what is involved we look at some simple examples first.

6.7.1 Introductory examples

The simplest transformation is a linear one. This is our first example.

EXAMPLE 6.7.1   We use the linear transformation

$$x = a_{11}u + a_{12}v, \tag{6.68}$$

$$y = a_{21}u + a_{22}v, \tag{6.69}$$

on the integral

$$\iint_{y>1} f(x,y)\, dx dy.$$

We assume that f is integrable and decompose the transformation (6.68), (6.69) as follows:

$$x = \left(a_{11} - \frac{a_{21}a_{12}}{a_{22}}\right)u + \frac{a_{12}}{a_{22}}y, \qquad u = u,$$

$$y = y, \qquad y = a_{21}u + a_{22}v.$$

For the substitutions in the repeated integrals we need the assumptions that $J_A = (a_{11}a_{22} - a_{21}a_{12}) \neq 0$ and $a_{22} \neq 0$. The case that $J_A < 0$ and $a_{22} > 0$ is more revealing. Hence we assume this, and have†

$$\iint_{y>1} f(x,y)\, dx dy = \int_1^\infty dy \int_{-\infty}^\infty f(J_A a_{22}^{-1}u + a_{12}a_{22}^{-1}y, y)|J_A|a_{22}^{-1}\, du$$

$$= \int_{-\infty}^\infty du \int_1^\infty f(J_A a_{22}^{-1}u + a_{12}a_{22}^{-1}y, y)|J_A|a_{22}^{-1}\, dy$$

$$= \int_{-\infty}^\infty du \int_{\alpha(u)}^\infty f(a_{11}u + a_{12}v, a_{21}u + a_{22}v)|J_A|\, dv,$$

where we have abbreviated $\alpha(u) = (1 - a_{21}u)a_{22}^{-1}$. Now we wish to convert the last repeated integral into a double integral over the set

$$G = \{(u,v)\,;\, u \in \mathbb{R}, v > \alpha(u)\}.$$

In order to use Fubini's theorem we assume that the function appearing in the last repeated integral is integrable on G. Abbreviating the transformation (6.68), (6.69) as $(x,y) = A(u,v)$ and noting that

$$\{(x,y)\,;\, x \in \mathbb{R}, y > 1\} = A(G)$$

we finally obtain

$$\iint_{A(G)} f(x,y)\, dx dy = \iint_G (f \circ A)(u,v)|J_A|\, du dv \tag{6.70}$$

---

† The absolute values in these calculations appear because we need to keep the upper limits of integration greater than the lower limits.

or denoting $A(G)$ by S we have

$$\iint\limits_{S} f(x,y)\,dxdy = \iint\limits_{A_{-1}(S)} (f \circ A)(u,v)|J_A|\,dudv. \qquad (6.71)$$

These formulae are already very similar to the general formula for change of variables. The factor $|J_A|$ is important; we note that

$$J_A = \begin{vmatrix} \frac{\partial x}{\partial u} & \frac{\partial x}{\partial v} \\ \frac{\partial y}{\partial u} & \frac{\partial y}{\partial v} \end{vmatrix}.$$

The determinant on the right-hand side is called the Jacobian of the transformation. If $A(u,v)$ is not linear, as it was in our example, J_A is a function of u and v but formulae (6.70) and (6.71) remain the same, as we shall see later. The assumption that $|J_A| \neq 0$ appears also in the general theorem. The assumption that both integrals in (6.70) exist seems reasonable. Since in practice one integral is obtained by means of the other, it is desirable to have a theorem which not only asserts the equality of the integrals but also guarantees the existence of the former from the latter. In order to obtain such a theorem we strengthen our assumption in Theorem 6.7.7 below to *absolute* integrability.

It can indeed happen that one integral exists and the other does not. To see this we consider the transformation

$$(x,y) = A(u,v) \equiv \begin{cases} x = \frac{1}{\sqrt{2}}(u-v) \\ y = \frac{1}{\sqrt{2}}(u+v) \end{cases}$$

and the integral $\iint_{A(D)} f$ where f is the function from Exercise 6.3 and $A(D) = \{(x,y)\,; x^2 + y^2 \leq 4\}$. The function f is KH-integrable on $A(D)$ since it is KH-integrable on $[0,1] \times [0,1]$ and is zero outside this interval. So the integral $\iint_{A(D)} f$ exists and is finite. On the other hand

$$\iint\limits_{D} f(A(u,v))\,dudv$$

cannot exist because† $K = [0,1] \times [0,-1] \subset D$ and

$$\iint\limits_{K} f(A(u,v))\,dudv = \sum_{1}^{\infty} \frac{1}{n} = \infty.$$

† Note that $D = A(D)$ and $J_A = 1$.

EXAMPLE 6.7.2 We consider polar coordinates

$$x = r \cos \varphi$$
$$y = r \sin \varphi.$$

The Jacobian of this transformation is

$$\begin{vmatrix} \frac{\partial x}{\partial r} & \frac{\partial x}{\partial \varphi} \\ \frac{\partial y}{\partial r} & \frac{\partial y}{\partial \varphi} \end{vmatrix} = r.$$

This leads to the integrals

$$\iint_{\mathbb{R}^2} f(x,y)\, dx\, dy \text{ and } \int \int_{\substack{-\frac{\pi}{2} < \varphi < \frac{\pi}{3} \\ 0 < r < \infty}} f(r \cos \varphi, r \sin \varphi) r\, dr\, d\varphi. \qquad (6.72)$$

We assume that both integrals exist and are finite. Then by the Fubini theorem and by the substitution theorem for one-dimensional integrals

$$\iint_{\mathbb{R}^2} f(x,y)\, dx\, dy = \int_{-\infty}^{0} dx \int_{-\infty}^{\infty} f(x,y)\, dy + \int_{0}^{\infty} dx \int_{-\infty}^{\infty} f(x,y)\, dy,$$

$$\int_{0}^{\infty} dx \int_{-\infty}^{\infty} f(x,y)\, dy = \int_{0}^{\infty} dx \int_{-\frac{\pi}{2}}^{\frac{\pi}{2}} f(x, x \tan \varphi) \frac{x\, d\varphi}{\cos^2 \varphi},$$

$$\int_{0}^{\infty} \int_{-\frac{\pi}{2}}^{\frac{\pi}{2}} f(r \cos \varphi, r \sin \varphi) r\, dr\, d\varphi = \int_{-\frac{\pi}{2}}^{\frac{\pi}{2}} d\varphi \int_{0}^{\infty} f(x, x \tan \varphi) \frac{x}{\cos^2 \varphi}\, dx.$$

Consequently

$$\int_{0}^{\infty} \int_{-\infty}^{\infty} f(x,y)\, dx\, dy = \int_{0}^{\infty} \int_{-\frac{\pi}{2}}^{\frac{\pi}{2}} f(r \cos \varphi, r \sin \varphi) r\, dr\, d\varphi. \qquad (6.73)$$

A similar argument can be given for the set $\{(x,y); x < 0\}$. Hence we have established that the integrals in (6.72) are equal.

In this example it was necessary to split the domain of integration into two, since we were unable to decompose, on all of $\mathbb{R}^2$, the transformation by polar coordinates into two transformations each of which changed only one variable. We shall encounter a similar situation in the general change of variables in a more complicated form. It will be necessary to split the domain of integration into a finite number of smaller domains and use our method on each of them separately.

EXAMPLE 6.7.3 (Viviani's problem) The problem consists in finding the (three-dimensional) measure of the set V described by the inequalities

$$x^2 + y^2 + z^2 \le a^2, \quad x^2 + y^2 \le ax.$$

The set is clearly an intersection of a ball and a cylinder. The existence of all multiple integrals in the following calculations is obvious. First we have

$$m_3(V) = \iint\limits_{x^2+y^2 \le ax} \left(\int\limits_{-\sqrt{a^2-x^2-y^2}}^{\sqrt{a^2-x^2-y^2}} dz \right) dx\,dy = 2 \iint\limits_{x^2+y^2 \le ax} \sqrt{a^2-x^2-y^2}\,dx\,dy.$$

Now we employ polar coordinates and have

$$m_3(V) \quad = \quad 4 \int\limits_0^{\frac{\pi}{2}} \left(\int\limits_0^{a\cos\varphi} \sqrt{a^2-r^2}\,r\,dr \right) d\varphi$$

$$= \quad \frac{4}{3} \int\limits_0^{\frac{\pi}{2}} (1 - \sin^3\varphi)\,d\varphi = \frac{2}{9}(3\pi - 4).$$

6.7.2 Notation, lemmas

From now on until the end of this chapter G will denote an open set. Partial derivatives will be denoted by subscripts and commas like this

$$\varphi_{,i} = D_i\varphi = \frac{\partial\varphi}{\partial x_i}.$$

For the change of variables theorem we need a number of lemmas. The first is

LEMMA 6.7.4 *If φ has continuous partial derivatives in G and $\varphi_{,n}(a) \ne 0$ then there exist numbers h and a function ψ defined on*

$$K(a) = (a^1 - h, a^1 + h) \times \cdots \times (a^{n-1} - h, a^{n-1} + h)$$
$$\times (\varphi(a) - h, \varphi(a) + h)$$

with the following properties:

(i) *for every $(x^1, x^2, \ldots, x^n) \in K(a)$ we have*

$$x^n = \varphi(x^1, x^2, \ldots, x^{n-1}, \psi(x^1, \ldots, x^n));$$

(ii) *ψ has continuous partial derivatives in $K(a)$.*

We postpone the proof to the appendix.

LEMMA 6.7.5 *If J is a compact interval and L_i with $i \in \mathbb{N}$, $i \leq m$ are open intervals covering J then there exist disjoint intervals j_k with $k \in \mathbb{N}, k \leq p$ such that each j_k is contained in some L_k and $J = \bigcup_{k+1}^p j_k$.*

Proof The intervals L_i define a gauge δ on J; indeed every $x \in J$ lies in some L_i and one can set $\delta(x) = L_i$. Let $\{(x, K)\}$ be a δ-fine partition of J. Now applying Lemma 6.2.3 to J and the K's yields the intervals j_k. ●

A map $\varphi : O \mapsto \mathbb{R}^n$, $\varphi(x) = (\varphi^1(x), \ \blacksquare \ )$ is called *regular* in O if

(i) O is an open set in $\mathbb{R}^n$;
(ii) the partial derivatives $\varphi^i_{,j}$ are continuous in O
(iii) the Jacobian of the mapping

$$
J_\varphi(x) = \begin{vmatrix} \varphi^1_{,1}(x) & \varphi^1_{,2}(x) & \cdots & \varphi^1_{,n}(x) \\ \varphi^2_{,1}(x) & \varphi^2_{,2}(x) & \cdots & \varphi^2_{,n}(x) \\ \cdots\cdots\cdots\cdots\cdots\cdots\cdots\cdots\cdots\cdots \\ \cdots\cdots\cdots\cdots\cdots\cdots\cdots\cdots\cdots\cdots \\ \varphi^n_{,1}(x) & \varphi^n_{,2}(x) & \cdots & \varphi^n_{,n}(x) \end{vmatrix} \neq 0 \qquad (6.74)
$$

for every $x \in O$.

In the rest of this chapter we reserve the letter φ for a map which is regular in G. It can be shown that for a regular map the set $\varphi(G)$ is open, but rather than get involved in a long proof we just make it an additional assumption which we keep for the rest of this chapter.

LEMMA 6.7.6 *If φ is a regular map in G then for every $x \in G$ there is an index i such that $\varphi^n_{,i}(x) \neq 0$.*

This lemma is obvious; if $\varphi^n_{,i} = 0$ for all i, $1 \leq i \leq n$ then $J_\varphi(x) = 0$.

6.7.3 The theorem

We already encountered the formula for change of variables in equation (6.71). For a map φ and an n-dimensional integral it takes the form

$$
\int \cdots \int_S f(x)\, dx = \int \cdots \int_{\varphi_{-1}(S)} (f \circ \varphi)(u)|J_\varphi(u)|\, du
$$

or briefly

$$\int_S f(x)\,dx = \int_{\varphi_{-1}(S)} (f \circ \varphi)(u)|\mathrm{J}_\varphi(u)|\,du \qquad (6.75)$$

with $x = (x^1, \bullet)$ and $u = (u^1, \bullet)$. Our main theorem reads

THEOREM 6.7.7 (Main theorem) *Assume that*

(i) *the regular map φ is one-to-one on G;*

(ii) *the function f is absolutely KH-integrable on a measurable set $S \subset \varphi(G)$.*

Then the function $(f \circ \varphi)|\mathrm{J}_\varphi|$ is absolutely KH-integrable on $\varphi_{-1}(S)$ and formula (6.75) holds.

It is desirable to have also a theorem which allows us to go from right to left in the formula. We return to this after we have dealt with this theorem first.

We have already seen that the assumption of absolute integrability is essential. Since absolute integrability is equivalent to Lebesgue integrability, Theorem 6.7.7 is essentially a change of variable theorem for the Lebesgue integral. The proof is always long, but the reader might like to see another proof. Change of variables theorems in Lebesgue theory which do not require anything more than what we already know can be found in [41] and [2]. We prove the theorem by several lemmas, each of which is is again proved in a number of steps. Fortunately many of these steps are easy. The main idea of the proof is to start with simple sets and functions and move successively to more complicated ones.

LEMMA 6.7.8 *The main theorem is valid if $f = 1$, the set $S \subset \varphi(G)$ is a bounded interval and φ merely interchanges two variables.*

Proof It follows from elementary rules of evaluation of determinants that $\mathrm{J}_\varphi = \pm1$, since φ in this simple case carries intervals into intervals of the same content there is nothing more to prove.   •

LEMMA 6.7.9 *The main theorem is valid if $f = 1$, the set S is a bounded interval, $S \subset \varphi(G)$ and $n = 1$.*

Proof Let $S = [a, b]$ with $a = \varphi(\alpha)$ and $b = \varphi(\beta)$. It is obvious that

$$\int_a^b dx = b - a = \varphi(\beta) - \varphi(\alpha) = \int_\alpha^\beta \varphi'(u)\,du\,.$$

The derivative φ' is continuous so it is either positive or negative for all $x \in S$. If it is positive then

$$\int_\alpha^\beta \varphi'(u)\, du = \int_{\varphi_{-1}([a,b])} |\varphi'(u)|\, du,$$

and if it is negative then

$$\int_\alpha^\beta \varphi'(u)\, du = -\int_\beta^\alpha \varphi'(u)\, du = \int_{\varphi_{-1}([a,b])} |\varphi'(u)|\, du \ . \qquad \bullet$$

The proof now proceeds by induction and we may assume that the theorem holds for $n-1$ if $S = I$ is a bounded interval $I \subset \varphi(G)$ and $f = 1$. In this particular case equation (6.75) reads

$$m_n(I) = \int_{\varphi_{-1}(I)} |\mathsf{J}_\varphi|. \qquad (6.76)$$

LEMMA 6.7.10 *If the map φ leaves one variable unchanged then (6.76) holds.*

Proof In view of Lemma 6.7.8 we can assume that the first variable remains unchanged, i.e. φ is of the form

$$\begin{aligned}
u^1 &= x^1, \\
u^2 &= \varphi^2(x^1, x^2, \ldots, x^n), \\
&\cdots \\
u^n &= \varphi^n(x^1, x^2, \ldots, x^n).
\end{aligned}$$

Let $I = I_1 \times \tilde{\ }$, $\tilde{I} = I_2 \times \cdots \times I_n$ and

$$\tilde{\varphi}^{x^1}(x^2, \ldots, x^n) = (\varphi^2(x^1, \ldots, x^n), \ldots, \varphi^n(x^1, \ldots, x^n)).$$

It is obvious that $\mathsf{J}_\varphi = \mathsf{J}_{\tilde{\varphi}^{x^1}}$ and also

$$m_n(I) = \int_I dx = \int_{I_1} dx^1 \int_{\tilde{I}} dx^2 \ldots dx^n.$$

In the $(n-1)$-dimensional integral we are allowed to change the variables and have

$$\int_I dx = \int_{I_1} dx^1 \int_{\tilde{\varphi}_{-1}^{x^1}(\tilde{I})} du^2 \ldots du^n.$$

The formula

$$\varphi_{-1}(I) = I_1 \times \tilde{\varphi}^{x^1}_{-1}(\tilde{I})$$

becomes obvious when one thinks what it means. It remains to show that the following conversion of the repeated integral into a multiple integral is valid:

$$\int_{I_1} dx^1 \int_{\tilde{\varphi}^{x^1}_{-1}(\tilde{I})} du^2 \ldots du^n = \int_{\varphi_{-1}(I)} |J_\varphi| \, .$$

This follows from Fubini's theorem as soon as the measurability of the set $\varphi_{-1}(I)$ is established. If I is an open interval then since φ is continuous $\varphi_{-1}(I)$ is open and hence measurable. If I is any bounded interval then it is easy to see that there is a sequence of bounded open intervals $\{O_i\}$ such that $I = \bigcap_{i=1}^{\infty} O_i$. Consequently $\varphi_{-1}(I) = \bigcap_{i=1}^{\infty} \varphi_{-1}(O_i)$ and is measurable.   •

The next lemma could be of some independent interest, so we call it a theorem.

THEOREM 6.7.11   *Assume that*

  (i) *the map φ is regular in G;*
  (ii) *there is a continuous non-negative function Δ with the property that*

$$m_n(I) = \int_{\varphi_{-1}(I)} \Delta(u) \, du$$

  for every bounded interval I with $\bar{I} \subset \varphi(G)$;
  (iii) *the function f is absolutely KH-integrable on $\varphi(G)$.*

Then

  (a) *the function $(f \circ \varphi)\Delta$ is absolutely KH-integrable on G and*

$$\int_{\varphi(G)} f(x) \, dx = \int_{G} (f \circ \varphi)(u)\Delta(u) \, du; \qquad (6.77)$$

  (b) *if N is of measure zero and $N \subset \varphi(G)$ then $m_n(\varphi_{-1}(N)) = 0$.*

REMARK 6.7.12   Condition (iii) is satisfied with $\Delta = |J_\varphi|$ if φ leaves one variable unchanged.

Proof of Theorem 6.7.11 It suffices to prove the theorem for non-negative f. We denote by $\mathfrak{F}$ the set of measurable functions f which are defined and non-negative *everywhere* in $\varphi(G)$ and for which (6.77) is true. We need the following implications:

I. If f_1 and f_2 are in $\mathfrak{F}$ then so is $f_1 + f_2$.

II. If $c \geq 0$ and $f \in \mathfrak{F}$ then $cf \in \mathfrak{F}$.

III. If f_i form an increasing sequence for every $x \in S$ or a decreasing sequence for every $x \in S$ and $f_i \in \mathfrak{F}$ for every $i \in \mathbb{N}$ then $\lim_{i \to \infty} f_i \in \mathfrak{F}$.

Statements I and II are obvious. To prove III we write

$$\iint\limits_{S} f_i(x)\, dx = \iint\limits_{\varphi_{-1}(S)} (f_i \circ \varphi)(u)\Delta(u)\, du.$$

Now we pass to the limit as $i \to \infty$.† If the sequence is increasing we use the monotone convergence theorem and if it is decreasing the Lebesgue dominated convergence Theorem, the majorant being f_1 or $(f_1 \circ \varphi)\Delta$, respectively.

Now we prove the theorem in eight steps.

Step 1. For every interval I there is an increasing sequence of bounded intervals I_k such that $\lim_{k \to \infty} \mathbf{1}_{I_k} = \mathbf{1}_I$. For each I_k (6.77) holds by assumption, and consequently $\mathbf{1}_I \in \mathfrak{F}$ by III.

Step 2. If $O \subset \varphi(G)$ is open then $\mathbf{1}_O \in \mathfrak{F}$. The open set O is a union of a sequence of disjoint intervals‡ $O = \bigcup\limits_{i=k}^{\infty} I_k$. Clearly

$$\mathbf{1}_O = \lim_{k \to \infty} \left(\mathbf{1}_{I_1} + \mathbf{1}_{I_2} + \cdots + \mathbf{1}_{I_k} \right)$$

and the proof is complete by appealing to III.

Step 3. We say that a set S is G_δ if it is the intersection of a sequence $\{O_k\}$ of open sets. If necessary this sequence can always be made decreasing. If S is bounded then S is a subset of an open interval I, if it is G_δ the sets $I \cap \varphi(G) \cap O_k$ are open, and by step 2 their characteristic functions are in $\mathfrak{F}$. By III for a decreasing sequence we have $\mathbf{1}_S \in \mathfrak{F}$. If S is any G_δ then the characteristic functions of $S_k = S \cap C(0, k)$ are in

† It is here that we need the functions f_i to be defined everywhere on $\varphi(G)$ in order to assert the convergence of the sequence $\{f_i \circ \varphi\}$ everywhere on $\varphi_{-1}(S)$.

‡ By the covering lemma O is the union of a sequence of non-overlapping closed intervals. Changing these to half open half closed intervals one obtains a disjoint sequence.

$\mathfrak{F}$ by what we have already proved. We can apply III to the increasing sequence $\mathbf{1}_{S_k}$ to obtain $\mathbf{1}_S \in \mathfrak{F}$ for a G_δ set S.

Step 4. Let $N \subset \varphi(G)$ be of measure zero. There are open sets O_k containing N with $m_n(O_k) < 1/k$. The intersection of all these sets S is a G_δ set of measure zero and by step 3 its characteristic function is in $\mathfrak{F}$, which means that

$$0 = \int_S dx = \int_G \mathbf{1}_S(\varphi(u))\Delta(u)\, du.$$

Since $\Delta > 0$ this implies that $\mathbf{1}_S \circ \varphi$ is zero almost everywhere. Obviously $\mathbf{1}_N \circ \varphi$ is also zero a.e. in G. Consequently

$$\int_G (\mathbf{1}_N \circ \varphi)\Delta = 0 = m_n(N)$$

and $\mathbf{1}_N \in \mathfrak{F}$. Moreover $\mathbf{1}_N \circ \varphi = \mathbf{1}_{\varphi_{-1}(N)}$ and it follows that

$$m_n(\varphi_{-1}(N)) = 0.$$

We have now proved part (b) of the theorem.

Step 5. Let S be measurable, $S \subset \varphi(G)$. Using the n-dimensional version of Theorem 3.11.15 we find a G_δ set $E \supset S$ such that $m_n(E \backslash S) = 0$. By steps 3 and 4 the characteristic functions of E and $E \backslash S$ are in $\mathfrak{F}$. So is their non-negative difference $\mathbf{1}_E - \mathbf{1}_{E \backslash S} = \mathbf{1}_S$.

Step 6. Let $f \geq 0$ be simple, finite and measurable in $\varphi(G)$. Then f is a linear combination with positive coefficients of characteristic functions of measurable sets and belongs to $\mathfrak{F}$ by step 5 and I and II.

Step 7. Let $f \geq 0$ be measurable in $\varphi(G)$. By Theorem 3.11.10 there exists an increasing sequence of simple non-negative measurable functions converging everywhere to f. It follows from step 6 and III that $f \in \mathfrak{F}$.

Step 8. If f is measurable and the set N where it is negative or not defined is of measure zero then redefining f to be zero on N does not change the integral on the left-hand side of (6.77). The integral on the right-hand side is not changed either because the function $f \circ \varphi$ is changed only on the set $\varphi_{-1}(N) = 0$ and by step 3 this set is of measure zero. •

LEMMA 6.7.13 *For every a in $\varphi(G)$ there is an open interval $K(a)$ such that for every interval $I \subset K(a)$ equation (6.76) holds.*

Proof By Lemma 6.7.6 one of the partial derivatives of φ^n is not zero. In view of Lemma 6.7.8 we can assume that $\varphi^n_{,n}(a) \neq 0$. Let $K(a)$ and $\psi = \psi_n$ be as in Lemma 6.7.4 with the rôle of φ being played by φ_n. We define maps ψ and σ by

$$\psi(x^1, \ldots, x^{n-1}, x^n) = (x^1, \ldots, x^{n-1}, \psi_n(x^1, \ldots, x^n))$$

and $\sigma = \varphi \circ \psi$. Denoting by Ψ the inverse of ψ we have

$$\Psi(x^1, \ldots, x^n) = (x^1, \ldots, x^{n-1}, \varphi^n(x^1, \ldots, x^n))$$

and $\varphi = \sigma \circ \Psi$. A theorem on Jacobians of composite mappings states that

$$\mathbf{J}_\varphi = \mathbf{J}_\sigma \mathbf{J}_\Psi = \mathbf{J}_\sigma \varphi^n_{,n}. \tag{6.78}$$

In this particular instance it can easily be verified. Obviously† $\Psi^k_{,j} = \delta^k_j$ if $j < n$ and $\Psi^n_{,n} = \varphi^n_{,n}$. Using the chain rule we obtain for $j < n$

$$\varphi^i_{,j} = (\sigma^i \circ \Psi)_{,j} = \sigma^i_{,j} \circ \Psi + \sigma^i_{,n} \varphi^n_{,j},$$

and

$$\varphi^i_{,n} = \sigma^i_{,n} \varphi^n_{,n}.$$

This shows $\mathbf{J}_\varphi$ and $\mathbf{J}_\sigma$ have the same columns except that the last column of $\mathbf{J}_\varphi$ is the last column of $\mathbf{J}_\sigma$ multiplied by $\varphi^n_{,n}$, hence we have (6.78). It follows that σ and Ψ are regular and by Lemma 6.7.10

$$m_n(I) = \int_{\sigma_{-1}(I)} |\mathbf{J}_\sigma|.$$

The mapping Ψ leaves one variable unchanged, so by Remark 6.7.12 assumption (ii) in Theorem 6.7.11 is satisfied with φ_{-1} replaced by ψ. Applying Theorem 6.7.11 to $f = \mathbf{1}_{\sigma_{-1}(I)}|\mathbf{J}_\sigma|$, abbreviating $K(a)$ to K and denoting $\psi(\sigma_{-1}(I))$ by T we have

$$m_n(I) = \int_K \mathbf{1}_{\sigma_{-1}(I)}|\mathbf{J}_\sigma| = \int_{\psi(K)} (\mathbf{1}_{\sigma_{-1}(I)} \circ \Psi)|\mathbf{J}_\sigma||\mathbf{J}_\Psi|$$

$$= \int_T |\mathbf{J}_\varphi| = \int_{\varphi_{-1}(I)} |\mathbf{J}_\varphi|. \qquad \bullet$$

LEMMA 6.7.14 *For every bounded interval I with $\bar{I} \subset \varphi(G)$ equation (6.76) holds.*

---

† δ^k_j is the Kronecker delta, i.e. $\delta^k_j = 0$ if $j \neq k$ and $\delta^j_j = 1$.

Proof For every $a \in \bar{I}$ we find the interval $K(a)$ according to the previous lemma. By Borel's theorem finitely many of these intervals, say $K(a_1), \dots, K(a_p)$, cover $\bar{I}$. Let j_k be the intervals according to Lemma 6.7.5; then the intervals $j'_k = j_k \cap I$ are disjoint and cover I and

$$m_n(j'_k) = \int_{\varphi_{-1}(j'_k)} |J(u)| \, du.$$

Adding for $k = 1, \dots, p$ gives (6.76).   •

We are now ready to give a short proof of the main result of this chapter.

Proof of Theorem 6.7.7 By the last lemma assumption (ii) of Theorem 6.7.11 is satisfied with $\Delta = |J_\varphi|$, hence

$$\int_{\varphi(G)} f(x) \, dx = \int_G (f \circ \varphi)(u) |J_\varphi| \, du.$$

Applying this to $\mathbf{1}_S f$ in place of f we obtain (6.75).   •

The inverse mapping theorem states that $\varphi(G)$ is open and if the regular map φ is one-to-one then ψ, the map inverse to φ, is also regular, moreover $J_\varphi J_\psi = 1$. This has important consequences. It makes the blanket assumption we made about $\varphi(G)$ superfluous. More importantly: If $f \geq 0$ and the integral on the right-hand side of (6.75) exists and is finite then making the substitution $u = \psi(x)$ in this integral proves this equation. Consequently we have

COROLLARY 6.7.15 (to the main theorem) *If assumption (i) of Theorem 6.7.7 is satisfied end either f is absolutely integrable on a measurable set S or $(f \circ \varphi)|J_\varphi|$ is absolutely integrable on a measurable set $\varphi_{-1}(S)$ then both integrals in equation 6.75 exist and this equation holds.*

6.8 Exercises

EXERCISE 6.1 ⓘLet $A = A_1 \times$ ▪, $B = B_1 \times$ ▪, $C = C_1 \times$ ▪ and $C = A \cup B$. Prove that if $A \neq \emptyset$ and $B \neq \emptyset$ then there is a subscript k such that $C_k = A_k \cup B_k$ and for $1 \leq j \neq k \leq n$ we have $A_j = B_j = C_j$.

EXERCISE 6.2 ⓘ Let J_k and $\hat{J}_k$ be the intervals $(\frac{1}{k+1}, \frac{1}{k}) \times (0, \frac{1}{k+1})$ and $(0, \frac{1}{k+1}) \times (\frac{1}{k+1}, \frac{1}{k})$, respectively. If $f = (k+1)^2$ on J_k and $f =$

$(-1)(k+1)^2$ on $\hat{J}_k$ and $f = 0$ otherwise show that f is not integrable on $[0, 1] \times [0, 1]$.

EXERCISE 6.3 ① *Let*

$$T_n = \left\{ (x, y); \frac{1}{n+1} < x < \frac{1}{n}, \frac{1}{n+1} < y < x \right\}.$$

If $f(x, y) = n(n+1)^2$ *for* $(x, y) \in T_n$, $n \in \mathbb{N}$ *and* $f(x, y) = -f(y, x)$ *and* $f(x, y) = 0$ *otherwise show that* f *is KH-integrable on* $[0, 1] \times [0, 1]$.

EXERCISE 6.4 ①① *Prove: If* $S(x, y) = [0, 1] \times [0, 1] \setminus (0, x) \times (0, y)$ *and* f *is KH-integrable on* $S(x, y)$ *for every* $x > 0$ *and every* $y > 0$ *then* f *is integrable on* $S(0, 0)$ *if and only if*

$$\lim_{(x,y)\to(0,0)} \int_{S(x,y)} f$$

exists and then $\int_{S(0,0)} f$ *is equal to this limit. [Hint: Use Henstock's lemma and the method of the proof of Theorem 2.8.3.]*

EXERCISE 6.5 ① *Give an example of a function* f *for which*

$$\int_{-1}^{1} dx \int_{-1}^{1} f(x, y) \, dy$$

exists but the other repeated integral does not. [Hint: $f(x, y) = y/x$.*]*

EXERCISE 6.6 *Let* $f : J \mapsto \overline{\mathbb{R}}$ *and* H *be an additive interval function on* J. *Assume that for every positive* ε *there is a gauge* δ *such that*

$$\left| \frac{\mathsf{H}(K)}{|K|} - f(x) \right| < \varepsilon$$

whenever $K \subset \delta(x)$. *Prove that* f *is KH-integrable and* $\mathsf{H}(J) = \int_J f$.

EXERCISE 6.7 ①① *Approximation by step functions. Prove that if* f *is absolutely KH-integrable on* J *then for every positive* ε *there exists a step function* φ *such that* $\int_J |f - \varphi| < \varepsilon$. *[Hint: Use Theorem 3.11.10 and Theorem 3.11.14.]*

EXERCISE 6.8 ①① *Prove: If* f *is absolutely KH-integrable on* J *then there exist two absolutely KH-integrable functions* f_1, f_2 *and two sequences of step functions* $\{\varphi_{1,n}\}$ *and* $\{\varphi_{2,n}\}$ *such that*

(i) $f = f_1 - f_2$ a.e.
(ii) both sequences $\{\varphi_{1,n}(x)\}$ and $\{\varphi_{2,n}(x)\}$ are increasing for every $x \in J$;
(iii) $\lim_{n\to\infty} \varphi_{i,n} = f_i$ a.e. for $i = 1, 2$.

[Hint: Use the previous exercise and the method of Subsection 3.13.1].

EXERCISE 6.9 Show that

$$A_1 = \int_0^\infty e^{-\alpha x^2} \cos \beta x^2 \, dx \;=\; \sqrt{\frac{\pi}{8(\alpha^2 + \beta^2)}} \sqrt{\alpha + \sqrt{\alpha^2 + \beta^2}}$$

$$A_2 = \int_0^\infty e^{-\alpha x^2} \sin \beta x^2 \, dx \;=\; \mathrm{sgn}\beta \sqrt{\frac{\pi}{8(\alpha^2 + \beta^2)}} \sqrt{-\alpha + \sqrt{\alpha^2 + \beta^2}}.$$

[Hint: Use the result and the same method as in Example 6.6.9. It is convenient to use complex valued functions and to consider $A_1 + \jmath A_2$.]

EXERCISE 6.10 ⓘⓘ Let S be a measurable subset of $\mathbb{R}^{n-1}$ and f a function defined and non-negative on S. Show that the set

$$E = \{(x, y); x \in S, \, 0 < y < f(x)\}$$

is measurable if and only if $\int_S f$ exists and then $m_n(E) = \int_S f$, where m_n denotes measure in $\mathbb{R}^n$. [Hint: Use Fubini's theorem.]

ⓘThe following exercises outline a different approach to Fubini's theorem. It is less demanding on the reader; however, it establishes the Fubini theorem under the additional assumption that the function is *absolutely* integrable.

EXERCISE 6.11 Let $\mathcal{F}$ be a set of non-negative KH-integrable functions on $\mathbb{R}^n$ for which formula (6.53) holds with $J = \overline{\mathbb{R}}^n$. Note that $cf \in \mathcal{F}$ if $c \in \mathbb{R}$ and $f \in \mathcal{F}$. Show that if f, $g \in \mathcal{F}$ then $f + g \in \mathcal{F}$. If moreover $g \le f$ then $f - g \in \mathcal{F}$ also.

EXERCISE 6.12 Show that the limit of an increasing sequence of functions from $\mathcal{F}$ is in $\mathcal{F}$, provided the sequence of integrals is bounded.

EXERCISE 6.13 Show that the characteristic function of an open set of finite measure is in $\mathcal{F}$.

EXERCISE 6.14 A set is called G_δ if it is the intersection of a countable number of open sets. Use Theorem 3.11.15 to show that for every measurable set S there is a G_δ set M such that $S \subset M$ and $M \setminus S$ is of measure zero.

EXERCISE 6.15 *Show that the characteristic function of a G_δ set S with $m_n(S) < \infty$ is in $\mathcal{F}$.*

EXERCISE 6.16 *Show that the characteristic function of a set of measure zero is in $\mathcal{F}$.*

EXERCISE 6.17 *Use Exercises 6.14 to 6.16 to show that the characteristic function of a measurable set of finite measure is in $\mathcal{F}$.*

EXERCISE 6.18 *Use part (IV) of Theorem 3.11.10 to prove that $f \in \mathcal{F}$ for any non-negative f which is KH-integrable on $\overline{\mathbb{R}}^n$.*

7

Some Applications

7.1 Introduction

In this last chapter we want to indicate some applications of the ideas presented in previous chapters. Applications of KH-integration are varied and many, and we just consider a few which are appropriate at our level of presentation. The origin and most important applications of the KH-integral lie in ordinary differential equations. We refer to the excellent monograph by Schwabik [38] and shall not deal with ordinary differential equations in this book.

In Section 7.2 we give the definition and basic properties of the so-called line integral of a function $F : G \mapsto \mathbb{R}^n$, $G \subset \mathbb{R}^n$, along a curve φ (we say path). This integral in physics describes the work of the field given by F when a particle moves along φ. In Theorem 7.2.8 we establish the existence of a function U with $dU = F(x)dx$ provided the integrability condition (7.23) is satisfied.

Section 7.3 deals with differentiation of series. The main theorem there is more general than the one usually given in analysis courses and has a simple proof. We could have proved it in Section 2.7 on applications of the Fundamental Theorem of calculus; the reason for this order is that we need it in the Section 7.4.

In Section 7.4 we solve the Dirichlet problem for the Laplace equation on a circle. This in itself is an important topic but we present it mainly as a motivation for Abel's summability of Fourier series considered in the following section 7.5.

The concluding section shows that a 2π-periodic function such that $\mathcal{KH} \int_0^{2\pi} f^2$ exists can always be represented by its Fourier series in a well defined sense which is described in that section. Also the important Riesz–Fisher Theorem is proved there.

7.2 A line integral

In this section the letter G will stand for an open set in $\mathbb{R}^n$ with $n \geq 2$.
A generic point in $\mathbb{R}^n$ will be denoted by x or $(x^1, x^2, \ldots, x^n)$, and the
norm $|x|_2$ is the usual Euclidean norm. For a mapping $F : G \mapsto \mathbb{R}^n$ we
have $F = (F^1, F^2, \ldots, F^n)$ with $F^i : G \mapsto \mathbb{R}^1$. Partial derivatives will
be denoted by subscripts with commas, for instance

$$\partial F^j / \partial x_i = F^j_{,i}. \tag{7.1}$$

The word *path* will be used for a continuous map of bounded variation
from an interval in $\mathbb{R}$ into $\mathbb{R}^n$. That is if

$$\varphi : [a, b] \mapsto \mathbb{R}^n \tag{7.2}$$

then φ is a path if it is continuous on $[a, b]$ and there exists a constant
K such that for any division D

$$\sum_D |\varphi(v) - \varphi(u)| < K. \tag{7.3}$$

The supremum of the sums in (7.3) over all divisions of $[a, b]$ will be
denoted by $\mathrm{Var}_a^b \varphi$ and is the length of the geometrical image of φ, i.e.
of the set $\{\varphi(t); t \in [a, b]\}$. We shall denote this set by $[\varphi]$. Similarly as
in (3.28) we have, for $a < c < b$,

$$\mathrm{Var}_a^b \varphi = \mathrm{Var}_a^c \varphi + \mathrm{Var}_c^b \varphi.$$

A path (7.2) is closed if $\varphi(a) = \varphi(b)$. Two *paths*

$$\varphi : [a, b] \mapsto \mathbb{R}^n \quad \text{and} \quad \psi : [c, d] \mapsto \mathbb{R}^n \tag{7.4}$$

are, by definition, *equivalent*† if there exists a continuous strictly increasing function θ mapping $[a, b]$ onto $[c, d]$ such that

$$\varphi(s) = \psi(\theta(s)) \quad \text{for every} \quad s \in [a, b]. \tag{7.5}$$

The path ψ is said to be a part of φ, in symbols $\psi \subset \varphi$, if $[c, d] \subset [a, b]$.
By a line integral $\int_\varphi F$ we understand the Kurzweil–Henstock limit of
the Riemann sums

$$\sum_k \sum_{i=1}^n F^i(\varphi(\xi_k))(\varphi^i(v_k) - \varphi^i(u_k)). \tag{7.6}$$

---

† It is easily seen that for two paths the relation of being equivalent is reflexive,
symmetric and transitive and therefore is a well defined equivalence.

More precisely the number $\int_\varphi F$ is the *line integral of F along φ* (as in (7.2)) if for every positive ε there exists $\delta : [a,b] \mapsto \mathbb{R}_+$ such that

$$\left| \sum_{k=1}^{N} \sum_{i=1}^{n} F^i(\varphi(\xi_k))(\varphi^i(v_k) - \varphi^i(u_k)) - \int_\varphi F \right| < \varepsilon,$$

for every δ-fine partition

$$\pi \equiv \{(\xi_k, [u_k, v_k]); \, k = 1, 2, \ldots, N\} \qquad (7.7)$$

of $[a, b]$. Sometimes we shall denote the sum in (7.6) briefly as†

$$\sum_\pi F(\varphi(\xi))(\varphi(v) - \varphi(u)).$$

We shall also denote the line integral $\int_\varphi F$ by $\int_\varphi F(x)\,dx$ and in concrete situations by $\int_\varphi \sum_1^n F^i(x)dx^i$. The line integral has the same value along two paths with the same geometrical image traversed it in the same direction. More precisely we have

THEOREM 7.2.1 *If φ and ψ are two equivalent paths then*

$$\int_\varphi F(x)dx = \int_\psi F(x)dx$$

as long as one of the above integrals exists.

This can be proved using the fact that the one-to-one correspondence between $[a, b]$ and $[c, d]$ by θ induces a one-to-one correspondence between Riemann sums of type (7.6). The usual theory of line integrals, given in advanced calculus courses, can be developed from this definition. However, just as the KH-integral has better properties than the Riemann integral so has 'our' line integral in comparison with the usual calculus line integral.‡ We specifically mention the following elementary properties of the line integral:

$$c \int_\varphi F(x)\,dx \;=\; \int_\varphi cF(x)\,dx, \qquad (7.8)$$

$$\int_\varphi F(x)\,dx + \int_\varphi G(x)\,dx \;=\; \int_\varphi (F(x) + G(x))\,dx, \qquad (7.9)$$

† The expression following the summation sign is to be understood as scalar product of two n-dimensional vectors.
‡ e.g. a modified version of the dominated convergence theorem is available for 'our' line integral; however, we shall not pursue this here.

If $|F| \le M$ then $\left| \int_\varphi F(x)\, dx \right| \le M \mathrm{Var}_a^b \varphi.$ (7.10)

If φ and ψ are as in (7.4) and $\varphi(b) = \psi(c)$ then $\varphi \oplus \psi$ is defined on $[a, b+d-c]$ by

$$(\varphi \oplus \psi)(t) = \begin{cases} \varphi(t) & \text{if } t \in [a, b] \\ \psi(t + c - b) & \text{if } t \in [b, b+d-c]. \end{cases}$$ (7.11)

The path $-\varphi$ is defined by

$$(-\varphi)(t) = \varphi(b - t + a)) \quad \text{for} \quad t \in [a, b].$$ (7.12)

It follows easily that

$$\int_{\varphi \oplus \psi} F(x)\, dx = \int_\varphi F(x)\, dx + \int_\psi F(x)\, dx,$$ (7.13)

$$\int_{-\varphi} F(x)\, dx = -\int_\varphi F(x)\, dx.$$ (7.14)

The Cauchy convergence principle applies to the line integral as well: The integral $\int_\varphi F$ exists if and only if for every positive ε there is a δ such that

$$\left| \sum_{\pi_1} F(\varphi(\xi))(\varphi(v) - \varphi(u)) - \sum_{\pi_2} F(\varphi(\eta))(\varphi(s) - \varphi(t)) \right| < \varepsilon$$ (7.15)

whenever $\pi_1 \ll \delta$ and $\pi_2 \ll \delta$. It is left to Exercise 7.7 to show that if $F \circ \varphi$ is continuous and φ a path then $\int_\varphi F$ exists. By the same method by which we prove the Fundamental Theorem of calculus it can be shown that

$$\int_\varphi F(x) dx = \mathcal{KH} \int_a^b (F \circ \varphi)(t) \varphi'(t)\, dt,$$ (7.16)

provided that the integral on the right-hand† side exists, φ^i are continuous and the derivatives of φ^i exist except on a countable set.‡

LEMMA 7.2.2 *If F is constant and φ closed then $\int_\varphi F = 0$.*

Proof This is obvious from the definition of the line integral. •

† $(F \circ \varphi)(t)\varphi'(t)$ is to be undersood as a scalar product of two vectors $(F \circ \varphi)(t)$ and $\varphi'(t)$.
‡ The weaker assumptions that φ^i are SL (or AC) and the derivatives exist a.e. on $[a, b]$ are also sufficient.

LEMMA 7.2.3 *If φ is closed and there is a constant k such that $|F(x) - k| \leq M$ then $\left|\int_\varphi F\right| \leq M\mathrm{Var}_a^b\varphi$.*

Proof This follows from the previous lemma and (7.10).   •

We call a path as in (7.2) *polygonal* if there is a division D of $[a, b]$ such that φ is linear on every subinterval of D.

THEOREM 7.2.4 (Approximation by a polygonal path) *Let φ be as in (7.2) and F continuous on an open set G containing $[\varphi]$. Then for every positive ε there exists a positive constant δ such that for every division*

$$\pi \equiv a = t_0 < t_1 < \cdots < t_n = b$$

with $n(D) < \delta$ the polygonal path defined by

$$\psi(t) = \frac{t_{k+1} - t}{t_{k+1} - t_k}\varphi(t_k) + \frac{t - t_k}{t_{k+1} - t_k}\varphi(t_{k+1}) \text{ for } t \in [t_k, t_{k+1}) \quad (7.17)$$

and $\psi(b) = \varphi(b)$ satisfies

$$\left|\int_\varphi F - \int_\psi F\right| < \varepsilon. \quad (7.18)$$

Proof Let $\varepsilon = 2\varepsilon_0\mathrm{Var}_a^b\varphi$. Since $[\varphi]$ is compact there is a positive r such that the set K of points whose distance from $[\varphi]$ does not exceed r is a part of G. Evidently K is compact. There exists a positive η such that

$$|F(x) - F(y)| < \varepsilon_0 \quad (7.19)$$

for $|x - y| < \eta$ and $x, y \in K$. Having obtained η we find δ such that

$$|\varphi(\bar{t}) - \varphi(\hat{t})| < \eta, \quad (7.20)$$

for $|\bar{t} - \hat{t}| < \delta$. Let $n(D) < \delta$ and ψ be defined by (7.17). Let γ_k be the closed path formed by the restrictions of φ and $-\psi$ to $[t_k, t_{k+1}]$. For every $x \in [\gamma_k]$ and $y = t_k$ we have (7.19) and by Lemma 7.2.3 we have

$$\left|\int_{\gamma_k} F\right| < 2\varepsilon_0\mathrm{Var}_{t_k}^{t_{k+1}}\varphi$$

and since

$$\int_\varphi F - \int_\psi F = \sum_\pi \int_{\gamma_\tau} F,$$

inequality (7.18) follows.   •

THEOREM 7.2.5 *If there exists a function $U : G \mapsto \mathbb{R}$ differentiable in G with $U_{,i} = F^i$ and the path φ lies in G, i.e. $[\varphi] \subset G$, then*

$$\int_\varphi F(x)dx = U(\varphi(b)) - U(\varphi(a)). \qquad (7.21)$$

In particular if φ is closed then

$$\int_\varphi F(x)dx = 0.$$

Proof For every positive ε and every $\xi \in [a, b]$ there exists by continuity of φ and differentiability of U a positive $\delta(\xi)$ such that

$$\left| U(\varphi(w)) - U(\varphi(\xi)) - \sum_{i=1}^n F^i(\varphi(\xi))(\varphi^i(w) - \varphi^i(\xi)) \right| < \varepsilon \left| \varphi(w) - \varphi(\xi) \right|$$
$$(7.22)$$

whenever $|w - \xi| < \delta(\xi)$. Let $\pi \equiv \{(\xi_k, [u_k, v_k])\}$ be a δ-fine partition of $[a, b]$. It follows from (7.22) that

$$\left| U(\varphi(v_k)) - U(\varphi(\xi_k)) - \sum_{i=1}^n F^i(\varphi(\xi_k))[\varphi^i(v_k) - \varphi^i(\xi_k)] \right|$$
$$< \varepsilon \left| \varphi(v_k) - \varphi(\xi_k) \right|,$$

$$\left| U(\varphi(\xi_k)) - U(\varphi(u_k)) - \sum_{i=1}^n F^i(\varphi(\xi_k))[\varphi^i(\xi_k) - \varphi^i(u_k)] \right|$$
$$< \varepsilon \left| \varphi(\xi_k) - \varphi(u_k) \right|.$$

Writing $U(\varphi(b)) - U(\varphi(a))$ as

$$\sum_\pi [(U(\varphi(v_k)) - U(\varphi(\xi_k))) + (U(\varphi(\xi_k)) - U(\varphi(u_k)))]$$

and replacing $\varphi^i(v_k) - \varphi^i(u_k)$ by

$$\varphi^i(v_k) - \varphi^i(\xi_k) + \varphi^i(\xi_k) - \varphi^i(u_k)$$

in (7.6) we obtain

$$\left| U(\varphi(b)) - U(\varphi(a)) - \sum_k \sum_{i=1}^n F^i(\varphi(\xi_k))(\varphi^i(v_k) - \varphi^i(u_k)) \right|$$
$$< \varepsilon \sum_\pi [|\varphi(v_k) - \varphi(\xi_k)| + |\varphi(\xi_k) - \varphi(u_k)|] \le \varepsilon \mathrm{Var}_a^b \varphi.$$

The last inequality is a consequence of the definition of $\mathrm{Var}_a^b \varphi$. •

It is an elementary result that if F has continuous partial derivatives in an n-dimensional interval K and

$$F^i_{,j}(x) = F^j_{,i}(x) \tag{7.23}$$

for all $x \in K$ there exists a function U with $U_{,i}(x) = F^i(x)$ for all $x \in K$ and the line integral is independent of the path. Moreover U is obtained by choosing an arbitrary point x_0 in K and integrating F from x_0 to a variable point x along any path in K. Our aim is to reduce the assumption of continuous derivatives to mere differentiability of F. We shall also allow more general sets than intervals.

In the proof of the next theorem we shall integrate over sides of triangles. To this end we need to make our terminology unambiguous and to say something about orientation of segments and triangles. If a and b are distinct points in $\mathbb{R}^n$ then the set

$$\overline{ab} = \{a + t(b-a); 0 \le t \le 1\}$$

is a segment and the points a, b are the endpoints of $\overline{ab}$. An oriented segment $\overrightarrow{ab}$ is the equivalence class of all paths equivalent to

$$\varphi : t \mapsto a + (b-a)t, \quad t \in [0,1].$$

For the oriented segment $\overrightarrow{ab}$, a is the initial point and b is the endpoint. The point $a + (b-a)t$ with $t \in [0,1]$ is also called a point of $\overrightarrow{ab}$ and if $0 < t < 1$ it is an interior point. Naturally, we regard $\int_{\overrightarrow{ab}} F(x)\,dx = \int_\varphi F(x)\,dx$ with φ as above and to simplify the notation we shall denote

$$\int_{\overrightarrow{ab}} F(x)\,dx = \int_a^b F(x)\,dx. \tag{7.24}$$

Clearly $\overline{ab} = \overline{ba}$ whereas $\overrightarrow{ab} = -\overrightarrow{ba}$ and

$$\int_a^b F(x)\,dx = -\int_b^a F(x)\,dx.$$

If $\overline{cd} \subset \overline{ab}$ we say that $\overrightarrow{cd}$ is consistently oriented with $\overrightarrow{ab}$ if $c \in \overline{ad}$. It is easy to see that $\overrightarrow{cd}$ is consistently oriented with $\overrightarrow{ab}$ if and only if $\overrightarrow{cd} \subset \overrightarrow{ab}$. We say that $\overrightarrow{cd}$ has the opposite orientation to $\overrightarrow{ab}$ if $d \in \overline{ac}$. In this situation

$$\int_a^b F(x)\,dx + \int_c^d F(x)\,dx = \int_a^d F(x)\,dx + \int_c^b F(x)\,dx. \tag{7.25}$$

A line is a set of the form $\{a + t(b - a); t \in \mathbb{R}\}$. If the points a, b, c do not lie on the same line then the set of all points x of the form

$$x = \lambda_x^a a + \lambda_x^b b + \lambda_x^c c, \tag{7.26}$$

$$\lambda_x^a + \lambda_x^b + \lambda_x^c = 1 \tag{7.27}$$

constitutes a plane. If moreover all λ are non-negative then the set of all x satisfying (7.26) and (7.27) becomes a triangle $\triangle abc$. The points a, b, c are the vertices of $\triangle abc$, and the segments $\overline{ab}, \overline{bc}, \overline{ca}$ are the sides of $\triangle abc$. A point $x \in \triangle abc$ is an *interior* point of $\triangle abc$ if all λ in (7.26) are positive. Two triangles belonging to the same plane are called non-overlapping if they have no interior point in common.

The *orientation of a triangle* is determined by an ordered triplet of its vertices and is the set of all ordered triplets which can be obtained from the determining triplet by cyclic permutations. Hence for a triangle $\triangle abc$, there are two distinct orientations

$$\{[a, b, c], \quad [c, a, b], \quad [b, c, a]\}$$

and

$$\{[a, c, b], \quad [c, b, a], \quad [b, a, c].\}$$

These orientations are *opposite* one to the other. A triangle to which one of its orientations has been assigned becomes an oriented triangle; the triangle $\triangle abc$ together with the orientation described by $[a, b, c]$ is an oriented triangle and we shall denote it by $\overrightarrow{abc}$. The *oriented boundary* ∂abc of $\overrightarrow{abc}$ is the path

$$\overrightarrow{ab} \oplus \overrightarrow{bc} \oplus \overrightarrow{ca}. \tag{7.28}$$

Intuitively we can think that the oriented boundary of a triangle describes a positive direction of moving around the triangle, for instance moving from $d = \frac{1}{2}(a + c)$ to b through a is moving 'positively' within ∂abc. See Figure 7.1.

For the line integral over ∂abc we have

$$\int_{\partial abc} F(x)\, dx = \int_a^b F(x)\, dx + \int_b^c F(x)\, dx + \int_c^a F(x)\, dx.$$

Obviously

$$\int_{\partial abc} F = \int_{\partial bca} F = \int_{\partial cab} F.$$

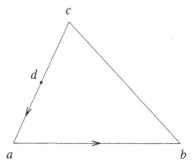

Fig. 7.1. An oriented triangle

We say that $\overrightarrow{xyz}$ is similarly oriented to $\overrightarrow{abc}$, in symbols $\overrightarrow{xyz} \sim \overrightarrow{abc}$, if

$$\begin{vmatrix} \lambda_x^a & \lambda_x^b & \lambda_x^c \\ \lambda_y^a & \lambda_y^b & \lambda_y^c \\ \lambda_z^a & \lambda_z^b & \lambda_z^c \end{vmatrix} > 0.$$

It is only a matter of simple linear algebra† show that

$$\overrightarrow{abc} \quad \sim \quad \overrightarrow{abc}. \tag{7.29}$$

$$\text{If } \overrightarrow{xyz} \sim \overrightarrow{abc} \quad \text{then} \quad \overrightarrow{abc} \sim \overrightarrow{xyz}. \tag{7.30}$$

$$\text{If } \overrightarrow{xyz} \sim \overrightarrow{abc} \quad \text{and} \quad \overrightarrow{abc} \sim \overrightarrow{pqr} \text{ then } \overrightarrow{xyz} \sim \overrightarrow{pqr}. \tag{7.31}$$

This means that the relation of being similarly oriented is reflexive, symmetric and transitive. Consequently all oriented triangles in the same plane belong to one and only one of the two distinct equivalence classes of similarly oriented triangles. It is easily checked that if $x \in \triangle abc$ but $x \notin \overline{ab}$ then $\overrightarrow{abc} \sim \overrightarrow{abx}$. In $\mathbb{R}^2$ the triangles similarly oriented to the triangle $\overrightarrow{oxy}$ with $o = (0,0)$, $x = (1,0)$ and $y = (0,1)$ are usually called positively oriented. Intuitively the oriented boundary ∂oxy describes an anticlockwise movement around $\triangle oxy$ and this is also true for all positively oriented triangles. Another intuitive way of describing it is to say that one moves along an oriented boundary of a positively oriented triangle in such a manner that the triangle stays on the left. Rather than rely on intuition, as most authors on our level of presentation do, we preferred to give a mathematically sound treatment of orientation of triangles and did not define orientation by means of a clock. If $\overrightarrow{wzc} \sim \overrightarrow{wzr}$ then these triangles have a common interior point, indeed, e.g.

$$x = (\frac{1}{2} - \varepsilon)w + (\frac{1}{2} - \varepsilon)z + 2\varepsilon c$$

† The theorem on multiplication of determinants is needed for the proof.

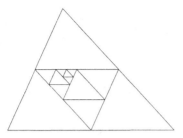

Fig. 7.2. Division by similar triangles.

for sufficiently small‡ positive ε. Let $\overrightarrow{abc}$ and $\overrightarrow{pqr}$ be two similarly oriented triangles with $\triangle abc$ and $\triangle pqr$ non-overlapping, $\overline{pq} \cap \overline{ab} = \overline{wz}$ and $w \in \overline{az}$. We are going to show that $\overrightarrow{zw} \subset \overrightarrow{pq}$. Assume contrary to what we want to prove that $w \in \overline{pz}$. Then

$$\overrightarrow{wzc} \sim \overrightarrow{abc} \sim \overrightarrow{pqr} \sim \overrightarrow{wzr}$$

and the triangles $\triangle wzc$, $\triangle wzr$ have a common interior point contrary to the assumption that $\triangle abc$ and $\triangle pqr$ are non-overlapping. Consequently $\overrightarrow{wz}$ is consistently oriented with $\overrightarrow{ab}$ but has the opposite orientation to $\overrightarrow{pq}$. Similarly as in (7.25) the integration over the common segment $\overline{wz}$ in the sum

$$\int_{\partial abc} F(x)\, dx + \int_{\partial pqr} F(x)\, dx \tag{7.32}$$

is cancelled.

We shall need the following generalization of Cousin's lemma: If $T \subset \mathbb{R}^n$ is a triangle and T_k, $k = 1, \ldots, r$ are non-overlapping triangles with $\bigcup_1^r T_k = T$, points y_k belong to T_k and $\delta : T \mapsto (0, \infty)$ then we say that *the set* $\{(y_k, T_k); k = 1, \ldots, r\}$ *is a δ-fine partition of the triangle T if* $d(T_k) < \delta(y_k)$.

LEMMA 7.2.6 (Cousin for triangles)   *For a triangle $T \subset \mathbb{R}^n$ there always exists a δ-fine partition consisting of triangles* **similar** *to T.*

See Figure 7.2. An indirect proof can be given which follows the usual pattern of the one-dimensional bisection argument in the proof of Theorem 2.3.1, except that now T would be divided into four similar triangles defined by the mid-points of sides of T.

‡ Note that $\lambda_r^c > 0$, so it is sufficient to have $\varepsilon < \mathrm{Min}(1/4, (|\lambda_r^w| + |\lambda_r^z| + \lambda_r^c)^{-1})$.

We shall say that assumption $\mathcal{D}$ is satisfied in G if F is continuous in G and there exists a countable set M such that F is differentiable and satisfies (7.23) in $G \setminus M$.

THEOREM 7.2.7 *If $\mathcal{D}$ is satisfied in G, the triangle T is oriented and $T \subset G$ then*

$$\int_{\partial T} F(x)dx = 0. \tag{7.33}$$

Proof For this proof we denote by $d(T)$, $A(T)$ and $p(T)$ the diameter, two-dimensional area and perimeter of a triangle T. The elements of M can be enumerated and for $w_m \in M$ and for arbitrary $\varepsilon > 0$ there exists a $\delta : T \mapsto \mathbb{R}_+$ such that

$$|F(x) - F(w_m)| < \frac{\varepsilon}{2^m} \tag{7.34}$$

whenever $|x - w_m| < \delta(w_m)$. For $y \in S \setminus M$ there is a $\delta > 0$ such that for $i = 1, 2, \ldots, n$

$$\left| F^i(x) - F^i(y) - \sum_{j=1}^{n} F^i_{,j}(y)(x^j - y^j) \right| < \varepsilon |y - x| \tag{7.35}$$

whenever $|x - y| < \delta(y)$. Let $\{y_k, T_k\}$ be a δ-fine partition of T with triangles similar to T. We give T a definite orientation and then we make the orientation of each T_k similar to T. In the sum

$$\sum_{k=1}^{r} \int_{\partial T_k} F(x)dx$$

integration over segments which are on common sides of adjacent triangles T_k cancels and only integration over segments which lie on the boundary remains. A moment's reflection reveals that these segments are consistently oriented with segments which form the oriented boundary ∂T. It follows that

$$\int_{\partial T} F(x)dx = \sum_{k=1}^{r} \int_{\partial T_k} F(x)dx. \tag{7.36}$$

If $y_k \in M$ then $y_k = w_m$ for some m and we obtain from (7.34)

$$\left| \int_{\partial T_k} F(x)dx \right| < \frac{\varepsilon}{2^m} p(T_k) \leq \frac{\varepsilon}{2^m} p(T). \tag{7.37}$$

If $y_k \notin M$ then because of (7.23)

$$\int_{\partial T_k} \sum_{i=1}^{n} \left(F^i(y_k) + \sum_{j=1}^{n} F^i_{,j}(y_k)(x^j - y_k^j) \right) dx^i = 0. \qquad (7.38)$$

This can be seen most easily by noting that the integrand in (7.38) has a 'primitive' U, where

$$U(x) = \sum_{i=1}^{n} F^i(y_k)x^i + \frac{1}{2} \sum_{i,j=1}^{n} F^i_{,j}(y_k)(x^j - y_k^j)(x^i - y_k^i),$$

and using Theorem 7.2.5. For $y_k \notin M$ we get from (7.35)† and (7.38) that

$$\left| \int_{\partial T_k} F(x)dx \right| < \varepsilon d(T_k)p(T_k). \qquad (7.39)$$

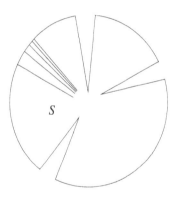

Fig. 7.3. Star-shaped set

The triangles T_k are similar to T, therefore there exists a constant C depending *only* on T and independent of k such that $d(T_k)p(T_k) < CA(T_k)$ for all k. Consequently (7.39) becomes

$$\left| \int_{\partial T_k} F(x)dx \right| < C\varepsilon A(T_k). \qquad (7.40)$$

Finally by (7.36), (7.37) and (7.40)

$$\left| \int_{\partial T} F(x)dx \right| < \varepsilon(CA(T) + p(T)). \qquad \bullet$$

† With $y = y_k$.

A set $S \subset \mathbb{R}^n$ is said to be *star-shaped* if there exists a point $c \in S$ (the centre of the star) such that for every $x \in S$ the whole segment joining x with c lies in S. See Figure 7.3.

THEOREM 7.2.8 (Path independence. Existence of a primitive) *If assumption $\mathcal{D}$ is satisfied in a star-shaped open set S then there is a **differentiable** function $U : S \mapsto \mathbb{R}$ such that*

$$dU(x) = \sum_{i=1}^{n} F^i(x)dx^i \qquad (7.41)$$

for all x in S.

If $\varphi : [a,b] \mapsto S$ is a path lying in S then

$$\int_\varphi F(x)dx = U(\varphi(b)) - U(\varphi(a)). \qquad (7.42)$$

Proof It suffices to prove (7.41); equation (7.42) then follows from Theorem 7.2.5. Let c be the centre of S and $\varepsilon > 0$. Given $x \in S$ we find $\eta > 0$ such that for $|h|_2 < \eta$ the point $x + h \in S$ and for $t \in [0,1]$

$$|F(x + th) - F(x)| < \varepsilon. \qquad (7.43)$$

We denote

$$U(x) = \int_c^x F(z)dz.$$

It follows from Theorem 7.2.7 that

$$U(x + h) - U(x) = \int_x^{x+h} F(z)dz. \qquad (7.44)$$

By combining (7.43) and (7.44) and using (7.10) we obtain

$$\left| U(x + h) - U(x) - \sum_1^n F^i(x)h^i \right| = \left| \int_x^{x+h} [F(z) - F(x)]\, dz \right| \le \varepsilon|h|_2. \quad \bullet$$

EXAMPLE 7.2.9 In $\mathbb{R}^2$ let us write x, y instead of x^1, x^2, respectively. If $\varphi(t) = (\cos t, \sin t)$ with $t \in [0, 2\pi]$ then

$$\int_\varphi \frac{-ydx}{x^2 + y^2} + \frac{xdy}{x^2 + y^2} = 2\pi. \qquad (7.45)$$

Theorem 7.2.8 is not applicable with $S = \mathbb{R}^2 \setminus \{(0,0)\}$ because this set is not star-shaped nor it is applicable to $\mathbb{R}^2$ because F is not continuous at $(0,0)$.

The line integral does not change its value when the path is continuously deformed. This is the content of the next theorem. First we give a precise meaning to the phrase 'continuously deformed' by using a function H below. We assume that the two paths φ and Φ for which we want to prove the formula $\int_\varphi F = \int_\Phi F$ are defined on $[0,1]$; this can be always achieved by passing to an equivalent path. We also assume that either the paths start and end at the same point, i.e. $\varphi(0) = \Phi(0) = a$ and $\varphi(1) = \Phi(1) = b$, or both paths are closed. Let $\Omega = [0,1] \times [0,1]$ and H a function† continuous on Ω with the following properties:

$$H(s,t) \in G \text{ for every } (s,t) \in \Omega; \tag{7.46}$$

$$H(0,t) = \varphi(t) \text{ and } H(1,t) = \Phi(t) \text{ for every } t \in [0,1]; \tag{7.47}$$

$$H(s,0) = a \text{ and } H(s,1) = b \text{ for every } s \in [0,1]; \tag{7.48}$$

or

$$H(s,0) = H(s,1) \text{ for every } s \in [0,1]. \tag{7.49}$$

THEOREM 7.2.10 *If*

- $G \subset \mathbb{R}^n$ *is open,*
- *assumption $\mathcal{D}$ is satisfied in G,*
- *there exists a continuous function H with properties* (7.46), (7.47) *and* (7.48) *or with properties* (7.46), (7.47) *and* (7.49)

then

$$\int_\varphi F = \int_\Phi F. \tag{7.50}$$

Proof Since $H(\Omega)$ is compact there exists a positive r such that the ball $B(s,t)$, centred at $H(s,t)$ and of radius r, lies in G for every $(s,t) \in \Omega$. The function H is uniformly continuous on Ω, so there is a positive δ such that

$$|H(s,t) - H(s',t')| < r$$

whenever $|s - s'| + |t - t'| < \delta$. Let $n \in \mathbb{N}$ with $n\delta < 1$. In this proof we shall integrate along a polygonal path φ_k joining $\varphi(0)$ with $\varphi(1)$ through the points $H(\frac{k}{n}, \frac{j}{n})$ for $j = 1, \ldots, n-1$. More precisely we set

$$\varphi_{k,j}(t) = H\left(\tfrac{k}{n}, \tfrac{j}{n}\right) + (nt - j)\left[H\left(\tfrac{k}{n}, \tfrac{j+1}{n}\right) - H\left(\tfrac{k}{n}, \tfrac{j}{n}\right)\right] \text{ for } t \in \left[\tfrac{j}{n}, \tfrac{j+1}{n}\right],$$

$$\varphi_k = \varphi_{k,0} \oplus \varphi_{k,1} \oplus \cdots \oplus \varphi_{k,n-1}.$$

† Commonly referred to as a homotopy.

We also denote

$$\psi_{k,j}(t) = H\left(\frac{k}{n}, \frac{j}{n}\right) + (nt - k)\left[H\left(\frac{k+1}{n}, \frac{j}{n}\right) - H\left(\frac{k}{n}, \frac{j}{n}\right)\right] \text{ for } t \in \left[\frac{k}{n}, \frac{k+1}{n}\right].$$

See Figure 7.4. First we note that $\int_{\varphi_0} F = \int_\varphi F$. This is because the line

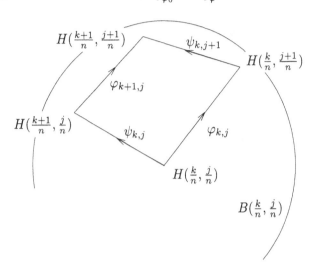

Fig. 7.4. Homotopy

segment joining $H(0, \frac{j}{n})$ with $H(0, \frac{j+1}{n})$ and the corresponding part of φ lie in a ball in which the integral is independent of its path. Similarly $\int_{\varphi_n} F = \int_\Phi F$. We now complete the proof by showing

$$\int_{\varphi_k} F = \int_{\varphi_{k+1}} F \quad \text{for} \quad k = 0, \ldots, n-1.$$

The integral along any closed path in $B(\frac{k}{n}, \frac{j}{n})$ is zero, consequently

$$\int_{\varphi_{k,j}} F - \int_{\varphi_{k+1,j}} F = \int_{\psi_{k,j}} F - \int_{\psi_{k,j+1}} F.$$

Finally we have

$$\int_{\varphi_k} F - \int_{\varphi_{k+1}} F = \sum_{j=0}^{n-1}\left(\int_{\varphi_{k,j}} F - \int_{\varphi_{k+1,j}} F\right)$$

$$= \sum_{j=0}^{n-1}\left(\int_{\psi_{k,j}} F - \int_{\psi_{k,j+1}} F\right) = \int_{\psi_{k,n}} F - \int_{\psi_{k,0}} F = 0.$$

If (7.48) holds the last two integrals are zero because $\psi'_{k,0} = \psi'_{k,n} = 0$; if (7.49) holds then $\psi_{k,n} = \psi_{k,0}$ and the last two integrals are equal. •

7.2.1 Green's theorem

The material presented so far in this section is intimately connected with the so-called Green–Stokes theorem. We restrict our attention to $\mathbb{R}^2$ and in this special case it is simply called Green's theorem after the British mathematician G. Green. The theorem, however was known before Green to C.F. Gauss and J.L. Lagrange. Green's theorem establishes the formula

$$\int_\phi P(x,y)\,dx + Q(x,y)\,dy = \iint_{K(\phi)} (Q_x(x,y) - P_y(x,y))\,dx dy. \quad (7.51)$$

In this formula ϕ is a positively oriented path and $K(\phi)$ is the domain enclosed by ϕ. The formula is often given a slightly different form

$$\int_{\partial K} P(x,y)\,dx + Q(x,y)\,dy = \iint_K (Q_x(x,y) - P_y(x,y))\,dx dy. \quad (7.52)$$

Here K is a set in $\mathbb{R}^2$ and ∂K is its positively oriented boundary.

In general there are considerable difficulties in proving or even rigorously stating the theorem. In equation (7.51) the problem lies in the phrase 'domain enclosed by', in equation (7.52) in the phrase 'positively oriented boundary'. To indicate what these difficulties are we offer the following examples. In Figure 7.5 the set K is the unit circle without a

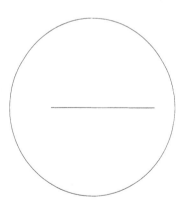

Fig. 7.5.

segment in the middle. In Figure 7.6 L_n is the segment joining the point $(n^{-1}, 0)$ with (n^{-1}, x_n) where $x_n \to 1$ and $K = [0,1] \times [0,1] \setminus \bigcup_1^\infty L_n$. In either case it is not clear what the positively oriented boundary of K should be. Moreover, in the second example, the 'length' of the boundary cannot be finite since the length of part of the comb $\bigcup_1^N L_n$

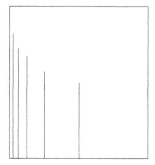

Fig. 7.6.

Fig. 7.7. Jordan curve

is $N - 1 + 2^{-N}$. The simple closed path $[\varphi]$ in Figure 7.7 is so complicated that it is not clear whether or not it encloses a domain and even admitting that it does it is difficult to decide what its interior is; if a fly (point) lands somewhere in the maze it is hard to say whether or not it is inside $[\varphi]$. These difficulties are easily overcome in special cases needed in applications, e.g. $[\varphi]$ is a circle or K is a rectangle. In this case the boundary is clearly a geometric image of a path, and the positive direction is the anticlockwise one, which can be explicitly described without any reference to a clock. We trust that it is clear what is meant by the positively oriented boundary ∂K of a rectangle K. See Figure 7.8. Let us prove the theorem for a rectangle first.

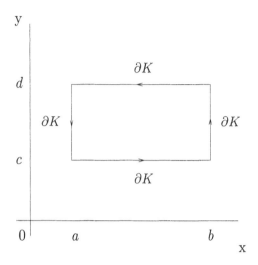

Fig. 7.8. Oriented boundary of K

THEOREM 7.2.11 *Let ∂K be the positively oriented boundary of*

$$K = [a,\, b] \times [c,\, d]$$

and K° its interior, i.e. $K^\circ = (a,\, b) \times (c,\, d)$. If

(i) *P, Q are continuous on K,*

(ii) *the set $S \subset K$ is countable, the functions P, Q are differentiable in $K^\circ \setminus S$,*

(iii) *the function $Q_x - P_y$ is absolutely KH-integrable on K*

then (7.52) is valid.

Proof We prove the theorem first under the additional assumption that P and Q are differentiable in an open set containing $K \setminus S$. Since the values of a function on a countable set influence neither the existence nor the value of the integral, we shall, during the proof, regard the expression $Q_x - P_y$ at the points of S as if it were zero. Let $\varepsilon > 0$. There is a positive δ_1 such that

$$|P(x,y) - P(\xi,\eta) - P_x(\xi,\eta)(x-\xi) - P_y(\xi,\eta)(y-\eta)|$$
$$\varepsilon[\,|x-\xi| + |y-\eta|\,], \qquad (7.53)$$
$$|Q(x,y) - Q(\xi,\eta) - Q_x(\xi,\eta)(x-\xi) - Q_y(\xi,\eta)(y-\eta)|$$
$$\varepsilon[\,|x-\xi| + |y-\eta|\,], \qquad (7.54)$$

for

$$|x-\xi| + |y-\eta| < \delta_1 \qquad (7.55)$$

and $(\xi,\eta) \in K \setminus S$. The elements of S can be ordered in a sequence and if (ξ_k,η_k) is the k-th element of this sequence then there exists δ_1 such that

$$|P(x,y) - P(\xi_k,\eta_k)| + |Q(x,y) - Q(\xi_k,\eta_k)| < \frac{\varepsilon}{2^k} \qquad (7.56)$$

for (x,y) satisfying (7.55). Finally there is a positive $\delta \le \delta_1$ such that for every δ-fine partition of K the Riemann sum for $Q_x - P_y$ differs by less than ε from the double integral in (7.52). Denote $p = 2(b-a+d-c)$ and $\lambda = (b-a)/(d-c)$. Let $\{((\xi,\eta),\, I)\}$ be a δ-fine partition of K with each I similar to K. It is fairly obvious that the integral

$$\sum_\pi \int_{\partial I} P\,dx + Q\,dy$$

is an additive function of intervals I and we have by Lemma 6.2.4

$$\int_{\partial K} P\,dx + Q\,dy = \sum_\pi \int_{\partial I} P\,dx + Q\,dy. \qquad (7.57)$$

If $(\xi,\eta) \in S$ then by (7.56)

$$\left| \int_{\partial I} P\,dx + Q\,dy \right| < \frac{\varepsilon p}{2^k}, \qquad (7.58)$$

for some natural k. If $(\xi,\eta) \notin S$ then by (7.53) and (7.54)

$$\left| \int_{\partial I} (P\,dx + Q\,dy) - [Q_x(\xi,\eta) - P_y(\xi,\eta)]|I| \right| < \frac{(1+\lambda)^2}{\lambda}|I|\varepsilon. \qquad (7.59)$$

Combining (7.58) and (7.59) with the definition of δ we have

$$\left| \int_{\partial K} P(x,y)\,dx + Q(x,y)\,dy \;-\; \iint_K (Q_x(x,y) - P_y(x,y))\,dxdy \right|$$
$$< \;\; \varepsilon p + \varepsilon \frac{(1+\lambda)^2}{\lambda}|K| + \varepsilon.$$

Since ε is arbitrary this establishes the theorem under the additional assumption. In the general case we apply the already proved part to the rectangle $[a+h, b-h] \times [c+h, d-h]$ and pass to the limit as $h \downarrow 0$. The limit passage in the line integral is easy because P and Q are continuous; the limit passage in the double integral follows from absolute continuity.●

We shall call a set a Green domain if the Green theorem can be applied to it. More precisely, we say that a compact set $K \subset \mathbb{R}^2$ with a non-empty interior K° is a Green domain if there exists a path ∂K with $[\partial K] = K \setminus K^\circ$ such that formula (7.52) holds for every pair of functions P, Q satisfying conditions (i), (ii) and (iii) of Theorem 7.2.11. The Lebesgue two-dimensional measure of a path is zero (see Exercise 7.9) and consequently K can be replaced by K° in equation (7.52). So far we know only one type of Green domain, namely a rectangle. The next theorem widens the class of Green domains considerably.

THEOREM 7.2.12 Let $\emptyset \neq K^\circ \subset K \subset O \subset \mathbb{R}^2$, with K compact and O open. If

(i) $H = (H^1, H^2)$ is a one-to-one mapping of K onto $H(K)$;

(ii) H^1 and H^2 have second order partial derivatives continuous in O;

(iii) $H^1_{,1} H^2_{,2} - H^1_{,2} H^2_{,1} > 0$ in K°;

then $H(K)$ is a Green domain if K is.

Proof The first order derivatives of H are bounded in K, i.e. there is a constant C such that $|H^i_{,j}(x^1, x^2)| \leq C$ for $i, j = 1, 2$ and all (x^1, x^2) in K. For brevity we denote $\varphi = \partial K$ and $\psi = H \circ \varphi$. It is clear that ψ is a path. The set $H(K)$ being a continuous image of a compact set is compact. It follows from the inverse mapping theorem that $H(K)^\circ = H(K^\circ)$ and consequently

$$H(K) \setminus H(K)^\circ = H(K \setminus K^\circ) = H([\varphi]) = [\psi].$$

It remains to be shown that the Green formula can be applied to ψ and $H(K)$. The idea of the proof is to transform the line integral first, then use the Green theorem on the domain K and conclude the argument by change of variables in the double integral. In order to save some cumbersome writing we denote derivatives by commas as in (7.1) and write F^1 and F^2 instead of P and Q. We need

LEMMA 7.2.13

$$\int_\psi \sum_{i=1}^2 F^i(y^1,y^2)\, dy^i = \int_\varphi \sum_{j=1}^2 \sum_{i=1}^2 F^i(H^1(x^1,x^2),H^2(x^1,x^2))H^i_{,j}\, dx^j.$$

(7.60)

Proof of the lemma Let δ be a positive number and $\pi \equiv \{(t,[r,\,s])\}$ a δ-fine partition of $[a,\,b]$. Denote by φ_t the restriction of φ to $[r,s]$ and

$$\varepsilon(\delta) = \sup\left\{|F \circ H \circ \varphi(t_2) - F \circ H \circ \varphi(t_1)|\,;\, |t_2 - t_1| \le \delta\right\}.$$

Denote

$$\sigma = \sum_\pi \sum_{i=1}^2 F^i \circ \psi(t)[\psi^i(s) - \psi^i(r)].$$

Using Theorem 7.2.5 we have

$$\psi^i(s) - \psi^i(r) = \int_{\varphi_t} \sum_{j=1}^2 H^i_{,j}\, dx^j$$

and consequently

$$\sigma = \sum_\pi \int_{\varphi_t} \sum_{i=1}^2 \sum_{j=1}^2 F^i \circ H(\varphi(t))H^i_{,j} dx^j.$$

The integral on the right-hand side of (7.60) exists. Let us denote it by A. For brevity set $W^i(x^1,x^2) = F^i \circ H(x^1,x^2)$. Then

$$A - \sigma = \sum_\pi \int_{\varphi_t} \sum_{i=1}^2 \sum_{j=1}^2 \left([W^i(x^1,x^2) - F^i \circ H(\varphi(t))]H^i_{,j}(x^1,x^2)\right) dx^j.$$

Using the boundedness of derivatives of H, the estimate (7.10) and the definition of $\varepsilon(\delta)$ we have

$$|A - \sigma| \le 2C\varepsilon(\delta)\mathrm{Var}_a^b\varphi.$$

This completes the proof of the lemma, since it shows that the integral on the left of (7.60) exists and equals A. •

Continuing with the **proof of the theorem** we have†

$$A = \int_{\varphi} \left(\sum_{i=1}^{2} F^i H^i_{,1} \right) dx^1 + \left(\sum_{i=1}^{2} F^i H^i_{,2} \right) dx^2$$

$$= \iint_{K^{\circ}} \sum_{i=1}^{2} \left[\sum_{k=1}^{2} \left(F^i_{,k} H^k_{,1} H^i_{,2} + F^i H^i_{,2,1} \right) - \sum_{k=1}^{2} \left(F^i_{,k} H^k_{,2} H^i_{,1} + F^i H^i_{,1,2} \right) \right]$$

$$= \iint_{K^{\circ}} \left(F^2_{,1} - F^1_{,2} \right) \left(H^1_{,1} H^2_{,2} - H^1_{,2} H^2_{,1} \right) dx^1 dx^2$$

$$= \iint_{H(K)^{\circ}} \left(F^2_{,1} - F^1_{,2} \right) dy^1 dy^2. \qquad \bullet$$

EXAMPLE 7.2.14 We are going to show that a domain D characterized by the inequalities

$$a \leq x^1 \leq b, \quad \varphi^1(x^1) \leq x^2 \leq \varphi^2(x^1),$$

(or a domain characterized by similar inequalities with the rôles of x^1 and x^2 interchanged) is a Green domain if $\varphi^1(x^1) < \varphi^2(x^1)$ on $[a, b]$, and the second derivatives of φ^i, $i = 1, 2$ are continuous in (a, b) with continuous extensions to $[a, b]$. Indeed, the domain D is an image of the rectangle $[a, b] \times [0, 1]$ under the mapping

$$
\begin{aligned}
H^1(x^1, x^2) &= x^1, \\
H^2(x^1, x^2) &= (1 - x^2)\varphi^1(x^1) + x^2 \varphi^2(x^1);
\end{aligned}
$$

and H satisfies the assumptions of Theorem 7.2.12. This can be further generalized: it suffices to require that there is a division $a = u_0 < u_1 < \cdots < u_n = b$ and φ^1 and φ^2 satisfy the above assumptions only on each $[u_i, u_{i+1}]$ for $i = 0, \ldots, n - 1$. By a limit passage it is possible to allow $\varphi^1(a) = \varphi^2(a)$ or $\varphi^1(b) = \varphi^2(b)$. It now follows easily that a disc, an ellipse, a triangle or a convex polygon is a Green domain.

Theorem 7.2.12 widens the class of Green domains sufficiently for applications within and outside mathematics. We now wish to make a few comments about even more general results. A path as in (7.2) is said to be *simple* if $\varphi(t_1) \neq \varphi(t_2)$ for $t_1 < t_2$ with the possible exception of $t_1 = a$ and $t_2 = b$. A simple closed path φ is by defintion positively

---

† Omitting the arguments x^1, x^2 in F^i, $H^i_{,j}$ etc. and leaving some routine calculations to the reader.

oriented if

$$\int_\varphi -ydx + xdy > 0. \tag{7.61}$$

Since $\int_\varphi -y_0 dx + x_0 dy = 0$ for any (x_0, y_0) the path φ is positively oriented if for some (x_0, y_0) the expression $-(y - y_0)dx + (x - x_0)dy$ is positive on $[\varphi]$. This happens if (x_0, y_0) is the centre of a circle or of a rectangle and φ is travelled anticlockwise. We see that defining 'positively oriented' by (7.61) is in accord with what our intuition tells us is 'anticlockwise'. A simple closed path φ divides the plane into two open and connected sets, one of which is bounded and called the interior of $[\varphi]$, while the other is unbounded and is called the exterior of $[\varphi]$. The common boundary† of the exterior and the interior of $[\varphi]$ is $[\varphi]$ itself. This intuitively obvious statement is called the *Jordan curve theorem* after the French mathematician C. Jordan who was the first to realize that the statement requires proof. Despite its plausibility the Jordan curve theorem is a deep theorem with a difficult proof. A relatively simple proof for a piecewise smooth curve can be found in [35]. We shall neither prove nor use the Jordan curve theorem. The Green theorem can be proved for a simple closed path in the form (7.51); the domain enclosed by $[\varphi]$ is simply the interior of $[\varphi]$. A proof of the Green theorem for a simple closed path within the framework of Riemann integration can be found in [1] but this proof can be extended to 'our' integrals as well.

REMARK 7.2.15 Assumption (iii) in Theorem 7.2.11 is general enough but it is desirable to remove it altogether. The situation here is similar to that of the Fundamental Theorem where *the integrability of f' is part of the assertion*. Unfortunately in this higher-dimensional case the KH-integration process is not good enough and yet another integral is needed for a Green-type theorem in which the integrability of $Q_x - P_y$ is asserted. This theme is a topic of current research and we refer to [22] and the bibliography there.

7.2.2 The Cauchy theorem

Theorems 7.2.8 and 7.2.10 contain as special cases the famous theorems of complex analysis, namely the Cauchy theorem and its homotopy version. In this special case it is however simpler to prove these results

---

† In the topological sense, the topological boundary of a set $S \subset \mathbb{R}^2$ is $\bar{S} \cap (\overline{\mathbb{R}^2 \setminus S})$.

directly. The set of complex numbers is denoted by $\mathbb{C}$. For $z \in \mathbb{C}$ we write $z = x + \jmath y$, i.e. $x = \Re z$ and $y = \Im z$. Functions in this subsection are complex valued, if $\Gamma \subset \mathbb{C}$ and $F : \Gamma \mapsto \mathbb{C}$ then we write $F(z) = U(x,y) + \jmath V(x,y)$. A path is similarly as before a continuous map of bounded variation from an interval in $\mathbb{R}^1$ into $\mathbb{C}$. Equivalence of paths is defined as in (7.4) and (7.5) with $\mathbb{R}^n$ replaced by $\mathbb{C}$. The opposite path $-\varphi$ and $\varphi \oplus \psi$ are defined exactly as in (7.12) and (7.11), respectively. The number $\int_\varphi F(z)\,dz$ is† the *line integral of* F *along* φ if for every positive ε there exists $\delta : [a,b] \mapsto \mathbb{R}_+$ such that if (7.7) is a δ-fine partition of $[a, b]$ then

$$\left| \sum_{k=1}^{N} F(\varphi(\xi_k))(\varphi(v_k) - \varphi(u_k)) - \int_\varphi F(z)\,dz \right| < \varepsilon. \qquad (7.62)$$

This complex valued integral has the properties (7.8) to (7.10), (7.13) and (7.14); moreover Theorem 7.2.1 holds. It is also easily checked that

$$\int_\varphi F(z)\,dz = \int_\varphi U(x,y)\,dx - V(x,y)\,dy + \jmath \int_\varphi V(x,y)\,dx + U(x,y)\,dy.$$

For a continuous F the integral always exists along a path. If φ' exists on $[a, b]$ except on a countable set the integral can be evaluated by the formula

$$\int_\varphi F(z)\,dz = \mathcal{KH} \int_a^b (F \circ \varphi)(t)\,\varphi'(t)\,dt. \qquad (7.63)$$

We rephrase and strengthen Theorem 7.2.5 as

THEOREM 7.2.16   *Let* $M \subset \Gamma \subset \mathbb{C}$ *and let* Γ *be an open set. If*

(i) *there exists a function* $\mathbf{P}$ *continuous in* Γ *and such that* $\mathbf{P}'(z) = F(z)$ *for all* $z \in \Gamma \setminus M$,

(ii) φ *is a path with* $[\varphi] \subset \Gamma$

(iii) $\varphi_{-1}(M)$ *is countable‡*

then

$$\int_\varphi F(z)\,dz = \mathbf{P}(\varphi(b)) - \mathbf{P}(\varphi(a)). \qquad (7.64)$$

In particular if φ *is closed*

$$\int_\varphi F(z)\,dz = 0. \qquad (7.65)$$

---

† With $F(z) \in \mathbb{C}$ and $[\varphi] \subset \mathbb{C}$.
‡ As usual $\varphi_{-1}(M) = \{t; \varphi(t) \in M\}$.

REMARK 7.2.17 If M is countable then assumption (iii) is satisfied if $\varphi'(t) \neq 0$ for every $t \in [a, b]$. (See Exercise 7.10.) It is also satisfied if φ is a polygonal path. Generally speaking, however, $\varphi_{-1}(M)$ need not be countable even if M is.

Proof of Theorem 7.2.16 Let $\varphi_{-1}(M) = \{t_1, t_2, \dots\}$ with $t_i \neq t_j$ for $i \neq j$. Since both $\mathbf{P} \circ \varphi$ and φ are continuous at t_k there exists $\delta(t_k) > 0$ such that

$$|\mathbf{P}(\varphi(v)) - \mathbf{P}(\varphi(u)) - F(\varphi(t_k))(\varphi(v) - \varphi(u))| < \frac{\eta}{2^k} \qquad (7.66)$$

whenever $t_k - \delta(t_k) < u \leq t_k \leq v < t_k + \delta(t_k)$. For $t \notin \varphi_{-1}(M)$ there is a positive $\delta(t)$ such that

$$|\mathbf{P}(\varphi(v)) - \mathbf{P}(\varphi(u)) - F(\varphi(t))(\varphi(v) - \varphi(u))| < \eta \, |\varphi(v) - \varphi(u)| \quad (7.67)$$

whenever $t - \delta(t) < u \leq t \leq v < t + \delta(t)$. If

$$\{(\tau_k, [u_k, v_k]); \ k = 1, 2, \dots, N\}$$

is a δ-fine partition then

$$\left| \mathbf{P}(\varphi(b)) - \mathbf{P}(\varphi(a)) - \sum_1^N F(\varphi(\tau_k)(\varphi(v_k) - \varphi(u_k)) \right|$$

$$\leq \ \sum_1^N \left| \mathbf{P}(\varphi(v_k)) - \mathbf{P}(\varphi(u_k)) - \sum_1^N F(\varphi(\tau_k)(\varphi(v_k) - \varphi(u_k)) \right|$$

$$\leq \ \eta \sum_1^n \frac{1}{2^k} + \eta \sum_1^N |\varphi(v_k) - \varphi(u_k)| \leq \eta + \eta \operatorname{Var}_a^b \varphi,$$

by inequalities (7.66) and (7.67). If $\varepsilon > 0$ is given it suffices to choose $\eta > 0$ such that $(1 + \operatorname{Var}_a^b)\eta < \varepsilon$ to show that (7.62) is satisfied with

$$\int_\varphi F(z) \, dz = \mathbf{P}(\varphi(b)) - \mathbf{P}(\varphi(a)). \qquad \bullet$$

THEOREM 7.2.18 (The Cauchy theorem) *If f is continuous in an open star-shaped set $S \subset \mathbb{C}$ and there is a countable set E such that f is differentiable in $S \backslash E$ then there exists a function F with $F'(z) = f(z)$ for every $z \in S$. If $\varphi : [a, b] \mapsto S$ is a path then*

$$\int_\varphi f = F(\varphi(b)) - F(\varphi(a)).$$

In particular, if φ is closed

$$\int_\varphi f = 0. \tag{7.68}$$

Proof The proof follows similar lines to the proofs of Theorems 7.2.7 and 7.2.8. The points of E can be enumerated, $E = \{z_1, z_2, \dots\}$, and for every positive ε there is a $\delta(z_k)$ such that

$$|f(z) - f(z_k)| < \frac{\varepsilon}{2^k} \tag{7.69}$$

for $|z - z_k| < \delta(z_k)$. For $\zeta \in S \setminus E$ there is a $\delta(\zeta)$ such that

$$|f(z) - f(\zeta) - f'(\zeta)(z - \zeta)| < \varepsilon |z - \zeta| \tag{7.70}$$

for $|z - \zeta| < \delta(\zeta)$. Let $T \subset S$ be a triangle, and $A(T)$, $p(t)$ and $d(T)$ its area, perimeter and diameter, in that order. We denote by ∂T the (positively) oriented boundary of T and by $\{(\zeta_i, T_i)\}$ a δ-fine partition of T into triangles similar to T (see Lemma 7.2.6). By inequality (7.69)

$$\left| \sum_{\zeta_i \in E} \int_{\partial T_i} f \right| \le \sum_{k=1}^{\infty} \frac{\varepsilon}{2^k} p(T). \tag{7.71}$$

Using Theorem 7.2.16 we have

$$\int_{\partial T_i} (f(\zeta_i) + f'(\zeta_i)(z - \zeta_i))\, dz = 0$$

and consequently by inequality (7.70)

$$\left| \sum_{\zeta_i \notin E} \int_{\partial T_i} f \right| < \varepsilon d(T_i) p(T_i). \tag{7.72}$$

The triangles T_i are similar to T, therefore there exists a constant C depending *only* on T and independent of i such that $d(T_i)p(T_i) < CA(T_i)$ for all i. Consequently

$$\int_{\partial T} f(z)\, dz = \sum_i \int_{\partial T_i} f(z)\, dz \le p(T)\varepsilon + A(T)\varepsilon.$$

This proves equation (7.68) in case φ is the oriented boundary of a triangle. Let c be the centre of S and $F(z) = \int_c^z f(\zeta)\, d\zeta$. If the segment joining z with $z + h$ lies in S and $|f(z + th) - f(z)| < \varepsilon$ for $0 \le t \le 1$

then by what we have already proved†

$$|F(z+h) - F(z)| = \left| \int\limits_{z}^{z+h} f(\zeta)\, d\zeta \right| \le \varepsilon |h|_2. \qquad \bullet$$

We now turn our attention to the homotopy version of the Cauchy theorem. Similarly as before we denote by Ω the square $[0,1] \times [0,1]$ and by H a function continuous on Ω with the following properties:

$$H(s,t) \;\in\; \Gamma \subset \mathbb{C} \text{ for every } (s,t) \in \Omega; \qquad (7.73)$$

$$H(0,t) \;=\; \varphi(t) \text{ and } H(1,t) = \Phi(t) \text{ for every } t \in [0,1]; \quad (7.74)$$

$$H(s,0) \;=\; a \text{ and } H(s,1) = b \text{ for every } s \in [0,1]; \qquad (7.75)$$

or

$$H(s,0) \;=\; H(s,1) \text{ for every } s \in [0,1]. \qquad (7.76)$$

The homotopy version of the Cauchy Theorem reads:

THEOREM 7.2.19 (The Cauchy theorem. Homotopy) *If*

   (i) $\Gamma \subset \mathbb{C}$ *is open,*
   (ii) $f'(z)$ *exists for every* $z \in \Gamma$
   (iii) *there exists a continuous function* H *satisfying conditions* (7.73), (7.74) *and* (7.75) *or conditions* (7.73), (7.74) *and* (7.76)

then

$$\int\limits_{\Phi} f(z)\, dz = \int\limits_{\varphi} f(z)\, dz.$$

The **proof** of this theorem follows from Theorem 7.2.18 in the same fashion as Theorem 7.2.10 followed from Theorem 7.2.8.

There is a very simple proof of Theorem 7.2.19 under the additional assumption that H has continuous partial derivatives of the second order. In this case

$$\int\limits_{\Phi} f(z)\, dz - \int\limits_{\varphi} f(z)\, dz = \int\limits_{\partial\Omega} [f(H(s,t))H_s(s,t)ds + f(H(s,t))H_t(s,t)\, dt].$$

$$(7.77)$$

Since

$$\frac{\partial(fH_s)}{\partial t} = f'H_tH_s + fH_{s,t} = f'H_sH_t + fH_{t,s} = \frac{\partial(fH_t)}{\partial s},$$

† We use the same notation as in equation (7.24).

the Green theorem is applicable and the right-hand side of (7.77) is zero. It is possible to obtain the full strength of the homotopy version of the Cauchy theorem by this proof, one has to employ an approximation argument to move from a twice continuously differentiable H to just a continuous one.

7.3 Differentiation of series

One cannot expect term by term differentiation of sequences (or series) to be permissible without some assumptions on the convergence of the sequence of *derivatives*, as the example of $F_n(x) = n^{-1}\sin nx$ shows. The sequence F_n is *uniformly* convergent on $\mathbb{R}$ yet the sequence $\{F_n'(0)\}$ is divergent. Exercise 7.12 provides an even more dramatic failure of term by term differentiation.

Differentiation and integration are inverse processes: under very general assumptions one cancels the other. It is therefore no surprise that theorems on integration of series lead naturally to theorems on differentiation of series.

THEOREM 7.3.1 *Let us assume that*

(A) *For every $n \in \mathbb{N}$ the functions F_n are SL and the derivatives F_n' exist almost everywhere on (a,b);*
(B) *for some $c \in (a,b)$, the sequence $n \mapsto F_n(c)$ converges, say to $F(c)$;*
(C) *the sequence $n \mapsto F_n'$ converges uniformly almost everywhere† to g, say.*

Then the sequence $n \mapsto F_n$ converges uniformly on (a,b) and the limit function F is differentiable at every x_0 for which $\lim_{n\to\infty} F_n'(x_0)$ exists and then

$$F'(x_0) = g(x_0). \tag{7.78}$$

Proof Firstly

$$F_n(x) = F_n(c) + \int_c^x F_n'. \tag{7.79}$$

By the almost uniform convergence of F_n'

$$F(x) = F(c) + \int_c^x g. \tag{7.80}$$

† i.e. there is a set S of measure zero and F_n' converges uniformly outside S.

Moreover the convergence of F_n to F is clearly uniform. Assume now the existence of

$$\lim_{n \to \infty} F_n'(x_0).$$

For a positive ε there is a natural n such that $F_n'(x_0)$ exists and

$$|F_n'(x_0) - g(x_0)| < \varepsilon \quad \text{and} \quad \left| \frac{1}{h} \int_{x_0}^{x_0+h} (g(t) - F_n'(t))dt \right| < \varepsilon.$$

Since

$$\frac{F(x_0 + h) - F(x_0)}{h} - g(x_0) = \frac{1}{h} \int_{x_0}^{x_0+h} (g(t) - F_n'(t))dt$$

$$+ \frac{F_n(x_0 + h) - F_n(x_0)}{h} - F_n'(x_0) + F_n'(x_0) - g(x_0),$$

we obtain

$$\left| \frac{F(x_0 + h) - F(x_0)}{h} - g(x_0) \right| < 2\varepsilon + \left| \frac{F_n(x_0 + h) - F_n(x_0)}{h} - F_n'(x_0) \right|.$$

By the choice of x_0 and n there is a positive δ such that the last term is less than ε for $0 < |h| < \delta$. Consequently

$$\left| \frac{F(x_0 + h) - F(x_0)}{h} - g(x_0) \right| < 3\varepsilon. \qquad \bullet$$

REMARK 7.3.2 Assumption (A) is satisfied if $F_n'(x)$ exists for every $x \in (a, b)$.

COROLLARY 7.3.3 *If*

(A1) *For every $n \in \mathbb{N}$ the derivative F_n' exists everywhere on (a, b);*

(B) *for some $c \in (a, b)$, the sequence $n \mapsto F_n(c)$ converges, say to $F(c)$;*

(C1) *the sequence $n \mapsto F_n'$ converges uniformly everywhere on (a, b) to g, say.*

then the function F is differentiable everywhere on (a, b) and (7.78) holds for every $x_0 \in (a, b)$.

COROLLARY 7.3.4 *If the series $\sum_1^\infty h_n'$ converges uniformly on (a, b) and the series $\sum_1^\infty h_n(c)$ converges for some $c \in (a, b)$ then*

$$\left(\sum_1^\infty h_n(x) \right)' = \sum_1^\infty h_n'(x)$$

for every $x \in (a, b)$.

Assumption (C) in Theorem 7.3.1 is indispensable but is very restrictive. We therefore state the next theorem although it is an almost trivial consequence of Theorem 3.5.17.

A KH-integrable function is said to be C-continuous at x if

$$\lim_{h \to 0} \frac{1}{h} \int_x^{x+h} f = f(x).$$

The more standard notation is $(C, 1)$ but we have no need for writing the additional 1. Recall that Theorem 2.6.8 asserts that f is C-continuous at every point at which it is continuous and that Theorem 3.8.2 says that a KH-integrable function is C-continuous a.e.

THEOREM 7.3.5 *Let us assume that*

(A) *For every $n \in \mathbb{N}$ the functions F_n are SL and the derivatives F_n' exist almost everywhere on (a,b);*

(B) *for some $c \in (a,b)$, the sequence $n \mapsto F_n(c)$ converges, say to $F(c)$;*

(C2) *the sequence $n \mapsto F_n'$ converges almost everywhere to g, say;*

(D) *there are KH-integrable functions h, H such that for all $n \in \mathbb{N}$ the inequalities $h(x) \le F_n'(x) \le H(x)$ hold almost everywhere on (a,b).*

Then the sequence $n \mapsto F_n$ converges on (a,b), the limit function F is differentiable at every x_0 at which g is C-continuous, and equation (7.78) holds.

Proof We clearly have (7.79) and (7.80) and consequently

$$\frac{F(x_0 + h) - F(x_0)}{h} - g(x_0) = \frac{1}{h} \int_{x_0}^{x_0+h} g - g(x_0).$$

Equation (7.78) follows by the definition of C-continuity. ●

COROLLARY 7.3.6 *If assumptions (A), (B), $(C2)$ and (D) are satisfied then F is differentiable a.e. and (7.78) holds for almost all x_0.*

REMARK 7.3.7 Assumption (D) can be replaced in Theorem 7.3.5 or Corollary 7.3.6 by any other assumption which guarantees the interchange of limit and integration.

7.4 Dirichlet's problem and the Poisson integral

In hydrodynamics, electromagnetic theory, heat conduction and in many other applications as well as in pure mathematics one encounters the Laplace equation:

$$\sum_1^n \frac{\partial^2 U}{\partial x_i^{\,2}} = 0.$$

Continuous solutions to this equation are called harmonic functions. An important boundary value problem, the so-called Dirichlet problem, is formulated as follows: *Given a bounded, open and connected set $G \subset \mathbb{R}^n$ and a function f defined on $\mathbf{b}G$, the boundary of G, find a function U satisfying the Laplace equation in G and equal to f on $\mathbf{b}G$.* An additional condition is required in the formulation of this problem: there is a need to connect the values of U on $\mathbf{b}G$ with the values of U inside G. Otherwise any function constant in G and equal to f on $\mathbf{b}G$ solves the problem. In the classical formulation of the Dirichlet problem it is required that U is continuous on $\bar{G}$, the closure of G. In this book we are not concerned with the theory of partial differential equations but as a motivation for the next section we prove the existence of a solution of the Dirichlet problem in its classical formulation when G is a two-dimensional disc†. Without loss of generality we assume this disc to be $G = \{(x, y);\ x^2 + y^2 < 1\}$. The boundary function f becomes a continuous 2π-periodic function of the polar angle φ. We seek the solution, at the first attempt purely formally, as an infinite series of harmonic functions. In elementary complex analysis it is shown, and is also easily directly verified (see Exercise 7.15), that functions $w_n :$ $(x, y) \mapsto \Re(x + \jmath y)^n$ with $n \in \mathbb{N}$ are harmonic in G. So we try‡

$$U(x, y) = \frac{a_0}{2} + \sum_1^\infty \Re[(a_i - \jmath b_n)(x + \jmath y)^n], \qquad (7.81)$$

i.e. we seek U as the real part of a power series. We set $x + \jmath y = r(\cos\varphi + \jmath \sin\varphi)$ and observe that owing to the factor r^n the series obtained by differentiating the right-hand side of (7.81) converges uniformly in any disc $\{0 \le r < a < 1\}$. Consequently§ U is harmonic. The boundary

---

† Uniqueness is proved in the Appendix.
‡ Denoting the coefficients by $a_0/2$ and $a_n - \jmath b_n$ is for convenience only, and has no material significance.
§ We allow a slight abuse of notation by writing U or $U(x, y)$ or $U(r, \varphi)$ interchangeably.

condition $U(1, \varphi) = f(\varphi)$ becomes

$$f(\varphi) = \frac{a_0}{2} + \sum_1^\infty (a_n \cos n\varphi + b_n \sin n\varphi). \qquad (7.82)$$

Our problem from partial differential equations leads naturally to another problem in mathematical analysis, to represent a given 2π-periodic function as a trigonometric series. Clearly this is in itself a question of significance for applications, for in science there are many situations when we would like to decompose a periodic phenomenon such as vibration (represented by f) into simple harmonic phenomena (represented by the terms of the series in (7.82)). It remains to determine a_n and b_n. If f were a trigonometric polynomial† then multiplying (7.82) by $\sin k\varphi$ or $\cos k\varphi$, integrating from $-\pi$ to π, and taking into account the formulae

$$\int_{-\pi}^{\pi} \sin k\varphi \cos n\varphi \, d\varphi = 0, \qquad (7.83)$$

$$\int_{-\pi}^{\pi} \sin^2 n\varphi \, d\varphi = \int_{-\pi}^{\pi} \cos^2 n\varphi \, d\varphi = \pi, \qquad (7.84)$$

$$\int_{-\pi}^{\pi} \cos k\varphi \cos n\varphi \, d\varphi = 0 \qquad \text{for } k \neq n, \qquad (7.85)$$

$$\int_{-\pi}^{\pi} \sin k\varphi \sin n\varphi \, d\varphi = 0 \qquad \text{for } k \neq n \qquad (7.86)$$

would give

$$a_n = \frac{1}{\pi} \int_{-\pi}^{\pi} f(\varphi) \cos n\varphi \, d\varphi \qquad \text{for } n = 0, 1, 2\ldots \qquad (7.87)$$

$$b_n = \frac{1}{\pi} \int_{-\pi}^{\pi} f(\varphi) \sin n\varphi \, d\varphi \qquad \text{for } n = 1, 2\ldots \qquad (7.88)$$

As long as they exist a_n and b_n determined by (7.87) and (7.88) are called the Fourier coefficients of f and the series

$$\frac{a_0}{2} + \sum_1^\infty (a_n \cos nt + b_n \sin nt)$$

---

† i.e. f would be equal to the right-hand side of (7.82) but with only a finite number of a_n and b_n distinct from zero.

with coefficients given by (7.87) and (7.88) is called the Fourier series associated with f. It now seems natural to try as a solution

$$U(r, \varphi) = \frac{a_0}{2} + \sum_1^\infty r^n (a_n \cos n\varphi + b_n \sin n\varphi), \qquad (7.89)$$

with a_n and b_n given by (7.87) and (7.88) for $n = 0, 1, \ldots$ However, an unpleasant surprise is waiting for us here: it was discovered already in the ninetennth century that for a continuous function its associated Fourier series can diverge†, which means that the series in (7.89) might diverge for $r = 1$. Nevertheless we are going to show that for $r < 1$ equation (7.89) does represent the solution of the Dirichlet problem. In order to show this we give the right-hand side a more compact form. Substituting from (7.87) and (7.88) into (7.89) and interchanging integration and summation, which is clearly permissible for $r < 1$, we obtain

$$U(r, \varphi) = \frac{1}{2\pi} \int_{-\pi}^{\pi} f(\psi) \left(1 + 2 \sum_1^\infty r^n \cos n(\psi - \varphi) \right) d\psi. \qquad (7.90)$$

Denoting $\varphi - \psi = \omega$ and summing a geometric series leads to:

$$1 + 2\sum_1^\infty r^n \cos n\omega \;=\; 1 + 2\Re \sum_1^\infty r^n e^{in\omega}$$

$$= \Re \frac{1 + re^{i\omega}}{1 - re^{i\omega}} \;=\; \frac{1 - r^2}{1 - 2r\cos\omega + r^2} \;.$$

Finally we have

$$U(r, \varphi) \;=\; \frac{1}{2\pi} \int_{-\pi}^{\pi} P(r, \varphi - \psi) f(\psi)\, d\psi \qquad (7.91)$$

$$= \; \frac{1}{2\pi} \int_0^{\pi} P(r, \omega) \left[f(\varphi + \omega) + f(\varphi - \omega) \right] d\omega, \qquad (7.92)$$

with

$$P(r, \omega) = \frac{1 - r^2}{1 - 2r\cos\omega + r^2} = 1 + 2\sum_1^\infty r^n \cos n\omega. \qquad (7.93)$$

† An example of such an f is given on page 405 of the book by Béla Sz.-Nagy [42].

The function P is called the Poisson kernel and has the following properties:

$$\int_0^\pi P(r,\omega)\,d\omega = \pi; \tag{7.94}$$

$$P(r,\omega) \geq 0 \quad \text{for} \quad 0 \leq r \leq 1,\ 0 < \omega \leq \pi; \tag{7.95}$$

$$\text{if } r \in (0,1) \text{ then } h : \omega \mapsto P(r,\omega) \text{ is decreasing on } (0,\pi). \tag{7.96}$$

These properties are easily verified from either the fraction form or the infinite series form of P.

We now show that for U given by (7.92)

$$\lim_{(r,\varphi)\to(1,\varphi_0)} U(r,\varphi) = f(\varphi_0). \tag{7.97}$$

In view of (7.94) it is sufficient to show: for every positive ε there are ρ and δ such that

$$\left| \int_0^\pi P(r,\omega)\left[f(\varphi+\omega) + f(\varphi-\omega) - 2f(\varphi_0) \right] d\omega \right| < \pi\varepsilon \tag{7.98}$$

whenever $\rho < r < 1$ and $|\varphi-\varphi_0| < \delta$. We choose such δ that $|t-\varphi_0| < 2\delta$ implies $|f(t) - f(\varphi_0)| < \varepsilon/4$, take $|\varphi - \varphi_0| < \delta$ and estimate

$$\left| \int_0^\delta P(r,\omega)\left[f(\varphi+\omega) + f(\varphi-\omega) - 2f(\varphi_0) \right] d\omega \right|$$

$$< 2\frac{\varepsilon}{4} \int_0^\pi P(r,t)\,dt \leq \frac{\pi\varepsilon}{2} \tag{7.99}$$

and

$$\left| \int_\delta^\pi P(r,\omega)\left[f(\varphi+\omega) + f(\varphi-\omega) - 2f(\varphi_0) \right] d\omega \right|$$

$$< 4\pi P(r,\delta) \sup\{f;\ [0,\pi]\}.$$

The right-hand side of the last inequality tends to zero as $r \to 1$, consequently there exists ρ such that

$$\left| \int_\delta^\pi P(r,\omega)\left[f(\varphi+\omega) + f(\varphi-\omega) - 2f(\varphi_0) \right] d\omega \right| < \frac{1}{2}\varepsilon\pi \tag{7.100}$$

for $r > \rho$. Combining (7.99) and (7.100) proves (7.98). We summarize our result as

THEOREM 7.4.1 (Dirichlet's problem for a disc) *If f is continuous 2π-periodic and*

$$U(r,\varphi) \;=\; \frac{1}{2\pi}\int_{-\pi}^{\pi} P(r,\varphi-\psi)f(\psi)\,d\psi \quad \text{for } 0\le r<1,\ 0\le\varphi<2\pi,$$
$$U(1,\varphi) \;=\; f(\varphi) \quad \text{for } 0\le\varphi<2\pi,$$

then

    (i) U *is continuous for* $0\le r\le 1$;
    (ii) U *is harmonic for* $0\le r<1$.

As a byproduct and a bonus we obtain easily the following classical theorem.

THEOREM 7.4.2 (Weierstrass' approximation theorem) *If f is continuous 2π-periodic then for every positive ε there is a trigonometric polynomial T such that*

$$|f(\varphi) - T(\varphi)| < \varepsilon$$

for all $\varphi \in \mathbb{R}$.

Proof By (i) of the previous theorem there is $r < 1$ such that

$$|f(\varphi) - U(r,\varphi)| < \frac{\varepsilon}{2}.$$

Using the infinite series form of U we find N such that

$$|U(r,\varphi) - \frac{a_0}{2} - \sum_{1}^{N} r^n(a_n\cos n\varphi + b_n\sin n\varphi))| < \frac{\varepsilon}{2}. \qquad \bullet$$

7.5 Summability of Fourier series

It was mentioned in the previous section that the Fourier series associated with a continuous function may diverge at some points. Nevertheless we shall show in this section that the Fourier series associated with a merely KH-integrable function represents this function in some well defined way. In order to describe this we need to attach rationally a 'sum' to an infinite series which need not be convergent. *An infinite series $\sum_{1}^{\infty} a_n$ is said to be Abel summable if the function $S(r) = \sum_{1}^{\infty} a_n r^n$ has a limit as $r \uparrow 1$. The Abel sum $(A)\sum_{1}^{\infty} a_n$ of the series $\sum_{1}^{\infty} a_n$ is*

$$(A)\sum_{1}^{\infty} a_n = \lim_{r\uparrow 1} S(r).$$

For example the series $1-1+1-\cdots$ is divergent but $S(r)=1/(1+r)$, $\lim_{r\uparrow1}S(r)=1/2$, and consequently $(A)\sum_1^\infty(-1)^n=1/2$. The Abel sum of a series shares some basic properties with sums of convergent series.

$$(A)\sum_1^\infty a_n+(A)\sum_1^\infty b_n=(A)\sum_1^\infty(a_n+b_n),$$

$$c(A)\sum_1^\infty a_n=(A)\sum_1^\infty ca_n,$$

$$(A)\sum_1^\infty a_n=\sum_1^k a_n+(A)\sum_{k+1}^\infty a_n.$$

We also have the important consistency property:

if a series $\sum_1^\infty a_n$ is convergent then it is Abel summable and $\sum_1^\infty a_n=(A)\sum_1^\infty a_n$.
Consistency of the Abel summation method is a consequence of the Abel theorem proved as Theorem A.5.2.

Theorem 7.4.1 of the previous section contains the following:

THEOREM 7.5.1 *The Fourier series of a continuous and 2π-periodic function f is Abel summable to $f(x)$ for every x.*

Recall that a KH-integrable function is C-continuous at x if

$$\lim_{h\to0}\frac1h\int_x^{x+h}f=f(x),$$

while Theorem 3.8.2 says that a KH-integrable function is C-continuous a.e. The main theorem of this section is

THEOREM 7.5.2 *The Fourier series of a KH-integrable function f is Abel summable to $f(x)$ at every point of C-continuity of f.*

The next two corollaries are immediate consequences of this theorem.

COROLLARY 7.5.3 *The Fourier series of a KH-integrable function f is Abel summable a.e. to f.*

COROLLARY 7.5.4 *If the Fourier series of a KH-integrable function f is convergent a.e. then it converges to $f(x)$ a.e.*

Proof of Theorem 7.5.2 It suffices to show (see (7.89), (7.92) and (7.93)) that

$$\lim_{r\uparrow 1} \int_0^\pi P(r,t)[f(t+x) + f(t-x) - 2f(x)]\,dt = 0 \qquad (7.101)$$

at every point x where f is C-continuous. At such a point x for every positive ε there is a positive δ such that

$$\left| \int_0^t [f(s+x) - f(s-x) - 2f(x)]\,ds \right| < t\varepsilon$$

for $0 < t < \delta$. Denoting $F_x(t) = \int_0^t [f(s+x) + f(s-x) - 2f(x)]\,ds$ we rewrite the last inequality as $|F_x(t)| < t\varepsilon$. In the following integration by parts and estimates we employ the property of the Poisson kernel expressed in (7.94)—(7.96).

$$\int_0^\delta P(r,t)[f(t+x) + f(t-x) - 2f(x)]\,dt$$

$$= P(r,\delta)F_x(\delta) + \int_0^\delta (-P_t(r,u))F_x(u)\,du, \qquad (7.102)$$

$$|P(r,\delta)F_x(\delta)| < \delta\varepsilon P(r,\delta), \qquad (7.103)$$

$$\left| \int_0^\delta (-P_t(r,u))F_x(u)\,du \right| \le \varepsilon \int_0^\delta (-P_t(r,u))u\,du \qquad (7.104)$$

$$= -\delta\varepsilon P(r,\delta) + \varepsilon \int_0^\pi P(r,s)\,ds \le -\delta\varepsilon P(r,\delta) + \varepsilon\pi. \qquad (7.105)$$

Combining (7.102)—(7.105) we have

$$\left| \int_0^\delta P(r,t)[f(t+x) + f(t-x) - 2f(x)]\,dt \right| < \pi\varepsilon. \qquad (7.106)$$

The function P is decreasing in t on $[\delta,\pi]$, and

$$\lim_{r\uparrow 1} P(r,t) = 0$$

for every $t \in [\delta,\pi]$. Theorems 3.7.5 and 3.7.6 are applicable to

$$\lim_{r\uparrow 1} \int_\delta^\pi P(r,t)[f(t+x) + f(t-x) - 2f(x)]\,dt = 0. \qquad (7.107)$$

Combining (7.106) and (7.107) gives

$$\limsup_{r\uparrow 1} \left| \int_0^\pi P(r,t)[f(t+x) + f(t-x) - 2f(x)]\,dt \right| \le \varepsilon\pi$$

and since ε is arbitrary we have (7.101).   •

We finish this section with a very important theorem, the so-called theorem on the completeness of the system of trigonometric functions. It is an immediate consequence of Theorem 7.5.2.

THEOREM 7.5.5 *If the Fourier coefficients of a KH-integrable function f are all zero then $f(x) = 0$ a.e.*

7.6 Fourier series and the space $\mathcal{L}^2$

The set of all measurable functions f for which the integral

$$\int_a^b f^2 < \infty$$

is called $\mathcal{L}^2$. We denote

$$\| f \| = \sqrt{\int_a^b f^2} \ .$$

If $\| f - g \| = 0$ then $f = g$ almost everywhere. In this section we shall identify functions which differ only on a set of measure zero. This really means that we are not working with individual functions but classes of functions, each class consisting of functions equal one another a.e. With a slight abuse of language we shall call these classes functions and the reader is welcome to think of them as functions defined up to a set of measure zero. We then have $\| f \| = 0$ if and only if $f = 0$.

The definition of the space $\mathcal{L}^2$ makes sense also for an infinite $[a, b]$, but in this book we shall restrict our attention to the case where the underlying interval $[a, b]$ is bounded. We also write $\mathcal{L}^2[a, b]$ to indicate that the domain of definition of functions in $\mathcal{L}^2$ is $[a, b]$. If $f \in \mathcal{L}^2$ and $g \in \mathcal{L}^2$ then the function $|fg|$ (and consequently fg itself) is KH-integrable. This follows from the inequality

$$|fg| \leq \frac{1}{2}(f^2 + g^2).$$

Taking $g = 1$ shows that every $f \in \mathcal{L}^2$ is absolutely KH-integrable. We denote $\int_a^b fg = (f, g)$ and call it the scalar product of f and g. One can think of (f, g) as a generalization of the scalar product of two vectors in $\mathbb{R}^n$. By the definition of the KH-integral (f, g) can be approximated by $\sum_1^n f(\xi)g(\xi)(v - u)$, i.e. by a scalar product of two finite dimensional vectors. The Cauchy–Schwartz inequality for vectors

$$\left(\sum_1^n f(\xi)g(\xi)(v-u) \right)^2 \le \left(\sum_1^n f^2(\xi)(v-u) \right) \left(\sum_1^n g^2(\xi)(v-u) \right)$$

implies the Cauchy–Schwartz inequality for integrals

$$\left| \int_a^b fg \right|^2 \le \int_a^b f^2 \int_a^b g^2,$$

or equivalently

$$|(f,g)| \le \| f \| \| g \| .$$

For $g = 1$ this gives

$$\int_a^b |f| \le \sqrt{b-a} \, \| f \| . \tag{7.108}$$

We also have the triangle inequality

$$\| f + g \| \le \| f \| + \| g \| . \tag{7.109}$$

It is proved as follows:

$$\| f + g \|^2 = (f+g, f+g)$$
$$= \| f \|^2 + 2(f,g) + \| g \|^2 \le \| f \|^2 + 2 \| f \| \| g \| + \| g \|^2$$
$$\le (\| f \| + \| g \|)^2 .$$

A sequence $\{f_n\}$ is said to be $\mathcal{L}^2$-Cauchy if for every positive ε there is a natural N such that

$$\| f_n - f_m \| < \varepsilon$$

for $n, m > N$. A function f is said to be the $\mathcal{L}^2$-norm limit† of $\{f_n\}$ if

$$\| f_n - f \| \to 0.$$

The $\mathcal{L}^2$-norm limit is obviously uniquely determined. We might say briefly norm limit or $\mathcal{L}^2$-limit instead of $\mathcal{L}^2$-norm limit. If f is the $\mathcal{L}^2$-norm limit of $\{f_n\}$ then $\lim_{n \to \infty} f_n(x)$ need not exist for any $x \in [a, b]$. This shows Example 3.5.23. The concept of $\mathcal{L}^2$-norm limit, or as we may also say $\mathcal{L}^2$-convergence, is far more important in analysis than pointwise convergence or pointwise convergence a.e. The following lemma is easy to prove.

---

† The term f_n is *mean convergent* to f is also often used.

LEMMA 7.6.1 *If $\{f_n\}$ is $\mathcal{L}^2$-Cauchy and there exists a subsequence $\{f_{n_i}\}$ which has a $\mathcal{L}^2$-norm limit f then the sequence $\{f_n\}$ itself has a $\mathcal{L}^2$-norm limit f.*

THEOREM 7.6.2 (Riesz–Fisher) *If $\{f_n\}$ is $\mathcal{L}^2$-Cauchy then there exists $f \in \mathcal{L}^2$ such that*

  (i) *there exists a subsequence $\{f_{n_i}\}$ convergent a.e. to f;*

  (ii) *f is $\mathcal{L}^2$-norm limit of the sequence $\{f_n\}$.*

Proof To prove (i) we choose n_i such that

$$\| f_{n_i} - f_m \| < \frac{1}{2^i} \tag{7.110}$$

for $m \geq n_i$. We also choose these n_i increasing, $n_i < n_{i+1}$. By inequality (7.108)

$$\int_a^b \left| f_{n_i} - f_{n_{i+1}} \right| < \frac{\sqrt{b-a}}{2^i}$$

and the series

$$f_{n_1} + \sum_{i=1}^{\infty} (f_{n_{i+1}} - f_{n_i})$$

is absolutely convergent a.e. on $[a, b]$ by Beppo Levi's theorem. Let $f = \lim_{i \to \infty} f_{n_i}$.

For the proof of (ii) it is sufficient to show, by Lemma 7.6.1, that

$$\lim_{i \to \infty} \| f_{n_i} - f \| = 0. \tag{7.111}$$

Let us denote $h_j = \left| f_{n_i} - f_{n_{i+j}} \right|$ and estimate $\| h_j \|$ by using (7.110).

$$\| h_j \| \leq \sum_{k=i}^{i+j-1} \| f_{n_k} - f_{n_{k+1}} \| < \frac{1}{2^{i-1}}.$$

We obtain successively by using Fatou's lemma on the sequence $\{h_j\}$ firstly

$$f_{n_i} - f \in \mathcal{L}^2$$

then $f \in \mathcal{L}^2$ and finally (7.111).   ●

Two functions f and g are said to be *orthogonal* if $(f, g) = 0$. A sequence of functions w_i with $i \in \mathbb{N}$ is *orthonormal* or forms an orthonormal system in $\mathcal{L}^2$ if w_i are mutually orthogonal, i.e. $(w_i, w_j) = 0$

for $i \neq j$ and $\| w_i \| = 1$ for $i \in \mathbb{N}$. A fairly simple† orthonormal system is $\omega_i = 2^{\frac{n+1}{2}}$ on $[2^{-(n+1)}, 2^{-n}]$ and zero otherwise. We know from (7.85) to (7.84) that the system of trigonometric functions

$$\frac{1}{\sqrt{2\pi}}, \quad \frac{\cos x}{\sqrt{\pi}}, \quad \frac{\sin x}{\sqrt{\pi}}, \quad \frac{\cos 2x}{\sqrt{\pi}}, \quad \frac{\sin 2x}{\sqrt{\pi}}, \quad \ldots \qquad (7.112)$$

is orthonormal.

An *orthonormal system* w_i, $i = 1, 2, \ldots$ is said to be *complete in* $\mathcal{L}^2$ if the only function orthogonal to all w_i is zero. We know from Theorem 7.5.5 that the system of trigonometric functions is complete. The sequence $\{\omega_i\}$ is not complete since the function $f = 1_{[1/2, 3/4]} - 1_{[3/4, 1]}$ is orthogonal to all ω_i.

The space $\mathcal{L}^2$ is a natural generalization of the n-dimensional Euclidean space $\mathbb{R}^n$ but it is infinite dimensional. A complete orthonormal sequence plays in $\mathcal{L}^2$ similar rôle as an orthonormal basis plays in $\mathbb{R}^n$. (See also equation (7.117).)

EXAMPLE 7.6.3 An interesting example of a complete orthonormal system in $\mathcal{L}^2[0,1]$ is the so-called Haar sequence. We define $h_1 = 1_{(0,1]}$. Every $n \in \mathbb{N}$, $n \geq 2$ can be uniquely represented as $n = 2^k + p$ with $k, p \in \mathbb{Z}$, $k \geq 0$ and $1 \leq p \leq 2^k$. For $n \geq 2$ we define h_n as follows:

$$h_n(x) = \begin{cases} \sqrt{2^k} & \text{if } \dfrac{p-1}{2^k} < x \leq \dfrac{2p-1}{2^{k+1}} \\[2mm] -\sqrt{2^k} & \text{if } \dfrac{2p-1}{2^{k+1}} \leq x < \dfrac{p}{2^k} \\[2mm] 0 & \text{otherwise.} \end{cases}$$

Two h_n with same k and distinct p are orthogonal because their product is zero on $[0,1]$. If k are distinct then the product is either also zero or is up to a constant factor equal to the h_n with the smaller n. The orthogonality then follows from $\int_0^1 h_n = 0$. Let $(g, h_n) = 0$ for all natural n and $G(t) = \int_0^t g$. If we show that

$$G\left(\frac{p}{2^k}\right) = 0 \quad \text{for} \quad k = 0, 1, 2, \ldots ; \ p = 0, 1, \ldots, 2^k, \qquad (7.113)$$

then completeness of the Haar sequence is established, since by continuity $G = 0$ on $[0,1]$ and consequently $g = 0$ a.e. For $k = 0$ equations (7.113) follow from (g, h_1). We also have

$$(g, h_n) = G\left(\frac{p}{2^k}\right) - G\left(\frac{p-1}{2^k}\right) + 2G\left(\frac{2p-1}{2^{k+1}}\right) = 0.$$

† And practically useless.

From this (7.113) follows by induction on k.   ●

If $f \in \mathcal{L}^2[a,b]$ and w_i, $i = 1, 2, \ldots$ is an orthonormal sequence then the numbers $c_i = (f, w_i)$ are called *the Fourier coefficients of f with respect to the sequence* $\{w_i\}$ and, if no confusion can arise, simply the Fourier coefficients of f. The series

$$\sum_{i=1}^{\infty} c_i w_i \tag{7.114}$$

is called *the Fourier series associated with f with respect to the sequence* $\{w_i\}$ or simply the Fourier series of f. The Fourier coefficients have the following minimizing property.

THEOREM 7.6.4   *If $f \in \mathcal{L}^2$ and $\{w_i\}$ is an orthonormal sequence of elements in $\mathcal{L}^2$ then among all functions of the form $\sum_{i=1}^{n} \alpha_i w_i$ the minimum of*

$$\| f - \sum_{i=1}^{n} \alpha_i w_i \|^2$$

is attained for $\alpha_i = c_i$, $i = 1, 2, \ldots, n$. Moreover

$$\sum_{i=1}^{n} c_i^2 \leq \| f \|^2 . \tag{7.115}$$

This inequality is called *Bessel's inequality*.
Proof The following calculations are fairly straightforward.

$$0 \leq \| f - \sum_{i=1}^{n} \alpha_i w_i \|^2 \quad = \quad (f - \sum_{i=1}^{n} \alpha_i w_i, \ f - \sum_{i=1}^{n} \alpha_i w_i)$$

$$= \| f \|^2 - 2 \sum_{i=1}^{n} \alpha_i c_i + \sum_{i=1}^{n} \alpha_i^2 \quad = \quad \| f \|^2 + \sum_{i=1}^{n} (\alpha_i - c_i)^2 - \sum_{i=1}^{n} c_i^2.$$

The minimum of the right-hand side is clearly achieved for $\alpha_i = c_i$. If this holds then Bessel's inequality is obvious from the above calculations.

●

COROLLARY 7.6.5   *If $f \in \mathcal{L}^2$ and $\{w_i\}$ is an orthonormal sequence in $\mathcal{L}^2$ then the series*

$$\sum_{i=1}^{\infty} c_i^2 = \sum_{i=1}^{\infty} (f, w_i)^2$$

converges.

The main theorem of this section is

THEOREM 7.6.6 *Let* $f \in \mathcal{L}^2$ *and let* $\{w_i\}$ *be an orthonormal se-quence in* $\mathcal{L}^2$, $\sigma_n = \sum_{i=1}^{n} (f, w_i) w_i$. *The following statements are equivalent:*

(i) *For every* $f \in \mathcal{L}^2$ *the Fourier series of* f *is* $\mathcal{L}^2$-*norm convergent to* f, *i.e.*

$$\lim_{n \to \infty} \| \sigma_n - f \| = 0. \tag{7.116}$$

(ii) *For every* f *in* $\mathcal{L}^2$ *the following Parseval's equation holds:*

$$\| f \|^2 = \sum_{i=1}^{\infty} (f, w_i)^2. \tag{7.117}$$

(iii) *The sequence* $\{w_i\}$ *is complete.*

Proof (i)$\Rightarrow$(ii). Firstly we have

$$\left| (f, \sum_{i=1}^{n} c_i w_i) - (f, f) \right| \leq \| f \| \, \| \sigma_n - f \|$$

and consequently

$$\| f \|^2 = \lim_{n \to \infty} (f, \sum_{i=1}^{n} c_i w_i) = \sum_{i=1}^{\infty} c_i^2.$$

(ii)$\Rightarrow$(iii) is obvious.

(iii)$\Rightarrow$(i). By Corollary 7.6.5 the sequence $\{\sigma_n\}$ is $\mathcal{L}^2$-Cauchy and by the Riesz–Fisher theorem (Theorem 7.6.2) there is a $g \in \mathcal{L}^2$ such that $\| \sigma_n - g \| \to 0$. Further

$$|(g, w_i) - (\sigma_n, w_i)| \leq \| g - \sigma_n \|$$

and therefore

$$(g, w_i) = \lim_{n \to \infty} (\sigma_n, w_i) = (f, w_i).$$

The function $f - g$ has all Fourier coefficients zero, and by (iii) we have $f - g = 0$.   ●

Since we know that the system (7.112) of trigonometric functions is complete by Theorem 7.5.5 we have

COROLLARY 7.6.7 *If* $f \in \mathcal{L}^2[-\pi, \pi]$ *then the Fourier series associated with* f *with respect to the orthonormal sequence of trigonometric functions is* $\mathcal{L}^2$-*norm convergent to* f.

We also have

COROLLARY 7.6.8 *If $f \in \mathcal{L}^2[0,1]$ then the Fourier series associated with f with respect to the Haar sequence is $\mathcal{L}^2$-norm convergent to f.*

We conclude this section with some general remarks about Fourier series. They form an important branch of mathematics; both set theory and functional analysis have their roots in the theory of convergence of Fourier series. It was in the nineteenth century that the German mathematician du Bois-Reymond produced a continuous periodic function whose Fourier series did not converge everywhere. The Hungarian mathematician Fejér much later produced a simpler example, which can be found in reference [42]. In 1926 the Russian mathematician Kolmogoroff created an absolutely KH-integrable function which had an everywhere divergent Fourier series. The question whether or not the Fourier series of a continuous function can diverge on a set of positive measure remained unsolved until 1966 when the Swedish mathematician Carleson proved that the Fourier series of a function $f \in \mathcal{L}^2$ converges a.e. to f. (Our Corollary 7.6.7 combined with the Riesz-Fisher theorem gives the much weaker result that a subsequence of the partial sums converges a.e. to f.)

7.7 Exercises

EXERCISE 7.1 *Prove relation* (7.16).

EXERCISE 7.2 *Let $o = (0,0)$, $a = (1,0)$, $b = (1,2)$, $c = (0,2)$. Evaluate the following line integral in $\mathbb{R}^2$: $\int_\varphi xydx - x^3dy$ if (1) $\varphi = \overrightarrow{ob}$, (2) $\varphi = \overrightarrow{oa} \oplus \overrightarrow{ab}$, (3) $\varphi = \overrightarrow{oc} \oplus \overrightarrow{cb}$, (4) $\varphi(t) = (t, 2t^2)$, $t \in [0,1]$.*

EXERCISE 7.3 *Evaluate the following line integral in $\mathbb{R}^2$:*

$$\int_\varphi \frac{xdx + ydy}{x^2 + y^2}, \quad \varphi(t) = (\cos^3 t, \sin^6 t), \; t \in [0, 2\pi].$$

EXERCISE 7.4 *If f is continuous show that the integral $\int_\varphi f(x+y+1)(dx + dy)$ is path independent.*

EXERCISE 7.5 *Evaluate the following line integral in $\mathbb{R}^3$: $\int_\varphi (y -$*

$z)dx + (z - x)dy + (x - y)dz$, $\varphi = (\cos t, \sin t, t)$, $t \in [0, 2\pi]$ (the screw line) and along the oriented segment joining the origin with $(0, 0, 2\pi)$.

EXERCISE 7.6 Prove the Cauchy convergence principle for line integrals. (See (7.15).)

EXERCISE 7.7 Prove that if F is continuous and φ a path then $\int_\varphi F$ exists.

EXERCISE 7.8 ①① If φ is a path and φ' exists except on a countable set then $\text{Var}_a^b \varphi = \int_a^b \sqrt{|\varphi'|^2}$. Prove this formula for the arclength.

EXERCISE 7.9 Prove: If φ is a path, $\varphi : [a, b] \mapsto \mathbb{R}^2$ then $m_2([\varphi]) = 0$ where m_2 denotes the two-dimensional Lebesgue measure. [Hint: Given ε let $D \equiv a = t_1 < t_2 \cdots < t_n = b$ be a division of $[a, b]$ such that $|\varphi(t_{i+1}) - \varphi(t_i)| < \varepsilon$ for $i = 1, \dots, n - 1$. The set $S = \bigcup_1^n B(\varphi(t_i), \varepsilon)$ covers $[\varphi]$ and $m_2(S) \leq n\pi\varepsilon^2 \leq \pi\varepsilon\text{Var}_a^b\varphi$.]

EXERCISE 7.10 Prove the statements from Remark 7.2.17. [Hint: Show that for $z_k \in M$ and $t \in \varphi_{-1}(z_k)$ there is an open interval I_t containing t such that $I_t \cap \varphi_{-1}(z_k) = \emptyset$.]

EXERCISE 7.11 Prove the Green formula under assumptions different from those in Theorem 7.2.11, namely that P, Q are continuous in G and each function $\dfrac{\partial Q}{\partial x}$ and $\dfrac{\partial P}{\partial y}$ is KH-integrable on G.

EXERCISE 7.12 ①Let $F_n(x) = n^{-\frac{1}{2}} \sin nx$. Show that F_n converges uniformly on $\mathbb{R}$ and $F_n'(x)$ diverges for every $x \in \mathbb{R}$. [Hint: For $n \in \mathbb{N}$ either $\cos nx \geq \frac{1}{2}$ or $\cos 2nx \geq \frac{1}{2}$.]

EXERCISE 7.13 Let $F_n(x) = nx^2$ for $x \in [0, \frac{1}{n}]$, $F_n(x) = 2x$ for $x > \frac{1}{n}$, $F_n(x) = -x^2$ for $x \in [-\frac{2}{n}, 0)$ and $F_n(x) = 4x$ for $x < -\frac{2}{n}$. What is the relevance of this example for Theorems 7.3.1, 7.3.5 and 3.14.6?

EXERCISE 7.14 Let $F_n(x) = nx^2 \exp(-n|x|)$. What is the relevance of this example for Theorems 7.3.1 and 7.3.5?

EXERCISE 7.15 Show directly without any use of complex analysis that the functions w_n with $w_n = \Re(x + \jmath y)^n$ are harmonic in $\mathbb{R}^2$ for any $n \in \mathbb{N}$. [Hint: Use the Binomial Theorem.]

EXERCISE 7.16 *Solve the Dirichlet problem if $U(1,\varphi) = 1$ for $0 < \varphi < \pi$ and $U(1,\varphi) = 0$ for $\pi < \varphi < 2\pi$.*

EXERCISE 7.17 *Prove the mean value theorem for harmonic functions. Namely: If U is harmonic in an open set $G \subset \mathbb{R}^2$ and the disc centred at (x_0, y_0) and of radius r lies in G then*

$$U(x_0, y_0) = \frac{1}{2\pi} \int_0^{2\pi} U(x_0 + r\cos\varphi, y_0 + r\sin\varphi)\, d\varphi.$$

[Hint: Use formula (7.91).]

EXERCISE 7.18 *Find the Abel sum of the series*

$$1 - 2 + 3 - 4 + 5 - \cdots$$

EXERCISE 7.19 ⓘ*Prove: If $(A) \sum_{n=0}^{\infty} a_n = a$ and $(A) \sum_{n=0}^{\infty} b_n = b$ then*

$$(A) \sum_{n=0}^{\infty} (a_0 b_n + a_1 b_{n-1} + \cdots + a_n b_0) = ab.$$

EXERCISE 7.20 ⓘⓛ *Prove Tauber's theorem. If*

$$(A) \sum_1^{\infty} a_n = s \quad \text{and} \quad \lim_{n\to\infty} na_n = 0$$

then the series $\sum a_n$ converges and $\sum_1^{\infty} a_n = s$. [Hint: see [44] Section 1.23.]

EXERCISE 7.21 ⓘ*Find the Fourier series of the following functions which are defined on $(-\pi, \pi)$ as (a) x; (b) $|x|$; (c) x^2. Use the result of (b) to prove*

$$\sum_1^{\infty} \frac{1}{n^2} = \frac{\pi^2}{6}$$

EXERCISE 7.22 ⓘ*Prove: If $f' \in \mathcal{L}^2$ and f is a continuous 2π-periodic function then the Fourier series of f converges uniformly to f.*

EXERCISE 7.23 *Let $\{w_n\}$ be a orthonormal sequence in $\mathcal{L}^2$. Prove that for a given $f \in \mathcal{L}^2$ equation (7.116) holds if and only if (7.117) is valid.*

EXERCISE 7.24 ⓣIt follows immediately from Corollary 7.6.8 that for every $f \in \mathcal{L}^2[a, b]$ there exists a sequence of step functions converging in $\mathcal{L}^2[a, b]$ to f. Give a direct proof without the use of the Haar sequence. [Hint: Use Theorem 3.10.2 to find a step function φ such that $\| f^N - \varphi \| < \varepsilon$ and $|\varphi| \leq N$, where f^N is as in (3.48).]

The next four exercises aim to establish Corollary 7.6.7 independently of Theorem 7.5.2 its corollaries and of Theorem 7.5.5.

EXERCISE 7.25 ⓣProve: Let $s(x) = \alpha_0/2 + \sum_1^\infty (\alpha_n \cos nx + \beta_n \sin nx)$ and $s_N(x) = \alpha_0/2 + \sum_1^N (\alpha_n \cos nx + \beta_n \sin nx)$. If there exists an $\mathcal{L}^2$-function G such that $|s_N(x)| \leq G(x)$ for all $N \in \mathbb{N}$ and all $x \in [0, 2\pi]$ then $\alpha_n = a_n$, $\beta_n = b_n$ and $\|s_N - s\| \to 0$ as $N \to \infty$.

EXERCISE 7.26 ⓣProve that

$$\sum_{n=1}^\infty \frac{1}{n} \sin nx = \frac{1}{2}(\pi - x)$$

for $x \in (0, 2\pi)$, that the partial sums are uniformly bounded and that the series converges in $\mathcal{L}^2$. [Hint: See reference [44] Section 1.75; for the $\mathcal{L}^2$-convergence use the previous exercise.]

EXERCISE 7.27 ⓣUse the previous exercise to show that the Fourier series of a characteristic function of an interval in $(0, 2\pi)$ converges in $\mathcal{L}^2[0, 2\pi]$ to it.

EXERCISE 7.28 ⓣCombine the results of the previous three exercises to prove Corollary 7.6.7.

Appendix

Supplements

A.1 The Cantor set

A.1.1 An uncountable set

We denote by $\mathbf{D}$ the set of sequences whose terms are zeros and ones only. A typical element of $\mathbf{D}$ is $x = \{\xi_1, \xi_2, \dots\}$ with $\xi_i = 0$ or 1. A sequence of elements from $\mathbf{D}$ is then

$$n \mapsto x_n \text{ with } x_n = \{\xi_1^n, \xi_2^n, \dots\}. \tag{A.1}$$

We show indirectly that $\mathbf{D}$ is not countable. If it were then there would be a sequence as in (A.1) containing all elements of $\mathbf{D}$. Let $\eta_k = 1$ if $\xi_k^k = 0$ and $\eta_k = 0$ if $\xi_k^k = 1$. Clearly $\{\eta_k\} \in \mathbf{D}$ but $\{\eta_k\}$ is not equal to x_n for any $n \in \mathbb{N}$.   •

The set $\mathbf{D}$ has the same cardinality as $\mathbb{R}$. This is usually shown in elementary set theory.

A.1.2 Cantor's discontinuum

Cantor's set, sometimes also called Cantor's discontinuum, is a set of measure zero and is not countable. We construct it from the interval $[0, 1]$, denoted by F_0, as follows: Firstly we remove from F_0 the middle third, more precisely the open interval $(\frac{1}{3}, \frac{2}{3})$, and denote the remaining set F_1, i.e. $F_1 = [0, \frac{1}{3}] \cup [\frac{2}{3}, 1]$. Then we remove from each interval of F_1 the middle third, denote the remaining set F_2 and continue this process. In the n-th step we obtain the set F_n which consists of 2^n closed intervals of length $(\frac{1}{3})^n$. The Cantor set C is (by definition) $\bigcap_1^\infty F^n$. It is clearly of measure zero because it is covered by F_n and the total length of all intervals† of F_n is $(\frac{2}{3})^n$. To show that C is not countable we put it into one-to-one correspondence with **D**. If $x \in C$ then x is in every F_n and we define a sequence $\{\xi_n\}$ as follows. If x lies in the left third of an interval of F_{n-1} then $\xi_n = 0$, if it is in the right third of an interval of F_{n-1} then $\xi_n = 1$. Now $x \to \{\xi_n\}$ is a map of C onto **D** and since the length of the intervals of F_n tends to zero it is one-to-one.

REMARK A.1.1 Many counterexamples in analysis use Cantor-like sets. These sets are constructed the same way as the Cantor sets, except the length of the removed middle intervals is smaller, e.g. in the n-th step it could be taken to be $(\frac{1}{6})^n$. The resulting Cantor like set is closed and bounded, its complement is dense and the set itself is not of measure zero. We used such a set in Example 1.4.5 of a derivative which was not Riemann integrable.

A.2 Dini's Theorem

The importance of uniform convergence stems from the theorem which asserts that a uniform limit of a sequence of continuous functions is continuous. Dini's Theorem is a partial converse.

THEOREM A.2.1 (Dini) *If, for every $x \in [a, b]$, the functions $\{f_n\}$ are continuous, the sequence $\{f_n(x)\}$ is monotonic and $\lim\limits_{n \to \infty} f_n(x) = f(x)$ is continuous then $\{f_n\}$ converges uniformly on $[a, b]$.*

Proof For the proof we assume that the sequence is decreasing. If need be we replace f_n by $f_n - f$ and can therefore further assume that $f \equiv 0$. Given $\varepsilon > 0$ there exists, for every $x \in [a, b]$, a natural number $N(x)$ such that $f_n(x) < \varepsilon$. Using continuity we find an open interval $J(x)$ containing x such that $f_n(t) < \varepsilon$ for all $t \in J(x)$. Applying the Heine–Borel covering theorem we find a finite number of intervals $J(x_i)$, $i = 1, 2, \ldots, s$ with $[a, b] \subset \bigcup_1^s J(x_i)$. Let $N = \text{Max}(N(x_1), N(x_2), \ldots, N(x_s))$. If $n > N$ and $t \in [a, b]$ then $t \in J(x_i)$ for some i and consequently

$$0 \le f_n(t) < \varepsilon. \qquad \bullet$$

It is important that the underlying interval is closed and bounded; neither $\{x^n\}$ nor $\{x - n + |x - n|\}$ converges uniformly on $(0, 1)$ or $\mathbb{R}_+$, respectively.

---

† The intervals are closed and in the definition of a set of measure zero the covering intervals were open, but this difference is irrelevant. See the end of the proof of Theorem 2.11.3, where K_n are replaced by J_n.

A.3 Sets in $\mathbb{R}^n$

In this section we review notation and summarize some rudiments of point sets in $\mathbb{R}^n$. A point x is an interior point of a set S if there exists a bounded open interval J such that $x \in J \subset S$. A set is open if all its point are interior points. The set S° of all interior points of S is the largest open set contained in S. The union of any system of open sets is open, the intersection of finitely many open sets is also open. Every open set in $\mathbb{R}$ is a disjoint union of open intervals. A point x is a limit point of S if every open interval containing x has at least one point of S distinct from x. A set is closed if it contains all its limit points. The smallest closed set containing S is called the closure of S and denoted $\bar{S}$. The intersection of any system of closed sets is closed, the union of finitely many closed sets is also closed. A set is closed if and only if its complement is open. A set is bounded if it is contained in a bounded interval. If a set is closed and bounded it is compact, which means that the following assertion holds.

THEOREM A.3.1 (The Borel Theorem) *If $S \subset \mathbb{R}^n$ is closed and bounded and G_λ with $\lambda \in \Lambda$ is a system of open sets such that*

$$S \subset \bigcup_{\lambda \in \Lambda} G_\lambda$$

then there are finitely many $\lambda_1, \lambda_2, \dots, \lambda_m$ such that

$$S \subset \bigcup_{i=1}^{m} G_{\lambda_i}.$$

For more about this topic see e.g. [4], [39] and [19].

A.4 Uniqueness of the Dirichlet problem

The so-called maximum principle plays an important rôle in the theory of elliptic and parabolic partial differential equations. Here we prove a simple version valid for harmonic functions. This simple version implies the uniqueness of the Dirichlet problem.

THEOREM A.4.1 (Maximum principle) *If U is harmonic in a bounded open set G, continuous on the closure of G and $U(x, y) \leq M$ on the boundary of G then*

$$U(x, y) \leq M \qquad \text{for every} \quad (x, y) \in G.$$

Proof of uniqueness If U_1 and U_2 are solutions then both $U_1 - U_2$ and $U_2 - U_1$ are non-positive on the boundary, by the maximum principle $U_1 - U_2$ and $U_2 - U_1$ are non-positive everywhere in G, hence we have $U_1 \equiv U_2$. •

Proof of the maximum principle Assume contrary to what we want to prove that at some point $(a, b) \in G$ and for some positive k

$$U(a, b) = M + k.$$

We denote by R the diameter of G and define an auxiliary function

$$v(x, y) = U(x, y) + k \frac{(x - a)^2 + (y - b)^2}{2R^2}.$$

v assumes its maximum at some point $(\tilde{x}, \tilde{y})$. This point is not on the boundary of G since there $v(x, y) \leq M + k/2$ and $v(a, b) = M + k$. At $(\tilde{x}, \tilde{y})$

$$\frac{\partial^2 v}{\partial x^2} \leq 0,$$

$$\frac{\partial^2 v}{\partial y^2} \leq 0.$$

This implies

$$\frac{\partial^2 U}{\partial x^2} < 0,$$

$$\frac{\partial^2 U}{\partial y^2} < 0.$$

U does not satisfy the Laplace equation. •

A.5 Abel's theorem

For the proof we need the following lemma:

LEMMA A.5.1 *For real numbers α_i, β_i with $i = 0, 1, \ldots, p$ and $S_k = \sum_{i=0}^{k} \alpha_i$ the following identity holds:*

$$\sum_{i=0}^{p} \alpha_i \beta_i = \sum_{1}^{p-1} S_i(\beta_i - \beta_{i+1}) + \beta_p S_p.$$

Proof An easy proof by induction is left to the reader. It is also possible to derive the lemma by writing $\alpha_i = S_i - S_{i-1}$ (with $S_{-1} = 0$), and rearranging the sum $\sum_{0}^{p}(S_i - S_{i-1})\beta_i$. •

THEOREM A.5.2 (Abel's theorem) *If the series $\sum_1^\infty a_n$ converges then*

(i) *the series* $\displaystyle\sum_1^\infty a_i x^i$ *converges uniformly on* $[0, 1]$

(ii) $\displaystyle\lim_{x \to 1} \left(\sum_1^\infty a_i x^i \right) = \sum_1^\infty a_i.$

Proof It suffices to prove (i). By the Cauchy convergence principle for every positive ε there is N such that for $n \ge N$ and natural p

$$\left| \sum_{i=0}^p a_{n+i} \right| < \varepsilon. \tag{A.2}$$

For $n \ge N$ and a natural p we have by the lemma

$$\left| \sum_{i=0}^p a_{n+i} x^{n+i} \right|$$

$$= \left| \sum_{i=0}^p \left(\sum_{k=1}^i a_{n+k} \right) (x^{n+i} - x^{n+i+1}) + x^{n+p} \sum_{i=0}^p a_{n+i} \right|$$

$$\le \varepsilon \sum_{i=n}^{n+p-1} (x^i - x^{i+1}) + \varepsilon x^{n+p} = \varepsilon(x^n - x^{n+p}) + \varepsilon x^{n+p} < \varepsilon. \qquad \blacksquare$$

A.6 Proof of Lemma 6.7.4

We choose $H > 0$ such that φ has continuous partial derivatives and $\varphi_{,n} \ne 0$ on $[a^1 - H, a^1 + H] \times \blacksquare$. For sake of definiteness let $\varphi_{,n} > 0$. The function F with $F(y) = \varphi(a^1, \dots, a^{n-1}, y)$ is strictly increasing on $[a^n - H, a^n + H]$ and therefore

$$\varphi(a) < \varphi(a^1, \dots, a^{n-1}, a^n + H),$$
$$\varphi(a) > \varphi(a^1, \dots, a^{n-1}, a^n - H).$$

By continuity of φ there is h, $0 < h < H$ such that

$$y < \varphi(x^1, \dots, x^{n-1}, a^n + H),$$
$$y > \varphi(x^1, \dots, x^{n-1}, a^n - H)$$

for $|x^i - a^i| \le h$ with $1 \le i < n$ and $|y - \varphi(a)| \le h$. By continuity of φ there exists $t \in [a^n - H, a^n + H]$ such that $y = \varphi(x^1, \dots, x^{n-1}, t)$. This t is unique since $\varphi_{,n} > 0$. Setting $\psi(x^1, \dots, x^{n-1}, y) = t$ proves (i).

Before proving differentiability of ψ we need to have continuity first.

Let us assume, contrary to what we want to prove, that there is a sequence $\{(x_k^1, \bullet)\}$ with $x_k^i \to b^i$ but $\psi(x_k^1, \bullet) \to c \neq \psi(b^1, \bullet)$. Then

$$\varphi(x_k^1, \ldots, x_k^{n-1}, \psi(x_k^1, \ldots, x_k^n)) \to \varphi(b^1, \ldots, b^{n-1}, c).$$

However

$$\varphi(x_k^1, \ldots, x_k^{n-1}, \psi(x_k^1, \ldots, x_k^n)) = x_k^n \to b^n.$$

By uniqueness of ψ it follows that $c = \psi(b^1, \bullet)$ – a contradiction.

We have by differentiability of φ

$$\varphi(x^1 + \Delta^1, \bullet) - \varphi(x^1, \bullet) = \sum_{i=1}^{n} A_i(\Delta^1, \bullet)\Delta^i, \qquad (A.3)$$

where A_i are continuous at $(0, \bullet)$ and $A_i(0, \bullet) = \varphi_{,i}(x^1, \bullet)$. We set

$$x^n = \psi(x^1, \ldots, x^{n-1}, y),$$
$$\Delta^n = \psi(x^1 + \Delta^1, \ldots, x^{n-1} + \Delta^{n-1}, y + t) - \psi(x^1, \ldots, x^{n-1}, y). \quad (A.4)$$

Substituting this into (A.3) leads to

$$\psi(x^1 + \Delta^1, \ldots, x^{n-1} + \Delta^{n-1}, y + t) - \psi(x^1, \ldots, x^{n-1}, y)$$

$$= -\sum_{i=1}^{n-1} \frac{A_i(\Delta^1, \ldots, \Delta^n)}{A_n(\Delta^1, \ldots, \Delta^n)}\Delta^i + \frac{1}{A_n(\Delta^1, \ldots, \Delta^n)}t.$$

It is understood that Δ^n here is given by (A.4). By the theorem on continuity of the composite function A_i/A_n and $1/A_n$ are continuous. This proves the differentiability of ψ and the formulae $\psi_{,i} = -\varphi_{,i}/\varphi_{,n}$, for $i < n$ and $\psi_{,n} = 1/\varphi_{,n}$. The continuity of the derivatives of ψ follows.

•

Bibliography

[1]   T.A. Apostol. *Mathematical Analysis*. Addison-Wesley, Reading MA and London, 1960.

[2]   Edgar Asplund and Lutz Bungart. *A First Course in Integration*. Holt, Rinehart and Winston, New York etc., 1966.

[3]   D. Austin. A geometric proof of the Lebesgue differentiation theorem. *Proc. Amer. Math. Soc.*, 16:220–221, 1965.

[4]   Ralph P. Boas. *A Primer of Real Functions*. The Mathematical Association of America, 1972.

[5]   P.S. Bullen and R. Výborný. Arzelà's dominated convergence theorem for the Riemann integral. *Bollettino U.M.I.*, 10-A:347–353, 1996.

[6]   Pierre Cousin. Sur les fonctions de n variables complexes. *Acta Matematica*, 19:1–62, 1895.

[7]   D. van Dalen and A.F. Monna. *Sets and Integration*. Wolters–Nordhoff, Groningen, 1972.

[8]   J.D. DePree and C.W. Swartz. *Introduction to Real Analysis*. Wiley, New York, 1988.

[9]   J.B. Diaz. Discussion and extension of a theorem of Tricomi concerning functions which assume all intermediate values. *J. Math. Mech.*, 18:617–628, 1968/69.

[10]   B.D. Gee. On Riemann integrability. *Math. Proc. Phil. Soc.*, 51:537–538, 1955.

[11]   C. Goffman. A bounded derivative which is not Riemann integrable. *Amer. Math. Monthly*, 84:205–206, 1977.

[12]   Russel A. Gordon. *The Integrals of Lebesgue, Denjoy, Perron, and Henstock*. AMS, New York and Amsterdam, 1991.

[13]   T. Hawkins. *Lebesgue's Theory of Integration. Its Origins and Development*. University of Wisconsin Press, Madison, 1970.

[14]   R. Henstock. Definitions of Riemann type of the variational integrals. *Proc. London Math. Soc.*, (3),11:401–418, 1961.

[15]   R. Henstock. *Theory of Integration*. Butterworths, London, 1963.

[16]   R. Henstock. *Linear Analysis*. Butterworths, London, 1967.

[17]   R. Henstock. A Riemann integral of Lebesgue power. *Canad. J. Math.*, 20:79–87, 1968.

[18] R. Henstock. *Lectures on the Theory of Integration*. World Scientific, Singapore, 1988.

[19] John L. Kelly. *General Topology*. Springer-Verlag, New York etc., 1955.

[20] J. Kurzweil. Generalized ordinary differential equations. *Czechoslovak Math. J.*, 7 (82):418–446, 1957.

[21] J. Kurzweil. *Nichtabsolute konvergente Integrale*. B.G. Teubner Verlagsgesellschaft, Leipzig, 1980.

[22] J. Kurzweil, J. Mawhin, and W.F. Pfeffer. An integral defined by approximating BV partitions of unity. *Czechoslovak Math. J.*, 41 (116):695–712, 1991.

[23] P.Y. Lee. *Lanzhou Lectures on Henstock Integration*. World Scientific, Singapore etc., 1989.

[24] P.Y. Lee and J.L. Garces. The Moore–Smith limit and the Henstock integral. *Real Analysis Exchange*, 24:447–455, 1998–99.

[25] P.Y. Lee and D. Zhao. Upper and lower Henstock integrals. *Real Analysis Exchange*, 22:734–739, 1996–97.

[26] J.W. Lewin. A truly elementary approach to the bounded convergence theorem. *Amer. Math. Monthly*, 93:395–397, 1986.

[27] W.A.J. Luxemburg. Arzelà's dominated convergence theorem for the Riemann integral. *Amer. Math. Monthly*, 78:970–979, 1971.

[28] J. Mawhin. *Introduction à l'Analyse*. Université de Louvain, Louvain, 1984.

[29] J. McCarthy. An everywhere continuous nowhere differentiable function. *Amer. Math Monthly,* reprinted in *Selected Papers on Calculus* by MAA, 60:709, 1953.

[30] Robert M. McLeod. *The Generalized Riemann Integral, Carus Math. Monograph #20*. Math. Assoc. Amer., Washington DC, 1980.

[31] E.J. McShane. A unified theory of integration. *Amer. Math. Monthly*, 80:349–358, 1973.

[32] E.J. McShane. *Unified Integration*. Academic Press, New York, 1983.

[33] F.A. Medvedev. *Razvitie Ponjatia Integrala*. Nauka, Moscow, 1974.

[34] I. P. Natanson. *Theory of Functions of a real variable, Vol 1*. Ungar, New York, 1974.

[35] R.N. Pederson. The Jordan curve theorem for piecewise smooth curves. *Amer. Math. Monthly*, 76:605–610, 1969.

[36] I.N. Pesin. *Classical and Modern Integration*. Academic Press, New York, 1970.

[37] Washek F. Pfeffer. *The Riemann Approach to Integration*. Cambridge University Press, Cambridge UK, 1993.

[38] Štefan Schwabik. *Generalized Differential Equations*. World Scientific, Singapore etc., 1992.

[39] G. F. Simmons. *Introduction to Topology*. McGraw-Hill, New York etc., 1963.

[40] M. Spivak. *Calculus*. W.A. Benjamin, New York and Amsterdam, 1967.

[41] Karl R. Stromberg. *An Introduction to Classical Real Analysis*. Wadsworth, Belmont, California, 1981.

[42] Béla Sz.-Nagy. *Introduction to Real Functions and Orthogonal Expansions*. Akadémiai Kiadó, Budapest, 1964.

[43] H.B. Thompson. Taylor's theorem using the generalized Riemann integral. *Amer. Math. Monthly*, 96:346–350, 1989.

[44]   E. Titchmarsh. *The Theory of Functions.* Oxford University Press, London, 1939.

[45]   V. Volterra. Sui principii del calcolo integrale. *Giorn. Mat. Battaglini,* 19:333–372, 1881.

[46]   R. Výborný. Applications of Kurzweil-Henstock integration. *Mathematica Bohemica,* 118:425–441, 1993.

[47]   R. Výborný. Kurzweil-Henstock absolute integrable means McShane integrable. *Real Analysis Exchange,* 20:363–366, 1994–95.

[48]   W. Walter. A counterexample in connection with Egoroff's theorem. *Amer. Math. Monthly,* 84:118–119, 1977.

Index

Printed in the United States
By Bookmasters